Evolution in Four Dimensions

Life and Mind: Philosophical Issues in Biology and Psychology
Kim Sterelny and Robert A. Wilson, editors

Cycles of Contingency: Developmental Systems and Evolution
Susan Oyama, Paul E. Griffiths, and Russell D. Gray, editors, 2000

Coherence in Thought and Action
Paul Thagard, 2000

Evolution and Learning: The Baldwin Effect Reconsidered
Bruce H. Weber and David J. Depew, 2003

Seeing and Visualizing: It's Not What You Think
Zenon Pylyshyn, 2003

Organisms and Artifacts: Design in Nature and Elsewhere
Tim Lewens, 2004

Molecular Models of Life: Philosophical Papers on Molecular Biology
Sahotra Sarkar, 2004

Evolution in Four Dimensions: Genetic, Epigenetic, Behavioral, and Symbolic Variation in the History of Life
Eva Jablonka and Marion J. Lamb, 2005

Evolution in Four Dimensions

Genetic, Epigenetic, Behavioral, and Symbolic Variation in the History of Life

Eva Jablonka and Marion J. Lamb
with
illustrations by Anna Zeligowski

A Bradford Book
The MIT Press
Cambridge, Massachusetts
London, England

MIT Press books may be purchased at special quantity discounts for business or sales promotional use. For information, please email special_sales@mitpress.mit.edu or write to Special Sales Department, The MIT Press, 5 Cambridge Center, Cambridge, MA 02142.

This book was set in Stone Sans and Stone Serif by SNP Best-set Typesetter Ltd., Hong Kong.
Printed and bound in the United States of America.

Library of Congress Cataloging-in-Publication Data

Jablonka, Eva.
Evolution in four dimensions : genetic, epigenetic, behavioral, and symbolic variation in the history of life / by Eva Jablonka and Marion J. Lamb with illustrations by Anna Zeligowski.
    p. cm.—(Life and mind)
Includes bibliographical references (p.).
ISBN 0-262-10107-6 (alk. paper)
1. Evolution (Biology) I. Lamb, Marion J. II. Title. III. Series.

QH366.2.J322 2005
576.8—dc22

                          2004058193

10 9 8 7 6 5 4 3 2 1

To our genetic, epigenetic, and cultural parents and offspring

Contents

Acknowledgments   ix

Prologue   1

I | **The First Dimension**   5

1 | **The Transformations of Darwinism**   9

2 | **From Genes to Characters**   47

3 | **Genetic Variation: Blind, Directed, Interpretive?**   79

II | **Three More Dimensions**   109

4 | **The Epigenetic Inheritance Systems**   113

5 | **The Behavioral Inheritance Systems**   155

6 | **The Symbolic Inheritance System**   193

Between the Acts: An Interim Summary   233

III | **Putting Humpty Dumpty Together Again**   239

7 | **Interacting Dimensions—Genes and Epigenetic Systems**   245

8 | Genes and Behavior, Genes and Language   285

9 | Lamarckism Evolving: The Evolution of the Educated Guess   319

10 | A Last Dialogue   355

Notes   385
Bibliography   417
Index   447

Acknowledgments

This book would not have been written without the encouragement and help of our friends, families, students and colleagues. We are grateful to all of them.

Part of the book was written while E.J. was a visitor at the Museum of Vertebrate Zoology, Berkeley, University of California, and we would like to thank David and Marvalee Wake and their colleagues for the good working environment they provided, and Martha Breed and the WW group for their company and the wonderful nature trips. We also want to thank everyone working at the Cohn Institute for the History and Philosophy of Science and Ideas at Tel Aviv University for their help and support. Our debt to the students in the Cohn Institute and the participants in the "Networks in Evolution" seminar at the European Forum Alpbach 2002 is a big one. Their comments and criticisms made us clarify many of our ideas and arguments, abandon some of them, and think deeply about how we should present the material in this book. We hope that they will enjoy the final product.

We have benefited from information and advice from many people, but our special thanks must go to those who have read and commented on various drafts of the book. Eytan Avital, Daniel Dor, Fanny Doljanski, Yehuda Elkana, Yehudit Elkana, Evelyn Fox Keller, James Griesemer, Revital Katznelson, Jawed Iqbal, Lia Nirgad, Christine Queitsch, Richard Strohman, Iddo Tavory, and Alan Templeton each read sections or chapters, and pointed out some of the errors and ambiguities in what we had written. Our long-suffering friends Lia Ettinger, Simona Ginsburg, and Joy Hoffman read drafts of the whole book, and their comments, criticism, and many valuable suggestions have made it a far better book than it would otherwise have been. Tom Stone and his colleagues at The MIT Press were helpful and encouraging throughout, and we thank them for their guidance and excellent editorial work.

We also want to thank Rami of the Ha'Shloshah restaurant in Jerusalem. His hamousta soup sustained us through many long days, and many problems were resolved at his tables.

Finally we need to acknowledge the contribution of Beauty-the-cat, who sat on every page of the manuscript, thereby delaying its completion by several weeks. As an ex-feral cat, she was a constant reminder of the power of learning, active niche construction, and the coevolution of humans and cats.

Evolution in Four Dimensions

Prologue

The content and format of this book are a little unusual, so we want to begin by explaining what it is about and how it is organized. Our basic claim is that biological thinking about heredity and evolution is undergoing a revolutionary change. What is emerging is a new synthesis, which challenges the gene-centered version of neo-Darwinism that has dominated biological thought for the last fifty years.

The conceptual changes that are taking place are based on knowledge from almost all branches of biology, but our focus in this book will be on heredity. We will be arguing that

- there is more to heredity than genes;
- some hereditary variations are nonrandom in origin;
- some acquired information is inherited;
- evolutionary change can result from instruction as well as selection.

These statements may sound heretical to anyone who has been taught the usual version of Darwin's theory of evolution, which is that adaptation occurs through natural selection of chance genetic variations. Nevertheless, they are firmly grounded on new data as well as on new ideas. Molecular biology has shown that many of the old assumptions about the genetic system, which is the basis of present-day neo-Darwinian theory, are incorrect. It has also shown that cells can transmit information to daughter cells through non-DNA (epigenetic) inheritance. This means that all organisms have at least two systems of heredity. In addition, many animals transmit information to others by behavioral means, which gives them a third heredity system. And we humans have a fourth, because symbol-based inheritance, particularly language, plays a substantial role in our evolution. It is therefore quite wrong to think about heredity and evolution solely in terms of the genetic system. Epigenetic, behavioral, and symbolic inheritance also provide variation on which natural selection can act.

When all four inheritance systems and the interactions between them are taken into account, a very different view of Darwinian evolution emerges. It is a view that may relieve the frustration that many people feel with the prevalent gene-centered approach, because it is no longer necessary to attribute the adaptive evolution of every biological structure and activity, including human behavior, to the selection of chance genetic variations that are blind to function. When all types of hereditary variation are considered, it becomes clear that induced and acquired changes also play a role in evolution. By adopting a four-dimensional perspective, it is possible to construct a far richer and more sophisticated theory of evolution, where the gene is not the sole focus of natural selection.

We have divided the book into three parts, each of which has a short introduction. Part I is devoted to the first dimension of heredity and evolution, the genetic system. In chapter 1 we outline the history of Darwin's theory and show how it became so gene-centered. Chapter 2 describes how molecular biology has changed the way biologists see the relation between genes and characters. In chapter 3 we examine the evidence suggesting that not all genetic changes should be seen as random, chance events.

Part II deals with the other dimensions of heredity. Chapter 4 is about the second dimension, epigenetic inheritance, through which different cells with identical DNA are able to transmit their characteristics to daughter cells. In chapter 5 we explore the ways in which animals transmit their behavior and preferences through social learning, which is the third dimension. We deal with the fourth dimension in chapter 6, which describes how information is transmitted through language and other forms of symbolic communication.

In part III of the book we put Humpty Dumpty together again. Having looked at each of the four dimensions of heredity more or less in isolation, we bring them together by showing how, in the long term, the systems of inheritance depend on each other and interact (chapters 7 and 8). In chapter 9 we discuss how they may have originated and how they have guided evolutionary history. Finally, in chapter 10, we summarize our position and put it into a wider perspective by considering some of the philosophical implications of the four-dimensional view, as well as some political and ethical issues.

Each chapter ends with a "dialogue," and the whole of chapter 10 takes this form. We use these dialogues as a device to enable us to reiterate some of the tricky points in our arguments, and to highlight areas of uncertainty and issues that are contentious. The participants in the dialogues are M.E. (who represents the authors, Marion Lamb and Eva Jablonka) and someone

Ifcha Mistabra.

who could have been called the devil's advocate, but who, in order to avoid the negative connotations of that term, we have chosen to call Ifcha Mistabra (I.M. for short). *Ifcha Mistabra* is Aramaic for "the opposite conjecture." It is a term that embodies the argumentative dialogue style used in the Talmud, in which arguments are countered and contradicted, and through this dialectic a better understanding of the subject is reached. The book can be read without the dialogues, but we think that readers may find them interesting and helpful, because they reflect many of the questions and concerns that our students and others have raised when we have spoken about our evolutionary views.

We hope that the book can be read not only by professional scientists but also by the many people who are interested in biological ideas, and are fascinated (and sometimes worried) by the current ways of thinking about biology, especially about modern genetics. To make it as reader-friendly as possible without compromising the science, we have relegated the more specialized material and the sources of information to endnotes, which are organized on a page-by-page basis. We use many examples and thought experiments to try to make our ideas clear, but we recognize that some chapters (particularly chapters 3, 4, and 7) may be a bit heavy going for nonbiologists. These chapters include quite a lot of molecular detail, which we need in order to make our case to skeptical biologists. Readers who do not wish to delve into the molecular nitty-gritty can skip the more technical parts of these chapters, and read the general discussions, although if they do so they will have to trust our intellectual honesty and judgment, rather than evaluate the data for themselves.

The book is intended to be both a synthesis and a challenge. It is a synthesis of the ideas about heredity that have come from recent studies in molecular and developmental biology, animal behavior, and cultural evolution. The challenge it offers is not to Darwin's theory of evolution through natural selection, but to the prevalent gene-based unidimensional version of it. There are four dimensions to heredity, and we should not ignore three of them. All four have to be considered if we are to attain a more complete understanding of evolution.

1 | The First Dimension

The first dimension of heredity and evolution is the genetic dimension. It is the fundamental system of information transfer in the biological world, and is central to the evolution of life on earth. For a century now, the genetic system has been studied intensely, and these studies have yielded rich dividends. Not only have they helped us to understand the natural world, they have also had significant practical effects in medicine and agriculture.

In the mid-twentieth century it became clear that the molecular basis of genetics was to be found in DNA and its replication, and from the mid-1970s, when genetic engineering got underway, knowledge about genetics began to expand at an unprecedented rate. With new technologies being invented almost daily, it was apparent by the early 1990s that the full DNA sequence of the human genome would soon be known. Molecular biologists were talking with prophetic certainty about the "book of life," which they would soon be reading; about the newly discovered "philosopher's stone"; about the Holy Grail they were uncovering. All of these metaphors referred to the sequencing of the human genome. Once the genome was sequenced, it was claimed, geneticists would be able to use the data to discover the hereditary weaknesses and strengths of an individual, and, where appropriate, benevolently intervene. Never before had biological knowledge seemed so powerful and so full of promise. And as the winter of 2001 drew to a close, the climax was at last reached—the draft sequence of the human genome was published. About 35,000 human genes (the number was later revised), scattered patchily on the twenty-three pairs of human chromosomes, had been identified, sequenced, and their locations made known. Newspapers were full of excited prophecies of a braver and healthier new world.

But the geneticists themselves, now in possession of the draft of the coveted "book of life," have shown a curious and almost schizophrenic response. On the one hand the excitement and sense of achievement are so overwhelming that prophecies about the newly revealed promised land have been even more daring. On the other hand there is a new sense of humility. And ironically, it is the achievements of molecular biology that are causing the humility. The discoveries that are being made show how enormously complicated everything is. Just as in an earlier century, when the telescope opened up new horizons for astronomers and the microscope revealed new worlds to biologists, the revelations of molecular biology cannot be neatly slotted into the existing framework of thought. They do not make the old genetics more complete; rather, they highlight the simplifying assumptions that have been made and reveal vast areas of unanticipated complexity. Genes and genetics can no longer be looked at in quite the same way as in the past.

One of the things that molecular studies have reinforced is something that had already been accepted by modern geneticists: the popular conception of the gene as a simple causal agent is not valid. The idea that there is a gene *for* adventurousness, heart disease, obesity, religiosity, homosexuality, shyness, stupidity, or any other aspect of mind or body has no place on the platform of genetic discourse. Although many psychiatrists, biochemists, and other scientists *who are not geneticists* (yet express themselves with remarkable facility on genetic issues) still use the language of genes as simple causal agents, and promise their audience rapid solutions to all sorts of problems, they are no more than propagandists whose knowledge or motives must be suspect. The geneticists themselves now think and talk (most of the time) in terms of genetic networks composed of tens or hundreds of genes and gene products, which interact with each other and together affect the development of a particular trait. They recognize that whether or not a trait (a sexual preference, for example) develops does not depend, in the majority of cases, on a difference in a single gene. It involves interactions among many genes, many proteins and other types of molecule, and the environment in which an individual develops. For the foreseeable future, predicting what a collection of interacting genes will produce in a certain set of circumstances is not going to be possible. But despite this awareness, the sense of power generated by the success of the genome project has often masked caution, sometimes creating great and unrealistic hopes, and great and unrealistic fears.

The contagious reactions of excited scientists and business people are fascinating and important, because they will influence where time and money are invested in the future, but in what follows we are going to focus on the more direct consequences of the molecular discoveries of the last two decades of the twentieth century. Not only have they made people think more deeply about what genes do, they have also challenged old ideas about what genes are. No longer can the gene be thought of as an inherently stable, discrete stretch of DNA that encodes information for producing a protein, and is copied faithfully before being passed on. We now know that a whole battery of sophisticated mechanisms is needed to maintain the structure of DNA and the fidelity of its replication. Stability lies in the system as a whole, not in the gene. Moreover the gene cannot be seen as an autonomous unit—as a particular stretch of DNA which always produces the same effect. Whether or not a length of DNA produces anything, what it produces, and where and when it produces it may depend on other DNA sequences and on the environment. The stretch of DNA that is "a gene" has meaning only within the system as a whole. And because the effect of a gene depends on its context, very often a change in a single gene does not have a consistent effect on the trait that it influences. In some individuals in some conditions it has a beneficial effect, in other individuals and other circumstances it is detrimental, and sometimes it has no effect at all.

The idea of the genome as a complex and dynamic system is not controversial among professional biologists, even though it does sometimes tend to be forgotten when the new genetics is presented to the public. However, the new ideas about genes and genomes have had only a very limited impact on evolutionary thinking. Yet, if a gene has meaning only in the context of the complex system of which it is a part, the standard way of thinking about evolution, in terms of changes in the frequency of one or more isolated genes, needs to be questioned. For example, it may be more appropriate to focus on changes in the frequency of alternative networks of interactions rather than on the frequencies of individual genes.

New knowledge of genes and genomes challenges the assumptions of current evolutionary theory in another way. If the genome is an organized system, rather than just a collection of genes, then the processes that generate genetic variation may be an evolved property of the system, which is controlled and modulated by the genome and the cell. This would mean that, contrary to long-accepted majority opinion, not all genetic variation is entirely random or blind; some of it may be regulated and partially directed. In more explicit terms, it may mean that there are Lamarckian mechanisms that allow "soft inheritance"—the inheritance of genomic changes induced by environmental factors. Until recently, the belief that acquired variations can be inherited was considered to be a grave heresy, one that should have no place in evolutionary theory.

By revealing the dynamic nature of the genome and the complexity of gene interactions, molecular biology is forcing a rethink of the genetic dimension of evolutionary theory. In part I we shall be looking at this dimension by describing (chapter 1) the origins of the conventional view, which is based on the perception of the gene as the unit of heredity, heritable variation, and evolution. We then go on to discuss (chapter 2) the complex relations between genes and developmental processes, and finally (chapter 3) look at the ways in which genetic variation is generated, and what this may mean for our view of evolution and heredity.

1 | The Transformations of Darwinism

No sphere of knowledge is free of controversy, and science is no exception. If anyone imagines that scientists are dispassionate and impartial people, discussing theories and ideas unemotionally in the cool clear light of reason, they have been seriously misled. Passion and fervor accompany all worthwhile scientific discussions. This is particularly evident when the discussion is about something like the theory of evolution, which bears directly on human history and our relationships with each other and the world around us. Because such discussions are tied up with ideas about "human nature," and impinge on moral judgments and ethical issues, they can be very emotional, as well as intellectually exciting.

We are not referring here to the arguments between people who accept evolutionary ideas and those who prefer to believe that the world was created by God in six real or metaphorical days. Such arguments have considerable sociological and political interest, but they are not really part of science, so we need say no more about them. What we are referring to are the heated discussions that have gone on and still go on among the evolutionary biologists themselves.

When you read popular accounts of new discoveries in biology, you often come across phrases such as "according to Darwin's theory of evolution . . . ," or "evolutionary biologists explain this as . . . ," or "the evolutionary explanation is. . . ." You get the impression that there is a tidy, well-established theory of evolution—Darwin's theory of natural selection—which all biologists accept and use in the same way. The reality is very different, of course. Ever since Darwin's book *On the Origin of Species* appeared in 1859, scientists have been arguing about whether and how his theory of evolution works. Can competition between individuals with heritable differences in their ability to survive and reproduce lead to new features? Is natural selection the explanation of all evolutionary change? Where does all the hereditary variation on which Darwin's theory depends come from? Can new species really be produced by natural selection? Darwin's book was crammed with observations that supported his theory, but there were some glaring gaps in his evidence. The biggest was that he could say little about the nature and causes of hereditary variation. Right from the outset, even those who accepted Darwin's evolutionary theory questioned its completeness and sufficiency, and struggled to try to find answers to the questions it raised about heredity and variation. In subsequent decades, as new discoveries were made and new theoretical approaches were developed, the debates continued. Existing ideas were constantly being challenged and revised, with the result that profound changes have occurred in the ways the concepts of evolution and heredity have been understood.

Today, most biologists see heredity in terms of genes and DNA sequences, and see evolution largely in terms of changes in the frequencies of alternative genes. We doubt that this will be the situation in twenty years' time. More and more biologists are insisting that the concept of heredity that is currently being used in evolutionary thinking is far too narrow, and must be broadened to incorporate the results and ideas that are coming from molecular biology and the behavioral sciences. We share this view, and in later chapters will explain why. But before doing so, we want to outline some of the history of evolutionary thinking over the last 150 years to see how the present gene-centered version of Darwinian theory came into being, and what it means for today's evolutionary biologists. Since we cannot even attempt to look at all of the many twists and turns in the pathway of ideas that led to the present position, we will focus on some of the major turning points and the arguments that influenced the direction taken.

Darwin's Darwinism

Darwin summarized his view of evolution in the last paragraph of *The Origin*. In what was for him an unusually poetic paragraph, he wrote:

It is interesting to contemplate an entangled bank, clothed with many plants of many kinds, with birds singing on the bushes, with various insects flitting about, and with worms crawling through the damp earth, and to reflect that these elaborately constructed forms, so different from each other, and dependent on each other in so complex a manner, have all been produced by laws acting around us. These laws, taken in the largest sense, being *Growth with Reproduction*; *Inheritance* which is almost implied by reproduction; *Variability* from the indirect and direct action of the external conditions of life, and from use and disuse; a Ratio of Increase so high as to lead to a *Struggle for Life*, and as a consequence to Natural Selection, entailing Divergence of Character and the Extinction of less-improved forms. Thus, from the war of nature, from famine and death, the most exalted object which we are capable of conceiving, namely, the production of the higher animals, directly follows. There is grandeur in this view of life, with its several powers, having been originally breathed into a few forms or into one; and that, whilst this planet has gone cycling on according to the fixed law of gravity, from so simple a beginning endless forms most beautiful and most wonderful have been, and are being, evolved. (Darwin, 1859, pp. 489–490)

The italics in the rather less poetical sentence in the middle are ours, not Darwin's. They are there to stress the "laws" to which Darwin pointed: the laws of reproduction, inheritance, variability between individuals, and a struggle for existence. By using these laws, it is possible to formulate Darwin's theory in a very general and abstract way, without referring to our own world or to the types of reproduction, inheritance, variation, and competition with which we are familiar. For example, in British evolutionary biologist John Maynard Smith's generalization, the properties that any group of entities and their world must have in order for evolution by natural selection to occur are the following:

- *Multiplication*—an entity can reproduce to give two or more others.
- *Variation*—not all entities are identical.
- *Heredity*—like begets like. If there are different types of entities in the world, the result of the multiplication of entity of type A will be more entities of type A, while the result of the multiplication of entity B will be more of type B.
- *Competition*—some of the heritable variation affects the success of entities in surviving and multiplying.

If all these conditions are met, evolution by natural selection is inevitable: the type of entity that has the greatest ability to survive and multiply will increase in frequency (figure 1.1). Eventually, evolution in this world will stop, because all the entities will be of the same type. However, if heredity is not always exact, so that from time to time new variants arise, then variations in a certain direction may accumulate and produce a complex functional system. Historically, the eye is the classic example of cumulative evolution in the living world, and the modern PC is a good example from the world of technology.

When formulated in Maynard Smith's way, Darwin's theory of evolution by natural selection is an extremely general theory. It says nothing about the processes of heredity and multiplication, nothing about the origin of

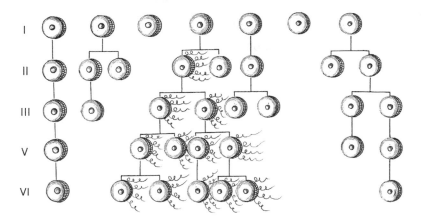

Figure 1.1
Universal Darwinism: the frequency of the hairy entity, which first appears in generation II, increases in subsequent generations because it survives better and multiplies more than its competitors.

the heritable variation, and nothing about the nature of the entity that is evolving through natural selection. Appreciating this is going to be crucial for the arguments that we develop in later chapters. Although we are not advocating it, we want it to be clear that it is possible to be a perfectly good Darwinian without believing in Mendel's laws, mutating genes, DNA codes, or any of the other accoutrements of modern evolutionary biology. That is why Darwin's theory can be and is so widely applied—to aspects of cosmology, economics, culture, and so on, as well as to biological evolution.

Darwin himself knew nothing about genes, Mendelian laws, and DNA, of course. These did not become part of evolutionary theorizing until the twentieth century. In fact, in Darwin's day, there was no good theory of heredity at all, and this was a problem. At that time, most people assumed that the characteristics of two parents blended in their young, so if you started out with a population with two types in it (say black and white), you would end up with a population in which everything was the same (gray). There would be no variability left. Yet Darwin's theory depends on the presence of heritable differences between individuals. Even without blending, if you continually selected one type (say black), the proportion of that type would increase until eventually all in the group would again be identical (this time black). So where does new variation come from? For the theory of natural selection to be believable, Darwin and his followers had to explain the origin and maintenance of variation.

As the quotation from the last paragraph of *The Origin* indicates, Darwin thought that heritable variation stems from the effects the conditions of life have on the organism, and from "use and disuse." Discovering that this is what Darwin thought surprises some people, because they associate the idea of evolutionary change through use and disuse with the name of Lamarck. Lamarck, they have been told, put forward a theory of evolution fifty years before Darwin did, but got the mechanism all wrong. Foolishly (somehow, Lamarck is always made to seem foolish), Lamarck believed that giraffes have long necks because their ancestors were constantly striving to reach the leaves on tall trees, stretching their necks as they did so. They passed on these stretched necks to their young, so that over many generations necks became longer and longer. Lamarck, the story goes, saw evolution as the result of the inherited effects of use (or disuse). His big mistake was to assume that "acquired characters"—changes in structures or functions that occur during an animal's life—could be inherited. Fortunately, the story continues, Darwin showed that natural selection, not use and disuse, is the cause of evolutionary change, so the idea that acquired characters can be inherited was abandoned.

This often repeated version of the history of evolutionary ideas is wrong in many respects: it is wrong in making Lamarck's ideas seem so simplistic, wrong in implying that Lamarck invented the idea that acquired characters are inherited, wrong in not recognizing that use and disuse had a place in Darwin's thinking too, and wrong to suggest that the theory of natural selection displaced the inheritance of acquired characters from the mainstream of evolutionary thought. The truth is that Lamarck's theory of evolution was quite sophisticated, encompassing much more than the inheritance of acquired characters. Moreover, Lamarck did not invent the idea that acquired characters can be inherited—almost all biologists believed this at the beginning of the nineteenth century, and many still believed it at its end. It was certainly part of Darwin's thinking, and his theory of natural selection certainly did not lead to the idea being abandoned. On the contrary, it led to endless acrimonious arguments (and even a few experiments) about whether or not acquired characters are inherited. For as long as there was no satisfactory and agreed theory of heredity, and no explanation of the origin of variation, the inheritance of acquired characters retained a place in evolutionary thinking.

The lack of a good theory of heredity and an explanation of variation were a constant frustration for Darwin and his followers, and Darwin tried to do something about it. From the 1840s onward, he collected together everything that was known about inheritance, and used it to develop his own heredity theory. He called it the "provisional hypothesis of pangenesis," and eventually described it in his massive book *The Variation of Animals and Plants under Domestication*. It wasn't very original, and was never very popular, but despite the criticism, Darwin never deserted it. It is worth describing Darwin's pangenesis theory, because most other heredity theories in the second half of the nineteenth century were quite similar. All were totally different from the theory of inheritance that we accept today.

What Darwin suggested was that every part of the body, at each developmental stage, sheds tiny particles, which he called "gemmules." These circulate in the body, sometimes multiplying as they do so. Some gemmules are used for regenerating damaged or missing parts, but most eventually aggregate in the reproductive organs. In asexual organisms, the gemmules in the egg, seed, spore, or whatever piece of the parent produces the next generation organize themselves and eventually each develops into the same type of part as that from which it originally came. In sexually reproducing organisms, the gemmules stored in the egg and sperm join together before development starts (figure 1.2). The offspring therefore become a blend of the parental characters, although sometimes, according to Darwin, gemmules are not used immediately, but remain dormant and reappear either later in life or in future generations.

Initially the gemmules present in a fertilized egg are not ordered in any particular way, but during development, as they grow and multiply, they are incorporated into the appropriate place at the appropriate time because they have certain special affinities for each other. Gemmules are therefore units of both heredity and development. According to this notion of heredity, what is inherited is the actual character itself. It is transmitted from one generation to the next in the form of its miniature representatives, the gemmules. In Darwin's words, "inheritance must be looked at as merely a form of growth" (Darwin 1883, vol. 2, p. 398).

Pangenesis could account for most of the things Darwin had found out about heredity, regeneration, hybridization, developmental abnormalities, and much else. But what about variation? Pangenesis should lead to blending and uniformity, so how did Darwin explain variation? First, he suggested, a change of nutrition or climate could affect growth and alter the proportions of the different gemmules in the reproductive organs; it could also reawaken dormant gemmules. Second, changed conditions or new experiences could at any stage lead to changes in the gemmules themselves. If parts of a parent were modified, for example through use or disuse, correspondingly modified gemmules would be produced. The new,

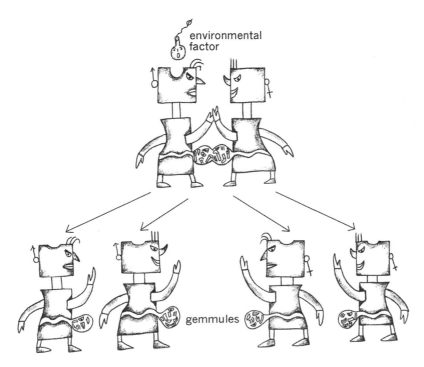

Figure 1.2
Sexual pangenesis. Representative particles (gemmules) from the male (left) and female (right) parents accumulate in their reproductive organs. Following insemination, these gemmules mix and together produce the next generation. An environmental effect (the bomb) induced a change in the male parent. This change is inherited, because the gemmules from the modified body parts are also modified, but the effect is diluted by gemmules from the unaffected female parent.

acquired character would be inherited, although it might not be expressed very strongly, because the modified gemmules would be mixed with those already present in the reproductive organs and with those contributed by the mate.

Obviously, accepting that the environment has a role in inducing variation in no way weakens Darwin's theory of evolution by natural selection. On the contrary, if new variation can arise in response to the conditions of life, it increases the amount of variation and the scope for natural selection. Darwin would no doubt have been amazed to hear that many biologists today think that Lamarckian views about the inheritance of acquired characters contradict the fundamental assumptions of his theory of natural selection. They do not. Darwin's pangenesis hypothesis shows that the theory of natural selection is really not very fussy. Gemmules turned out to be no more than fascinating figments of Darwin's imaginative mind, but as a cause of the heredity and variation needed for animals and plants to evolve through natural selection, they did very nicely. Darwin's theory of natural selection is a very general theory; it is not tied to any particular mechanism of heredity or cause of variation.

Weismann's Neo-Darwinian Theory: Acquired Characters Discarded

We tend to assume that the great increase in the rate of scientific progress began in the twentieth century, but imagine what it must have been like to be a biologist in the late 1850s. First, Rudolph Virchow propounds the theory that cells come only from other cells; they cannot arise from non-cellular matter. Soon after, Darwin tells the world that species arise only from other species; they are not produced by special creation, but by natural selection. Then Louis Pasteur reports his experiments showing that living things are not generated spontaneously; organisms come only from other organisms. Trying to keep up with all that was going on must have been as big a nightmare for scientists in the mid-nineteenth century as it is now. So it is not surprising that when Darwin was dealing with the finer points of his pangenesis theory, he left the question of the formation of cells rather vague, "as I have not especially attended to histology." Given how much else he was attending to, no one can really blame him for deciding that he didn't know enough to evaluate the various ideas about the origin of cells. It was left to others to try to relate the new cell biology to heredity and evolution. Among those who tried to do so was the German biologist August Weismann, one of the most profound and influential evolutionary thinkers of the nineteenth century.

Weismann's ideas about heredity and development changed over time, but the essentials were in place by the mid-1880s. By then it was generally recognized that organisms are made of cells, and that cells have a nucleus containing threadlike *chromosomes* (the word itself was not invented until 1888). It was known that ordinary body cells divide by *mitosis*, a process in which each chromosome doubles and then splits longitudinally, with one half going to each of the daughter cells. Once this rather precise method of allocating the nuclear material had been recognized, it became clear to Weismann and several other people that the chromosomes probably contain the hereditary substance that determines the characteristics of the cell and its descendants.

Weismann realized, however, that if the chromosomes of the nucleus contain the hereditary materials, then there is a problem when it comes to inheritance between generations. The link between generations of sexually reproducing organisms is through the eggs and sperm (or as we would now say, the gametes). Yet, if both egg and sperm have the same chromosomal content as other cells, the fertilized egg and the new organism it produces will have twice as much chromosomal material per cell as either parent had. Obviously, this cannot be what is happening. Weismann therefore concluded that during sperm and egg production, the cells in the reproductive system must undergo a different kind of division from that of other cells. It has to be a *"reduction division,"* he said, in which each daughter cell receives only half of the parent cell's chromosomal material. Then, when the nuclei of sperm and egg are united during fertilization, the two halves become a new whole with the same amount of nuclear material as other cells. When Weismann first suggested this, there was no real evidence for a reduction division, although it was known that odd things happen during the cell divisions that produce eggs. It took some years for people to unravel the nature of the process that was eventually called *meiosis*, and recognize its significance in inheritance. As Weismann guessed, the amount of chromosomal material is indeed halved, but there is a lot more to meiosis than that.

How did Weismann's deductions about cell division relate to his ideas about heredity and evolution? The first thing to be said is that Weismann emphatically rejected any possibility that acquired characters are inherited. The big muscles that the blacksmith develops through his hard work cannot be transmitted to his sons and daughters. If his sons want to be blacksmiths, they will have to go through the muscle-building process themselves, because they do not inherit their father's big muscles. There is no way, according to Weismann, in which properties that reside in the cells and tissues of the arms can be transmitted to the father's sperm cells. The same is true for circumcision. Although for three thousand years Jews have been circumcising their newborn boys, this has not resulted in their male babies being born without a foreskin. Eight-day-old baby boys still have to undergo the painful ritual operation. There is no route through which information about a cut-off foreskin can be passed to the sperm. Not only are there no empirical data of any kind to support the conjecture that acquired characters are inherited, claimed Weismann, there is no way in which it could happen.

Weismann's insistence that it is impossible to inherit acquired characters was tied up with the way in which he saw heredity and embryonic development. He devised a scheme that was founded on what he called "the continuity of the germ plasm," which we have shown in figure 1.3. It involved a division of labor between the elements that maintain individual life and the elements that are devoted to producing future generations—a division between the soma (the body) and the germ line. He argued that right from the beginning of development, a part of the chromosomal material, which he called the "germ plasm," is set aside for the production of the eggs, sperm, spores, or whatever else gives rise to the next generation. In many animals it is separated off into special gamete-producing cells—germ cells—very early in development. Sometimes the germ cells are the very first cells to form, but even if they form later, they still have germ plasm that is identical to that in the fertilized egg. According to Weismann, the other cells of the body, the somatic cells, do not.

Weismann's scheme for development was quite complicated and, as it turned out, quite wrong. It involved a whole hierarchy of units, each present in the chromosomes in multiple copies. In essence, what Weismann thought was that when embryonic cells divide, each daughter cell can receive different parts of the nuclear material—a different set of "determinants." That is why daughter cells develop into different cell types. Determinants move out of the nucleus to impose their characteristics on the cells, so the nuclear material gets simpler and simpler as cells continue to divide and produce the different tissues. Development therefore depends on gradual, regulated, qualitative changes in the nuclear substance. Only the germ plasm in the germ line retains the full hereditary potential—a full set of determinants. It is this unaltered and untainted

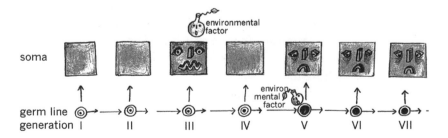

Figure 1.3
Weismann's doctrine: hereditary continuity is through the germ line. An environmentally induced change in the soma (bomb in generation III) does not affect the offspring, whereas a change in the germ line (bomb in generation V) affects all subsequent generations.

germ plasm that is used for the sperm and eggs that will produce the next generation.

If, as Weismann maintained, acquired characters cannot be inherited because bodily events do not affect the protected germ line, where did he think all the variation that Darwin's theory demanded came from? Here he had an important insight: it comes from sexual reproduction, he said. He reasoned that since the father's germ plasm in the sperm mixes with the mother's germ plasm in the egg, there are two mingled germ plasms in their offspring. In the next generation, the two mingled germ plasms in these and similar offspring's eggs will mix with two mingled germ plasms from their sperm to give offspring with four mingled germ plasms; and in the next generation four mingled germ plasms in the egg will mix with four more from the sperm to give eight. And so it will go on. Every individual is thus the product of a mixture of minute quantities of vast numbers of ancestors' germ plasms. Now, since the amount of nuclear material is kept constant by the reduction division that halves the amount of germ plasm during sperm and egg formation, what Weismann cleverly suggested was that the half of the germ plasm that is eliminated is not the same for every egg or sperm. In each a different group of ancestral germ plasms is retained. It is like a card game in which a deck of ancestral germ plasms is shuffled before a gamete is formed, and the gamete is then dealt half of the deck. Since there is an enormous number of possible combinations of ancestral germ plasms, no two gametes will be the same. Thus there is always a lot of variability in sperm and eggs, and even more in the offspring they produce when they fuse. There have been some wonderful words written about sex, but what Weismann rather prosaically said was "The object of this process [sex] is to create those individual differences which form the material out of which natural selection produces new species" (Weismann, 1891, p. 279).

Sex could provide endless variability by recombining the hereditary material from different ancestors, but Weismann still had to explain how ancestral germ plasms came to be different in the first place. The ultimate origin of variation, he said, was in changes in the quantity and qualities of the many growing and multiplying determinants for each character that are present in the germ line. From time to time, small random accidents would alter determinants. Some would survive and multiply better than others, so through natural selection among the determinants, the germ plasm would gradually change. Weismann called this process "germinal selection." Exactly which determinants were selected would depend on factors such as nutrition and temperature, said Weismann.

It is worth noting two things about Weismann's germinal selection. The first is that although he was adamant that environmental effects on the body cannot be inherited, he did accept that the conditions of life had heritable effects. They did so because they affected the determinants in the germ plasm *directly*. The second is that the germinal selection idea shows that Weismann appreciated how very general Darwin's theory is: he recognized that natural selection can occur between units other than individual organisms. As well as believing that it occurs between determinants in the germ plasm, he accepted that selection must also occur between the cells within a tissue. Like Darwin, he also recognized that natural selection must occur between groups of organisms, because this was the only satisfactory way of explaining the evolution of sterile worker ants and bees. We will come to the evolutionary problems with social insects later, but here we just want to point out that in applying Darwinian theory to other levels of biological organization, Weismann and some of his contemporaries were really way ahead of their time. It took another three-quarters of a century for the idea of multilevel selection to be incorporated into mainstream evolutionary biology.

In summary we can say that the key differences between Darwin's original theory and Weismann's version of it are as follows:

• Weismann gave natural selection an exclusive role. He completely excluded change through use and disuse, and every other form of the inheritance of acquired characters.
• Weismann's heredity theory was totally different from Darwin's. His heredity determinants were transmitted from generation to generation only through the germ line. In contrast to Darwin's gemmules, determinants were not derived anew in each generation, but were stable, replicating entities. Not only were they not derived from the parent's body structures, those retained in the germ line were totally immune from anything occurring in the body.
• For Weismann, the only source of new heritable variation was accidental or environmentally induced changes that directly affected the quantity or quality of determinants in the germ line itself.
• Weismann recognized that it was the sexual process, which brought together different combinations of the parent's determinants, that produced the heritable differences between individuals that were needed for evolution through natural selection.

Historically, one of the most interesting things about Weismann's theories is that although many of his contemporaries hated them, they were still extremely influential. His theory of heredity and development was far too speculative and complicated to gain much acceptance, yet elements of it were incorporated into the new science of genetics in the early twentieth century. Similarly, Weismann's version of Darwinism was seen as far too restricted, yet it had long-lasting effects on the direction ideas about evolution took in later years.

Doubts about Darwinism

By the 1880s, although most biologists accepted the idea of evolution, Darwin's theory of natural selection was thought to be on its deathbed. It didn't recover until well into the twentieth century. One reason for its decline was probably Weismann's dogmatic insistence that natural selection is the *only* mechanism of evolution. This hardened the attitudes of those who preferred Darwin's more pluralistic views, which included the inheritance of acquired characters. Some people rejected Darwin's natural selection almost completely, assigning to it the minor role of merely weeding out the oddities and mistakes. In place of natural selection, various "neo-Lamarckian" mechanisms were proposed.

The term *neo-Lamarckism* was invented in 1885, but was never well defined and meant different things to different people. A dominant element in neo-Lamarckism was the idea that adaptation could occur through the inherited effects of use and disuse. In addition, however, many neo-Lamarckians believed that there were internal forces that made evolution progressive and goal-directed, just like embryonic development. Ideas like these seemed to provide a better basis for adaptation and what was known of evolutionary history. They also fitted better with many peoples' deep-seated religious or moral beliefs. To some the idea of human beings improving as a result of experience was much more attractive than change through ruthless Darwinian competition.

People from both within and outside the scientific community attacked Weismann's ideas from all sides, not always in moderate language. Prominent figures such as Herbert Spencer, Samuel Butler, and later even George Bernard Shaw, ensured that the Lamarckian aspects of evolution were given the widest publicity. Herbert Spencer, one of the leading thinkers of the second half of the nineteenth century, was a believer in biological evolution even before *The Origin*. In fact he was the person who brought the term *evolution* into general currency, using it for all sorts of developmental processes that lead from the simple to the more complex. It was an explanatory concept that united events in the solar system, in society, in the development of mind and body during an individual's lifetime, and in structures and functions in lineages over generations. For Spencer, evolution extended beyond biology, and he assumed that all evolutionary change was fueled by similar mechanisms. He was convinced that the inheritance of acquired characters played a major role in both biological and social evolution, and battled publicly with Weismann about it in the pages of the widely read *Contemporary Review*.

Since Lamarckians rejected Weismann's ideas about inheritance, they needed a heredity theory that would allow the effects of use and disuse to be transmitted. Darwin's pangenesis hypothesis might have done, because it was compatible with the inheritance of acquired characters, but it never found favor, partly because of some work done by Darwin's cousin, Francis Galton. Galton tested pangenesis experimentally by making massive blood transfusions between rabbits with different-colored fur. If Darwin was right, he reasoned, then when blood from white rabbits is transfused into gray rabbits, white-fur gemmules should be transferred too, and some should reach the gray rabbits' reproductive organs. The offspring of these gray rabbits should therefore have some white fur. Unfortunately for Darwin, Galton found that they did not. Although Darwin tried to wriggle out of this embarrassment for his theory by pointing out that he had never said that gemmules circulate *in the blood*, Galton and many others saw it as evidence against pangenesis. However, the main reason why pangenesis-type theories fell from favor was probably not so much that there was no experimental evidence for them, but that they didn't fit with cell biology. As the cell theory became better established, it was impossible to reconcile gemmules or similar hereditary particles coming from all parts of the body with the idea that all cells, including the sperm and egg, come only from other cells. Increasingly, heredity theories had to be seen to be consistent with the growing knowledge of the behavior of cells.

Lamarckians suggested various ways in which what happened in the body could influence the hereditary material in the germ cells, but their theories were extremely speculative. They and their opponents also made many attempts to show experimentally that the inheritance of acquired characters did or did not occur, and such attempts continued until well into the twentieth century. It is not worth going into the details of these experiments and the arguments about them here, however, because in the long run they had little influence on the debate about Lamarckism. As Peter Bowler, one of the leading historians of biology of this period, has stressed, it was not the lack of experimental evidence that eventually led to the demise of Lamarckism, but the lack of a good theoretical model of inheritance.

Neo-Lamarckians were not the only people who were attacking Darwinism in the latter part of the nineteenth century. The idea of *gradual* evolution through the selection of *small* variations was also under attack. People began to argue that evolutionary change was saltatory—it occurred by big jumps, not through the selection of many little differences. Once again, Darwin's cousin Francis Galton was in the forefront of those who caused problems for Darwin's ideas. In an effort to understand human heredity better, Galton applied statistical reasoning to characters that show continuous variation. Continuous characters—characters such as height, for which there is a whole range of possible values—were those which Darwin believed were really important in evolution. According to Darwin, it was selection of small differences over many generations that led to gradual change. Galton, however, decided that this type of selection simply would not work. He did some calculations that suggested (incorrectly) that because you inherit not only from your parents but also from your grandparents and more distant ancestors, the average value of a character could never be permanently changed by selection. He concluded that for permanent change you needed a "sport"—a large, qualitative change in the hereditary material.

Galton's conclusions were fiercely contested by other biometricians, who said he had made a logical mistake in his mathematics. They claimed that selection *could* shift a population average, in exactly the way Darwin had suggested. However, support for the idea that evolution occurred through big jumps also came from a totally different direction. Hugo de Vries in Holland and William Bateson in England had both studied variation in nature, and recognized that a lot of it is discontinuous. Often there are just a few distinct, alternative types, with no intermediates. The same is true if you compare species—there are distinct differences between them; they do not grade into each other. Bateson and de Vries therefore agreed with Galton that discontinuous variation is of greatest importance in evolution, and that evolution occurred through sudden big jumps, not slow crawling. According to de Vries, the driving force in evolution was *mutation*, a process that suddenly and without cause irreversibly changed the germ plasm. Mutation produced a new type of organism in a single step.

De Vries and Bateson were to be significant figures in the development of Mendelism in the first decade of the twentieth century, and it is worth remembering that almost all of the pioneers of the new science of genetics were, like them, "mutationists." Although the term *mutation* didn't mean exactly the same then as it does now, it did relate to a quantum change in the hereditary material. Among most of the founders of modern genetics, both Lamarckism and Darwinism were deemed irrelevant to evolution—mutations were believed to be the important factor.

The Modern Synthesis: Development Vanishes

Debates about the relative importance of selection, mutation, and the inheritance of acquired characters continued until well into the 1930s, but during that decade a far more specific version of Darwin's theory began to be established. Biologists from several disciplines started to shape what became known as the "Modern Synthesis" of evolutionary biology. Weismann's ultra-Darwinism was combined with Mendelian genetics, which had adopted the concept of the gene as the hereditary unit of biological information. Using this framework, many aspects of comparative anatomy, systematics, population biology, and paleontology were explained in terms of natural selection. We are not going into all the details of this, but want to look quite closely at the theory of heredity that was incorporated into the Modern Synthesis, because it was this that began to bias many biologists' approach to evolution.

Mendel gave the world the laws that now bear his name in 1865, when he told the Brno Scientific Society about the hybrids he had made between varieties of the garden pea. His paper was published in the society's journal in the following year, but its significance was not appreciated until decades later. It was not until 1900 that three botanists—Hugo de Vries (the mutationist), the German Carl Correns, and the Austrian Erich von Tschermak—published results from their own breeding experiments which confirmed the validity of the laws that Mendel had established more than thirty years earlier. The year 1900 is now regarded as the birthdate of the discipline for which William Bateson a few years later coined the term *genetics*.

According to the formulation of Mendel's theory that was produced in the early years of the twentieth century, individuals contain hereditary units that determine the development of their characteristics. The crucial thing about these heredity units, which were called *genes*, is something that Weismann (and initially de Vries) had failed to recognize—they exist in pairs. One member of each pair is inherited from the male parent, the other from the female parent. The members of a pair can be identical or somewhat different, but both can affect the development of a particular trait, such as the color of pea seeds or the shape of human ear lobes. The different versions of a gene are known as *alleles*. When sperm or pollen and eggs are formed, they contain only one allele from each pair because, just as Weismann had said, the formation of gametes involves a reduction division, which halves the hereditary material. During fertilization, when the sperm and egg, or pollen and egg, unite, the full hereditary complement is reestablished, and there are once again two alleles for each character.

Mendel's "laws" describe the regularity of the distribution of alleles in the gametes and at fertilization. The "first law" asserts that during the formation of gametes, the two alleles of each pair separate. They have not been changed by being with their partner or by being in that particular body. They leave it in exactly the same condition as they entered it. The "second law" asserts that alleles that belong to different pairs segregate independently of each other. This means that if you are thinking about a lot of characters and a lot of pairs of alleles, there is a vast amount of hereditary variation in the gametes. The argument is basically the same as that which Weismann used—any two eggs or two sperm are very unlikely to get exactly the same combination of alleles. Mendel's laws also assume that which particular sperm and egg unite is not influenced by the alleles they carry, so even more variation is present in the fertilized eggs.

A crucial part of Mendel's findings was that, with the strains he chose to use (and he made his choice very carefully), hybrid offspring did not show intermediate characteristics; they resembled one or other of the parents. For example, in crosses between a pure breeding strain with yellow seeds and one with green seeds, all the offspring were yellow, not yellowish-green. In Mendelian jargon, yellow is dominant, green is recessive. The explanation is simple: if the allele that determines yellowness is given the symbol Y, and that determining greenness is given the symbol y, seeds have to have two copies of y to be green, but a single Y allele is enough for yellowness. So with parents that are pure yellow (YY) and pure green (yy), the offspring inherit a Y allele from the yellow parent and a y allele from the green parent, so they are Yy. The single Y allele is enough to make them yellow. When these are self-fertilized, you get the famous Mendelian ratio of three yellow to one green. The reason why can be seen in figure 1.4, which shows the behavior of characters in a typical Mendelian cross and its genetic interpretation.

Within a few years of the rediscovery of Mendel's laws, hundreds of crosses confirming them had been made using a variety of animals and plants. It was quickly realized that the behavior of the hypothetical hereditary units, the genes, which was deduced from breeding experiments, was

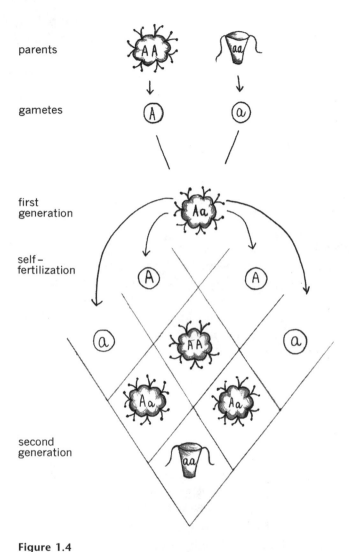

parents

gametes

first
generation

self –
fertilization

second
generation

Figure 1.4
A Mendelian cross between two strains with different alleles for a structural feature.
Crossing *AA* and *aa* produces *Aa*, which resembles the *AA* parent. *A* is therefore
dominant, *a* recessive. When individuals of type *Aa* are self-fertilized, three-quarters
of their offspring resemble the dominant parental type, and one-quarter resemble
the recessive type.

paralleled by the behavior of chromosomes during gamete formation and fertilization. Alleles come in pairs, and so do the chromosomes in body cells; in the gametes there is only a single allele of each gene, and there is only a single copy of each chromosome. From this starting point it did not take long to show that genes are linearly arranged on the chromosomes. This has some consequences when you are looking at more than one trait, but we do not have to worry about this at the moment. We just need to appreciate that genes were soon being regarded as discrete particles, organized rather like beads on a string.

Before moving on, we need to stress something that at first glance may seem rather trivial. It is that Mendelian genetics is based on the analysis of differences. When differences in alleles lead to differences in appearance, we can deduce something about the genetic constitution of parents and offspring. From the ratios of the different types of offspring, we can say which alleles the parents probably have. Conversely, if we know the parents' genetic constitution, we can predict the expected proportions of each type of offspring. But if there are no visible differences, we can say nothing about the genetic constitution, and we know nothing about inheritance.

At first, the distinct character differences that genetics dealt with so well—yellow or green, tall or short, long wings or vestigial wings—reinforced the mutationists' view that evolution depends on discrete qualitative jumps. Mendelism lent no support to Darwinism. Later, however, it was realized that genes can also explain the inheritance of characters such as height or weight, which show continuous variation. All that is necessary is to assume that the character is controlled by many genes, each having a small effect. When there are many genes involved, genetic differences between individuals can supply all the variation needed for adaptive evolution through Darwinian selection.

How genes brought about their effects was at first totally unknown, and for Mendelian analysis and evolutionary theorizing it seemed unimportant. Many of the pioneer geneticists made a conscious decision to ignore development. The newly formed departments of genetics concentrated on counting the different types of progeny obtained in crosses between plants or animals with visible differences, and from their numbers deducing the relationship of the underlying genes to each other and to the chromosomes. Thomas Hunt Morgan and his students at Columbia University launched the small, rapidly breeding, fruit fly *Drosophila* on its career as the geneticists' favorite experimental animal, and used it to produce a wealth of information about the transmission of genes and the chromosomes that carried them. It was the Mendelist-Morganist view of heredity that was later adopted by the architects of the Modern Synthesis of evolution. It was a view that was based on genes located firmly and exclusively in the nucleus, and ignored the surrounding cytoplasm.

The conceptual basis of the Morgan school's view of heredity was provided in the very early days of genetics by Wilhelm Johannsen, a Danish botanist. It was Johannsen who coined the term *gene* as part of his attempt to formulate a biological concept of heredity. Johannsen worked with pure lines of plants—strains that are initiated from a single individual, and maintained by repeated self-fertilization. They can differ from each other, but within any particular line there is very little variation among individuals, and any differences that there are, are not inherited. Johannsen found that if he selectively bred from the extremes—say the tallest and the shortest—it had absolutely no effect: the selected lines still had the same average height as those from which they came. This work led Johannsen to define two key concepts—genotype and phenotype. The genotype is an organism's inherited potential—the potential to have green seeds, green eyes, or to be tall. Whether or not this potential is realized depends on the conditions in which the organism is raised. For example, the height of a plant will depend on the quality of the soil, the temperature, how much water it gets, and so on. So even if a plant has the genotype to be tall, it will not manifest this potential tallness unless the conditions are right. How tall the plant actually is—its phenotype—depends on both its genotype and environmental conditions. Johannsen's interpretation of his pure-line work was simple: all individuals in a pure line have the same genotype. Because they all have the same genes, any differences in their phenotypes cannot be passed on. Differences in phenotype can be inherited and selected only if they are the result of differences in genotype.

The distinction between genotype and phenotype is fundamental to classical genetics. According to Johannsen, heredity does not involve the transmission of characters, but of the potential for characters. As early as 1911, he said quite clearly, "Heredity may then be defined as *the presence of identical genes in ancestors and descendants . . .* " (Johannsen, 1911, p. 159; the italics here are Johannsen's, not ours). His unit of heredity, the gene, was neither a part of the phenotype nor a representation of it. It was a unit of information about the potential phenotype. Genes are not affected by the way that the information is used. They are extremely stable, although occasionally an accident happens and a gene mutates to a new allele, which is then inherited.

The architects of the Modern Synthesis adopted these chromosomal genes as the foundation of the revised neo-Darwinian theory. They rejected both de Vries's type of mutationism and all forms of Lamarckism. By the late 1930s, the mathematical geneticists had shown theoretically how the frequencies of different alleles in a population would alter in response to changes in the mutation rate, the intensity of selection, or when migrants entered the population or its size was restricted. Laboratory experiments and natural populations were soon showing how, give or take a bit, when there are two genetically controlled alternative characters, they behaved as the mathematical geneticists' equations predicted. So, according to the Modern Synthesis:

- Heredity is through the transmission of germ-line genes, which are discrete units located on chromosomes in the nucleus. Genes carry information about characters.
- Variation is the consequence of the many random combinations of alleles that are generated by the sexual processes, with each allele usually having a small phenotypic effect. New variations in genes—mutations—are the result of accidental changes; genes are not affected by the developmental history of the individual.
- Selection occurs among individuals. Gradually, through the selection of individuals with phenotypes that make them more adapted to their environment than others, some alleles become more numerous in the population.

One of the major figures of the Modern Synthesis, the Russian-American geneticist Theodosius Dobzhansky, in 1937 described evolution as "a change in the genetic composition of populations" (Dobzhansky, 1937, p. 11). The genes he was thinking about were, at that time, entirely hypothetical units whose existence had been deduced from numerical data obtained in breeding experiments. What a gene was chemically, and what went on between the genotype and the phenotype, were entirely unknown.

The view of heredity that was taken into the Modern Synthesis did not go unchallenged. Many embryologists maintained that heredity involves more than the transmission of nuclear genes from generation to generation. They argued that the egg cytoplasm is crucial for the inheritance and the development of species characteristics. Moreover, some European biologists, particularly those making crosses between plant varieties, insisted that their results showed that the cytoplasm influences heredity and must carry hereditary factors of some kind. They rejected what was called the "nuclear monopoly" of the Morgan school. But in the English-speaking world their protests went largely unheeded. The influence of the Mendelist-Morganists spread as genetics was taken up by plant and animal breeders, and by the eugenicists, who wanted to "improve" human populations.

Molecular Neo-Darwinism: The Supremacy of DNA

Even though rumblings of dissent about the exclusively nuclear location of the hereditary material continued, the influence of the American and British schools of genetics grew. During the 1940s and 1950s, biochemistry developed rapidly, and many of the chemical processes that go on in cells and tissues were worked out. Geneticists began to recognize the value of microorganisms for their work, and adopted various bacteria and fungi to help them discover what genes are and what they do. Fungi have a few genetic quirks, many of which turned out to be useful, but their genetics can be studied by the classical methods of Mendelian analysis. Bacteria, on the other hand, have no proper nucleus and no pairs of chromosomes, so Mendel's rules do not apply to them. However, they do have a type of sexual process, so genetic analysis is possible. It showed that for the bacteria being studied, genes were linearly arranged on a single circular chromosome.

Through a combination of biochemical and genetic analyses using a variety of organisms, it became clear that genes are involved in the production of proteins. By the early 1950s, it was accepted that the hereditary substance was not the many chromosomal proteins, but a rather simple molecule, deoxyribonucleic acid (DNA). In 1953 Watson and Crick deciphered its structure, the famous double helix, and pointed to how it might do the job required of the genetic material. At amazing speed, molecular biology raced forward. The way DNA replicates was characterized, and the relationship between the DNA of genes and the production of proteins began to be worked out. We shall have to go into this in more detail later, but in essence what was discovered was that a DNA molecule consists of two strings of four different units, called nucleotides. Proteins are made up of one or more polypeptide chains, which are strings of another kind of unit, amino acids, of which there are twenty types. The sequences of nucleotides in DNA encode the sequences of amino acids in the polypeptide chains of protein molecules. However, the translation from DNA into proteins is not direct; the DNA sequence is first copied into mRNA (messenger ribonucleic acid, another linear sequence of nucleotides), and only then is it translated into proteins.

As the code and the way it is translated were worked out, it became clear that a change in the sequence of nucleotides in DNA often brings about a corresponding change in the sequence of amino acids in the protein it encodes. However, the way the process works seemed to offer no way in which a change in a protein could alter the corresponding nucleotides in DNA. "Reverse translation" was deemed impossible. In 1958 Francis Crick proclaimed this unidirectional flow of information from DNA to protein as the "central dogma" of molecular biology. As figure 1.5 shows, the central dogma is conceptually very similar to Weismann's doctrine, which says that somatic events cannot influence the germ line.

Up to this point, the discoveries of molecular biology had little effect on the Modern Synthesis of evolutionary biology that had been developed in the 1930s and 1940s. The gene was interpreted as a DNA sequence, which produced its phenotypic effects by coding for the proteins involved in cell structure and function. Mutations were random changes in the nucleotide sequences of DNA in the nucleus. And just as evolutionary biologists had believed for a long time, because of the central dogma there was no way in which induced phenotypic changes could have any effects on the genetic material. However, soon things began to change, and the Modern Synthesis version of neo-Darwinian evolution had to be updated.

The undercurrent of dissent about the hegemony of the nuclear gene that had been rumbling since the early days of genetics intensified.

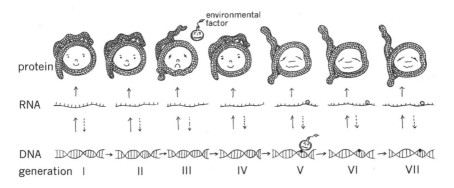

Figure 1.5
The central dogma. Induced changes in the protein product of DNA (bomb in generation III) do not affect the protein in the offspring, whereas changes in the DNA (bomb in generation V) affect the protein in all subsequent generations. Information flows from DNA to RNA to proteins (solid arrows), and possibly from RNA to DNA (dashed arrows), but never from protein to RNA or DNA.

Eventually, studies made in the 1960s confirmed what a few people had been saying for years—there are perfectly good hereditary units outside the nucleus. Genes, made of DNA, were identified in the cytoplasmic organelles known as mitochondria and chloroplasts. This meant that nuclear chromosomes could no longer be regarded as the sole repository of hereditary information.

Molecular studies also showed that there was much more variation in populations than had previously been thought. In fact, there was an embarrassingly large amount of it. It had generally been assumed that any new variant allele that cropped up in a population would either have beneficial effects, in which case it would spread through natural selection and eventually replace the original allele, or, more commonly, it would have detrimental effects and be selectively eliminated. It was recognized that occasionally two or more alleles might persist in a population, and theories about why and when this might happen had been worked out. But in the mid-1960s it was found that for many proteins there were often several allelic variants in a single population. As a result, a new spate of arguments erupted in the evolutionary community. Do all small differences in the amino acid sequence of a protein matter, as the selectionists claimed, or are most of them selectively irrelevant, and kept in the population by chance, as the neutralists said? It was not the first time that chance effects had been given a place in evolutionary theory: ever since the 1930s, Sewall Wright had maintained, somewhat controversially, that differences between small populations would arise by chance, not just by selection. His reasoning was mathematical, but now there were real biochemical data to argue about.

Eventually, after several years of heated debate, it was more or less agreed that many differences in proteins and alleles are, on average, selectively equivalent. In other words, if you think about a genetically diverse population over many generations, it will experience a lot of slightly different conditions, and a small difference in a protein will sometimes improve and sometimes reduce the survival chances or fertility of the organisms carrying it, but on average it will have no effect. As often happens, both sides in the controversy could claim to have been right.

Another cause of controversy was the result of the realization that most DNA in higher organisms does not code for proteins at all. What does this noncoding DNA do? Is it just "junk," or does it have a regulatory function? There has been a lot of argument about both the term and the idea that DNA can be "junk," and the matter is still being discussed. Some noncoding sequences are undoubtedly control sequences, which help regulate when and where the information in DNA is processed to form proteins, but it is also true that a vast amount of DNA has no obvious function. Part of it consists of sequences that are present in many copies, either clustered together or dispersed all over the genome. Some have been found to be similar in organization to the genomes of viruses, and can change their location, moving around the genome. We will say more about these "mobile elements" and "jumping genes" in later chapters, but here we just want to note that their discovery complicated the Modern Synthesis view of the causes of changes in genes and gene frequencies.

As it was recognized that a lot of DNA is concerned with regulating gene activities rather than coding for proteins, the way people thought about hereditary information changed. They began thinking in terms of a genetic program—a set of instructions, written in the genes, which guides the development of traits. The relationship between genotype and phenotype was transformed into the relationship between a plan and a product. John Maynard Smith, an aeronautical engineer by training, likened the genotype to a plan for building an airplane, and the phenotype to the actual plane. Another British biologist, Richard Dawkins, likened the genotype to a recipe for a cake, and the phenotype to the actual cake that is baked. Changes in the recipe or in the plan lead to changes in the product, but changes in the product do not affect the recipe or plan. If a cake is accidentally burnt during its baking, it does not change the recipe; modifications made while building an airplane do not change the written plan. Only changes in plans or recipes—the programs—are inherited, not changes in products.

The discoveries in molecular biology inevitably led to a partial revision of the Modern Synthesis version of Darwinian evolution:

• The gene, the unit of heredity in the Modern Synthesis, became a DNA sequence, which codes for a protein product or an RNA molecule.
• Inheritance became associated with DNA replication, a complex but precise copying process that duplicates chromosomal DNA.
• It was recognized that in higher organisms DNA-containing chromosomes are present in the cytoplasmic organelles as well as in the nucleus.
• Mutations were equated with changes in DNA sequence, which arise through rare mistakes during DNA replication, through chemical and physical insults to the DNA and imprecise repair of the damage, and through the movement of mobile elements from one DNA site to another. Some physical and chemical agents (mutagens) increase the rate of mutation, but since they do not increase specifically those variations that are adaptive,

these induced variations, like all others, were still considered to be random, or blind.

Selfish Genes and Selfish Replicators

While the molecular biologists were busy working out what genes are and what they do, some evolutionary biologists became preoccupied with another problem—the problem of the level at which selection acts. As we mentioned earlier, in the nineteenth century Weismann and others had recognized that natural selection can occur among units other than individuals, but interest in the subject had waned. It revived in the early 1960s, when people began looking more seriously at who benefits from certain types of behavior found in group-living animals. For years most biologists had been happy to accept that some behaviors were "for the good of the species," or "for the good of the group," because they were certainly (or so it seemed) of no benefit to the individual. The most famous and extreme examples are worker ants and bees, where females work for the good of other members of their colony, but do not themselves have young. There are other less extreme examples, such as the alarm notes of birds. The bird that calls out, thereby warning others when it sees potential danger, often is not doing itself any good; on the contrary, it may make it more likely that it will be spotted and killed. It was therefore argued that this type of "altruistic" action must have been selected because it benefits the group, rather than the individual.

Not everyone agreed. A few evolutionary biologists had been pointing out for some time that the for-the-good-of-the-group argument is beset with problems. The most obvious one is that if genes crop up that make individuals selfish—that turn a bird into a noncaller, for example—then those genes will spread in the population and replace the genes for altruistic behavior. Compared with altruistic callers, who keep drawing attention to themselves, noncallers are less likely to be caught, so they will, on average, produce more offspring. Noncaller genes will increase in frequency, and eventually the population will end up as all noncallers. The only way for the altruistic calling behavior to survive in spite of this is if groups of individuals with calling behavior do very much better than groups without it. The question that had to be asked, therefore, was could a behavior (or any other characteristic) be maintained because selection between groups overrides the effects of selection between individuals within the group?

At first the mathematical evolutionists said no. So compelling were their arguments that group selectionists tended to be derided and accused of mathematical illiteracy. Later, however, different equations with different assumptions showed that evolution through group selection was possible after all. Others took different approaches to the problem of why altruism and the genes underlying it do not disappear. Bill Hamilton, one of the most original evolutionary biologists of the second half of the twentieth century, provided an answer that was seen initially as a viable alternative to the idea of group selection. He realized that the beneficiaries of most altruistic behavior tend to be the altruist's own kin. The significance of this is that an animal and its kin are likely to have inherited copies of the same genes. How many genes family members have in common depends on the closeness of their genetic relationship: it is 50 percent among parents and children, 50 percent among brothers and sisters, 25 percent among grandparents and grandchildren, and the same among half brothers and sisters; cousins share only 12½ percent of their genes. The genes that relatives have in common include, of course, any gene or genes that underlie altruistic behavior. So, if altruistic behavior leads to a large increase in the number of offspring reared by members of the altruist's family, the genes underlying the behavior may increase in frequency, even if the altruist has fewer offspring than it would have had had it not helped its kin.

Whether or not altruism genes increase in frequency depends on first, how close the relationship is (and therefore the chances that relatives carry the genes for altruism); second, by how much the altruistic behavior decreases the number of the altruist's own offspring; and third, by how much it increases the number of offspring reared by the beneficiaries of its altruistic actions. It may sound complicated, but the basic idea is very simple. From the point of view of a gene for altruism, it can increase its representation in the next generation if it makes the animals carrying it help their kin to survive and reproduce, because kin are likely to carry copies of it.

Richard Dawkins took up Hamilton's approach, extended it, and popularized it. He suggested that taking a gene's-eye view can help us to understand the evolution of *all* adaptive traits, not just the paradoxical ones like altruism. He coined the term *the selfish gene*, which recognizes that the "interests" of a gene may not coincide with the interests of the individual carrying it. Metaphorically speaking, the gene is "selfish" because the effects it has on the well-being or the reproductive success of the individual carrying it do not matter so long as they enhance the chances that it, the gene, will have more representatives in the next generations. Adaptations are always "for the good of" the gene. They are all outcomes of competition between selfish genes.

According to Dawkins, thinking about evolution in terms of competition between rival genes, rather than between individuals or other units such as genomes, groups, or species, unifies many aspects of evolution. The gene is not just the unit that is inherited, it is also the unit that is ultimately selected. Genes have the stability and permanence that is required for units of selection, whereas most other potential units do not. If you think about individual bodies, then a child is really a rather poor copy of its parent: it does not inherit most of the features that the parent has acquired during its lifetime, and parental characters get separated and mixed up during sexual reproduction. So individual bodies are not faithfully inherited, whereas genes usually are. The living and breathing body is just a carrier—a vehicle—for selfish genes.

On the basis of his image of the selfish gene, Dawkins has constructed a unifying scheme in which he has generalized the molecular neo-Darwinian approach. He argues that genes belong to a category of entities (not necessarily made of DNA) that he calls "replicators." He defines the replicator as "anything in the universe of which copies are made" (Dawkins, 1982, p. 83). At first sight this definition seems very general, and capable of including many types of entities and processes, because "copying" is a conveniently vague word. But Dawkins immediately restricted what he meant by "copying." Bodies are not replicators, because an acquired feature, such as a scar, is not copied to the next generation. But a stretch of DNA or a sheet of paper that is photocopied is a replicator, because any change in DNA or the scribbles on a sheet of paper will be copied. "Copying" is thus restricted so that the term *replicator* cannot be applied to entities that are changed by their own development or product. To make this point Dawkins defined another entity, the "vehicle":

A vehicle is any unit, discrete enough to seem worth naming, which houses a collection of replicators and which works as a unit for the preservation and propagation of those replicators. (Dawkins, 1982, p. 114)

Individual bodies are therefore vehicles, not replicators.

The replicator concept fits the gene so well because it is a generalization of the properties of the classical gene. The distinction between gene and body, and more generally between replicator and vehicle, is derived from Johannsen's distinction between genotype and phenotype, which was built on Weismann's view that the inheritance of acquired characters is impossible. The gene-replicator in the germ line has a special status: it is the unit of heredity, of variation, of selection, and of evolution. It causes the vehicle-body to behave in a way that will increase its frequency, even at the price of sacrificing the body. The move is unidirectional: variations in genes affect corresponding variations in the body, while variation in the body, resulting from the history of the body and from the environment, do not cause corresponding variations in the gene. Development is a process that vehicles (bodies) undergo, and it is controlled by genes that replicate to ensure their own further propagation.

Notice that there is a claim here about the nature of the relationship between genes and development. According to Dawkins, heredity and variation cannot be influenced by adaptive processes that go on in individuals. There is therefore a big difference between this neo-Darwinian generalization and the version of Darwinism with which we started, which was not committed to any type of replicator-vehicle distinction or to assumptions about the origin of heritable variation. In addition to the gene, Dawkins discusses another type of replicator, the *meme*, which is a cultural unit of information that is passed among individuals and generations through cultural replication processes. We shall have more to say about this replicator in chapter 6.

Needless to say, Dawkins's selfish-gene view of evolution has not gone unchallenged. In fact it has been aggressively attacked (and defended) ever since *The Selfish Gene* was published in 1976. But as Hamilton, Dawkins, and others soon realized, a lot of the initial disagreements between those who went along with the selfish-gene view and those who insisted that individuals and groups are the focus of natural selection was the result of scientists talking past each other. The two ways of viewing evolution are not incompatible. Dawkins centers his evolution on the gene-replicator, a permanent unit whose frequency changes during evolutionary time. Other biologists center their evolutionary ideas on the targets of selection, the vehicles—the organism or groups of organisms that survive and multiply. But whatever the targets of selection—whether individuals, interacting groups of kin, or larger groups—biologists still assume that the underlying hereditary units that affect the properties of these targets are genes. Today's models of group selection are as gene-centered as any other models of natural selection, including Hamilton's explanation of the evolution of altruistic traits. Many biologists are now quite comfortable with the idea that kin selection is a form of group selection, in which the interacting kin group is the target of selection, and the unit whose frequency changes during selection is the gene.

One of Dawkins's most bitter critics was the American paleontologist Stephen Jay Gould, who insisted that focusing evolutionary ideas on genes is misleading. According to Gould, tracing the fate of genes through generations is no more than bookkeeping, because it can tell us little about evolution. It is individuals, groups, or species that survive or fail to survive, that reproduce or fail to reproduce, not genes. Moreover, said Gould, we cannot explain the varieties of animals and their adaptations solely in terms of natural selection, whether of genes, individuals, or anything else. We have to take into account historical events such as catastrophic climate changes; we have to think about accidents that affect the amount of genetic variation in populations and lineages; we have to appreciate the way evolutionary change is constrained by development, and remember the side effects that are an inevitable consequence of selection. Natural selection is just one of the many factors that have brought about the wonderful adaptations and patterns of evolution that we see in the living world. For Gould, the central focus of evolutionary studies had to be organisms, groups, and species, which are the targets of natural selection and the entities that develop. For Dawkins, it has to be the gene, the unit of heredity.

The controversy between Gould and Dawkins continued until Gould's death in 2002. Like many of the controversies that punctuated the earlier history of evolutionary thinking, it was bitter, venomous, and often unfair. Arguments were pushed *ad absurdum*, and the ambiguities of language were used and misused to erect and demolish straw men. We cannot and need not go into the details here, because for us what is important is not the disagreements, but what Gould's and Dawkins's ideas have in common. What is interesting for us is that although their different perspectives put them at opposite ends of the spectrum of views held by orthodox evolutionary biologists, they were in agreement when it came to the nature of hereditary variation. Gould and Dawkins were united in assuming that genes are the only units of heredity relevant to the evolution of organisms other than humans, and that acquired characters are not inherited.

The Transformations of Darwinism

Our account of the history of Darwinism has been sketchy, but we hope that we have said enough to show that Darwin's theory is not something set in stone. Ever since the publication of *The Origin*, the theory of natural selection has been the subject of intense debate, and its fortunes have

Table 1.1

Type of theory	Hereditary transmission	Unit of variation	Origin of variation	Target of selection	Unit of evolution
Darwin's Darwinism	Gemmules transferred from the soma to sex cells	Gemmule	Random + induced in the soma	Individual (sometimes also the group)	The population of individuals
Weismann's neo-Darwinism	Transfer of determinants through the germ line	Determinant	Random + induced in the germ line	Individual (mainly) + determinants, cells, organs	The population of individuals, cells, or determinants
Modern Synthesis neo-Darwinism	Transfer of genes in the germ line	Genes in the germ line	Random mutation	Individual	The population of individuals
Molecular neo-Darwinism	DNA replication	DNA sequence	Random DNA changes; rarely also directed changes (see chapter 3)	Mainly the individual (also the gene, the group, the lineage, and species)	Mainly the population of individuals
Selfish gene neo-Darwinism	DNA replication	DNA sequence	Random DNA changes	The gene, the individual, the group	The population of alleles of the gene

waxed and waned. Sometimes the predominant view has been that it has played only a minor role in evolution; at other times, it has been seen as the most important part of the evolutionary process.

Not only has opinion about the theory of natural selection as a whole changed over the years, there have also and inevitably been changes in the details. We have summarized the various historical transformations of Darwin's theory that we have described in table 1.1. It shows how ideas about the nature of the hereditary process, the unit of heritable variation, the origin of variation, the target of selection, and the units of evolution have changed. New facts and new scientific fashions, often promoted by powerful and persuasive voices, have molded Darwin's theory of evolution into its present form.

Today, the gene-centered view of evolution predominates. It certainly provides a tidy framework for evolutionary thinking, and biologists are generally comfortable with it. That does not mean, of course, that it is the final, correct, and complete interpretation of Darwin's theory. In fact, there is a growing feeling that Darwinism is due for another transformation. We shall be putting the case for this in subsequent chapters.

Dialogue

I.M.: I am not entirely comfortable with the implications of the characterization of evolution by natural selection that you borrowed from Maynard Smith. If I am not mistaken, both Maynard Smith and Dawkins see natural selection not only as the mechanism underlying adaptive evolution but also as a kind of litmus paper for life. The conditions for natural selection—multiplication, heritable variation, and competition—are the conditions for life itself. According to this view, if we ever make robots that are able to produce robots like themselves, you will have to define them as evolving and hence alive. This contradicts our intuitions. What is your position?

M.E.: The "definition of life" issue is a really messy subject. First of all, self-production is not sufficient for there to be evolution by natural selection. You also need a mechanism through which variation that is generated during the production of robots is transmitted. Only then can you have evolution by natural selection. You have to have heritable variation. And the variation has to affect the chances of self-production.

I.M.: Let's say my robots can produce themselves and also transmit some variants that occur during the production process. But let us also assume that the number of variations is very limited—let's say that four possible robot variants can arise, and each variant affects self-production in a different manner, which depends on the environment. Nothing very exciting can happen—you can have one of four possibilities reoccurring and changing in frequency as the environment cycles. But that's all. Would you call these robots "living"?

M.E.: John Maynard Smith and Eörs Szathmáry call these cases in which you have only a very few variants "limited heredity" systems. With them you can certainly have evolution by natural selection, but very restricted and boring evolution. Functional complexity and the evolution of functional complexity are the hallmarks of living organisms. Maybe we should be talking about different manifestations of life, rather than about whether there is a clear distinction between life and nonlife. Maybe there is no simple line of demarcation.

I.M.: Since you obviously accept the principle of natural selection, and seem to be prepared to generalize it even to self-producing and varying robots, why do you imply that Dawkins's generalization is insufficient and that Darwinism is due for another transformation? As you showed, Dawkins has suggested a unifying scheme, which allows us to understand the evolution of many different traits, both the straightforward ones and the seemingly paradoxical ones like altruism. It seems very logical to me. What is your problem with it?

M.E.: Our problem is with Dawkins's replicator/vehicle concepts. There are several difficulties. First, he assumes that a replicator has to have a high level of permanence to be a unit of evolutionary change. It has to be copied with very high fidelity. He rightly pointed out that a particular individual—Charles Darwin, for example—is unique and is never replicated, whereas his genes are. It is his faithfully replicated genes that are passed on and effect evolutionary changes. That is why, according to Dawkins, genes, not individuals, are the units of evolution. However, like many other people, we think this argument is misleading, because no one ever thought that individuals are units of heredity and selection in the sense implied by Dawkins. When looking at levels of organization above the gene, evolutionary biologists have focused on traits—for example, on Darwin's square jaw or the shape of his nose, or an aspect of his intelligence—not on whole individuals. So the alternative units should be genes or traits, not genes or individuals. Alternative traits can be traced from one generation to the next and their frequency may change. They have sufficient permanence through time to be units of evolution, even though many genes concurrently affect them and these genes are reshuffled in every generation through sex.

Our second difficulty is with Dawkins's assumption that the relation between replicator and vehicle is unidirectional—variations in the replicator (gene) affect the vehicle (body), but not vice versa. He assumes that development does not impinge on heredity, and we take issue with this assumption. Our third problem is that Dawkins assumes that the gene is the only biological (noncultural) hereditary unit. This simply is not true. There are additional biological inheritance systems, which he does not consider, and these have properties different from those we see in the genetic system. In these systems his distinction between replicator and vehicle is not valid. We will come to them in later chapters.

I.M.: So I shall wait for you to develop these arguments. Meanwhile, I want to ask you about your historical reconstruction. I realize that it is very sketchy, but you pictured the historical trend as one in which Darwinian thinking has become more and more specific about the nature of heredity and the origins of variation. Now that biology has gone so molecular, ideas about heredity and evolution are presented in ever more molecular terms. I see this as progress, and surely so do you. Yet there is a note of discontent in your story.

M.E.: Of course we welcome the molecular level of description. In fact some of the new ideas and the challenges to orthodoxy that we are going to describe in the next chapters are consequences of the new findings in molecular biology. But the molecular-genetic description does not come instead of other levels of description. We shall be making the case that some variations at the physiological and behavioral levels are heritable, and can lead to interesting processes of heredity and evolution even when there is no variation at the genetic level. At this point in time, as at most previous stages of the history of evolutionary ideas, certain findings in biology are being ignored or underplayed. That is why we decided to present today's standard view of Darwinian evolution and how it was reached historically.

I.M.: I have a question about this claim of yours that findings were underplayed or ignored at certain times in the history of evolutionary theory. It is not difficult to be wise in retrospect, and see imperfections and dogmatism, but what does it mean? It seems to me that the most important turning point in the history of twentieth-century evolutionary thinking was the formulation of the Modern Synthesis, so I'll focus on that. You mentioned the rumblings of disagreement about the importance of nuclear genes that came from certain Europeans, but there was nothing in your depiction of the Modern Synthesis to suggest that it did not accurately reflect the biology of the time. The biologists involved in the Synthesis had a certain concept of heredity and evolution, which was derived, I assume, from what they found. It was not as if it was an ideological decision, like it was with the Lysenko doctrine in the USSR, where there was only one politically correct genetic theory. Surely the Synthesis had a wide empirical basis? What was wrong or misleading in the Modern Synthesis? Are you claiming that the view of evolution that emerged was the consequence of scientific ideology?

M.E.: It depends on how you think about ideology. At a very basic level, there is no scientific activity that is totally free from ideology. You can't build a theory without assumptions, and some of them stem from a socio-political general worldview, and feed into that worldview. This doesn't mean that it is a cynical and conscious type of process—that scientists are just puppets in the hands of politicians, or that power-hungry and amoral scientists are recruited for the service of an explicit ideology. Of course this can happen, as the sad story of Russian genetics during the Stalinist era testifies. German eugenics also showed it in a dreadful manner. But in many and perhaps most cases, everything is rather more subtle. Even in nontotalitarian regimes, ideological considerations appear in various guises, and they are important in determining the route of science. This occurred in the United States. There is a fascinating book written in 1966 by Carl Lindegren, an American microbial geneticist. The book is called *The Cold War in Biology*. It describes the political attitudes that surrounded the study of genetics in the West, and the discussions about the nature of the gene and the gene-environment relationship that took place during the Cold War. Self-evidently there were also scientific-ideological presuppositions about the genetic research in which some of the architects of the Synthesis were engaged. They decided what were the important things, and what belonged to the unimportant fringe.

I.M.: And what, for example, did they decide?

M.E.: The Synthesis was based on genetic research that focused on traits that could be studied using the methods of Mendelian analysis. Mendelian analysis depended on discrete qualitative traits that showed fairly regular segregation. Traits that did not behave like that were pushed aside. It was easy to believe that they were the consequences of experimental mistakes, or the overcomplexity of the system. If there are a lot of genes and they interact, it was said that the trait is obviously too difficult to analyze. Extra genes, called "modifiers," which interact with the main gene, were readily evoked whenever there were problems of interpretation. As early as 1949, Lindegren was pointing out that in the bread mould *Neurospora*, two-thirds of the mutations he found did not show Mendelian segregation. But most scientists ignored these cases, even though they were in fact the majority. They were considered to be part of the "noise" in the system. When these deviant traits were acknowledged at all, they were excused, not studied. And even when there was agreement that there are indeed some strange phenomena—jumping genes in maize, for example, or strange inheritance of cortical structures in unicellular organisms—they were brushed under the carpet. At best they were considered to be eccentric cases that did not alter the general picture, and at worst they were simply ignored.

Animal geneticists worked mainly on the mouse or the fruit fly, and organisms that reproduced asexually were of little interest to them. They worked largely with traits that, in the jargon, show "strong developmental canalization." In other words, the organisms develop the same phenotype whatever the environmental conditions. Moreover, much of the genetics of the Synthesis was based on organisms in which the germ cells are separated off from the rest of the body early in development. In plants the germ line and soma do not separate early—you can often take a piece of stem or a leaf from a mature plant and grow another plant from it, and this plant can then produce pollen and eggs. There is no real segregation of germ line and soma in plants, and of course they are much less canalized. On the whole, the botanists were always much less dogmatic about heredity than the zoologists, but their influence on the Synthesis was not great.

I.M.: And do you think that the choices geneticists made were ideological? They seem to me to be good practical decisions.

M.E.: Of course they were not *just* ideological, and usually it was not a conscious and simple process. There was certainly an element of historical continuity. A lot of the early work was done with fruit flies, for example, and this no doubt led to a tendency to generalize from them and see all genetic phenomena in the light of this research. Again, it depends on what you mean by ideology and choice. There were conservatives, liberals, and communists participating in the Modern Synthesis. But there was also a commitment to the Mendelian view and the conception of heredity promoted by Johannsen, and a rejection of the possibility of the inheritance of acquired characters. And these views hardened as a result of the Cold War and the discovery of the charlatanism of Lysenko in the USSR, where the inheritance of acquired characters was fundamental and Mendelism was seen as a bourgeois perversion.

I.M.: What is wrong with generalizing from the genetics of the fruit fly to other species? I thought Mendel's laws were general.

M.E.: They are, but the fruit fly is really peculiar in many ways. Some of these peculiarities were a great help to genetic research, but some were a handicap to evolutionary theorizing. For example, in the fruit fly there is very early segregation between somatic cells and germ cells; the cells of the adult fly do not divide and, in general, development is very stable. So it is difficult to see the effects of the environment on the phenotype, especially any long-term, transgenerational effects. These are much clearer in plants, for example. But there was another more human element too—there was a struggle over the way heredity should be studied. People argued over what kind of research really yields the most significant results, about the status of nuclear genes relative to cytoplasmic factors, about the place of developmental research in the study of heredity. The Mendelist-Morganists, who focused on nuclear genes and on the transmission rather than the expression of characters, won this battle. There were others who took a different approach, especially in prewar Germany, but they lost the battle for various reasons, both scientific and extrascientific.

I.M.: Today biologists are excited about what is happening in genetic engineering and molecular biology, and I know that battles are going on about what work should and should not be done, because some of it has social implications. But whatever the ideologies and whatever the decisions made, isn't it inevitable that this emphasis on molecular biology will lead to a hardening of the gene-centered approach to evolution?

M.E.: We think not. There is a lot more to molecular biology than genes, and the current selfish-gene view does not fit easily with some of the things that molecular studies are turning up. In the next chapters we will look at what molecular biology is telling us about genes and development, and you will see that what has been found is not compatible with an exclusively gene-centered view of heredity and evolution. In addition, although molecular biology is hogging the limelight and the money at present, new facts and ideas are still coming from other areas of biology, and these too are having repercussions on evolutionary thinking.

The relationship between genes and development is one of the hottest topics in biology today. In 2001 the Human Genome Project delivered the first draft of the promised sequence of human DNA, and revealed that we have about 35,000 genes. This was far fewer than most geneticists had anticipated, and recent estimates suggest the figure may be even lower— as low as 25,000. The big question now is how these relatively few genes can be the basis of all the exotic and intricate events that occur during embryonic and postembryonic development. Finding out exactly what genes do has acquired a new urgency. Will it be possible to work out what each gene does? And if it is, what will it tell us about the inherited differences between us?

To answer these questions, we first need to say something about the molecular nature of the gene, and how it works as a unit of function and inheritance. In doing so, we shall be stressing what we see as the major properties of the genetic system, and try to explain why biologists think that this DNA-based inheritance system is so special. DNA is not the only thing that we inherit from our parents, of course. We inherit the other materials that are present in the egg, and also things such as our parents' food preferences, their ideas, and their real estate. Obviously, there are several routes through which materials and information can be transferred from parents to their young, and in later chapters we shall argue that all are potentially important in determining what happens in evolution. However, inheritance systems differ in what kind of information is transmitted, how it is transmitted, how much and how faithfully it is transmitted, and in the relationship between what is transmitted and the effects it has. This is why we are going to focus on these aspects of the genetic system.

From DNA to Proteins

We have already mentioned the discovery of the structure of DNA by Watson and Crick in 1953. One of the fascinating and surprising things about the structure they revealed was how it reflects in a relatively straightforward way the properties that are required by the genetic inheritance system. The way that DNA is organized immediately suggests how replication may occur, and hints at how the molecule might carry information for making proteins. Both are possible because DNA is a linear molecule, made up of strings of a small number of different components.

Figure 2.1 shows the DNA double helix, one of the icons of the twentieth century. The two strands of the helix are wound around the same axis and are held together by weak chemical bonds. Each strand is made of four different units, the nucleotides, which are joined together to form a long chain. A nucleotide consists of a sugar molecule, a phosphate group, and a nitrogenous base. The differences between the four nucleotides lie in their nitrogenous bases, which can be thymine, adenine, guanine, or cytosine, and the different nucleotides are always known by an abbreviation of the names of these bases—T, A, G, and C, respectively. The weak bonds between the two nucleotide chains are formed because a base in one strand is paired with a base in the other. The association is such that A is always paired with T, and C is always paired with G. There are good chemical reasons why the pairs are always A–T and C–G, but we need not go into these. What is important for us here is that the two strands are complementary: if you know the sequence of nucleotides in one strand, you can work out the sequence in the other. Apart from the fact that the two

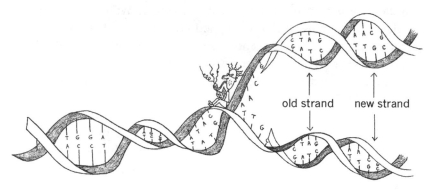

old strand   new strand

Figure 2.1
The structure and replication of DNA.

strands must be complementary, the sequence of nucleotides in DNA is not restricted in any way.

It was clear from the outset what the significance of the nucleotide pairs might be. In a calculated understatement at the end of their famous paper, Watson and Crick wrote, "It has not escaped our notice that the specific pairing we have postulated immediately suggests a possible copying mechanism for the genetic material." It was the complementary relationship between the nucleotides in the two strands that hinted at the nature of the copying process. If the strands separate, each single strand can be a template on which a complementary strand, made up of complementary nucleotides, is formed: a free A is attached to the T in the single strand, a free C is attached to the G in the single strand, and so on. In this way the structure of the original double helix is reconstructed, and two identical daughter molecules are produced from one parent molecule. Of course this replication does not occur spontaneously. There is no "self-replication" of DNA. The replication process requires many enzymes and other proteins to unravel the two parental strands, attach the nucleotides to the single strand, assemble the daughter molecules, and check that mistakes are not made. The ability to replicate is not the property of DNA, but of the cellular system.

From the structure of DNA, Watson and Crick deduced not only how it could be copied but also that "the precise sequence of the bases is the code which carries the genetical information." By this time, the early 1950s, biologists were comfortable with the idea that genes carry coded information that is deciphered in the cell. Work on information transmission and code-breaking had been among the top priorities during the Second World War, and communication technology continued to flourish in the postwar period, influencing the framework of thought in biology, as well as in other spheres. Ideas about the type of information that could be carried by genes stemmed mainly from the biochemical geneticists' work with the bread mold *Neurospora*, which had shown that genes can specify the production of enzymes, which are proteins. Proteins were known to be folded strings of amino acids, of which there were about twenty types. So once the structure of DNA had been revealed, the problem of how genes have their specific influences on the cell was quite rapidly boiled down to how a sequence of four types of DNA nucleotide can encode the specific sequence of the twenty or so types of amino acids in a protein. It took several years to come up with the answer to the problem, but after considering various theoretical possibilities, it was eventually shown experimentally that the nucleotide sequence in DNA is a triplet code. Successive groups of three

DNA nucleotides can be translated into the sequence of amino acids in the polypeptide chain of a protein.

During the 1960s and 1970s, some of the processes that enable the information in a sequence of DNA to be converted into the polypeptide chains of proteins were worked out. Most DNA sits in the nucleus, but most proteins, which are large molecules, often made up of several polypeptide chains, are found in the cytoplasm. It was predicted and shown that the information embodied in the DNA sequence of a gene is carried to the cytoplasm before it is decoded. The carrier molecule is another nucleic acid, ribonucleic acid (RNA). RNA is very similar to DNA in that it too consists of a string of nucleotides, although the sugar in the nucleotides is different. Three of the nucleotides in RNA have the same bases as in DNA, but instead of the thymine-containing nucleotide (T), RNA has a nucleotide containing uracil (U). Like T, U pairs with A. RNA molecules also differ from DNA in being basically single-stranded, and relatively short.

Figure 2.2 shows how a cell uses DNA to make polypeptides. The segment of DNA coding for a polypeptide is first copied into RNA. This process, known as transcription, involves only one of the two DNA strands. The RNA is then modified a bit (we will come to this later), and eventually messenger RNA (mRNA) is transported from the nucleus to the cytoplasm. In the cytoplasm, the mRNA associates with a ribosome—a huge molecular complex made of proteins and another type of RNA, ribosomal RNA (rRNA). Ribosomes enable mRNAs to be translated into polypeptide chains. Each triplet of nucleotides (known as a "codon") in mRNA encodes a specific amino acid or acts as a signal to begin or end the polypeptide chain. For example, UUU and UUC are both codons for the amino acid phenylalanine (Phe), GUU is one of the codons for valine (Val), GAA codes for glutamic acid (Glu), UAA is stop (end of message), and so on.

Translation from the sequence of nucleotides in mRNA to the sequence of amino acids in a polypeptide chain involves a whole battery of enzymes and other molecules, including several types of a smaller RNA—transfer RNA (tRNA). These small molecules act as adaptors, carrying amino acids to the ribosome and adding them to the growing polypeptide chain in the order dictated by the sequence of codons in the mRNA. Each type of tRNA has an attachment site for a specific amino acid at one end, and at the other end it has a recognition site—a triplet of nucleotides that recognizes the mRNA codons for that amino acid because they are more or less complementary.

Through the transcription and translation processes, information in DNA is used to produce the polypeptides of proteins. Some proteins are

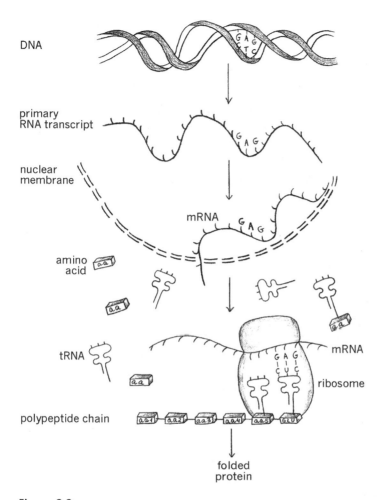

Figure 2.2
The interpretation of information in DNA: transcription within the nucleus (upper part) and translation in the cytoplasm (lower part).

enzymes that participate in chemical reactions, others are the building blocks of the structural elements of cells and tissues, and yet others function as regulatory molecules that affect translation and transcription. DNA therefore codes for many different proteins with many different functions. However, an awful lot of DNA does not code for proteins at all. In fact, most of it does not. It has been estimated that in the human genome at most 1.5 percent of the DNA codes for proteins. A little codes for tRNAs and other nontranslated RNAs, but most of it is never or hardly ever transcribed, let alone translated. So what role does it play?

As we indicated in chapter 1, a lot of DNA has no obvious function, and is commonly regarded as "junk." However, some nontranscribed sequences certainly do have a function—they are involved in the regulation of gene activity. Not every gene is active in every cell all the time. This is why cells can be different, even though most have exactly the same set of genes. Cells are able to respond to internal and external conditions, turning genes on and off when and if required, and nontranscribed DNA is an important part of the regulatory system that determines which coding sequences are being transcribed.

In figure 2.3 we have tried to show in a simple way how one type of gene regulation works. The coding sequence of each of two genes is associated with a regulatory region—a DNA sequence that is not transcribed into RNA. Gene P produces a protein that can bind to the regulatory region of gene Q. When it does so, the enzymes that are necessary for the coding sequence of Q to be transcribed cannot get at the DNA, so gene Q does not produce RNA. However, as the lower part of the figure shows, when the configuration and properties of the gene P protein are altered through its association with a second type of regulatory molecule, it can no longer bind to the regulatory region of gene Q. The transcription enzymes can therefore get at Q, and the protein that it encodes is produced. The second type of regulatory molecule in the system could be the product of another gene, or a direct or indirect result of nutritional or other environmental conditions. In this and many similar ways, environmental factors can participate in the processing of genetic information.

What Is Information?

Before looking at the relationship between the information in DNA and the phenotype in more detail, we need to digress and say something about the term "information," which we have already used a lot and will use a lot more. If someone says that a DNA sequence carries or contains infor-

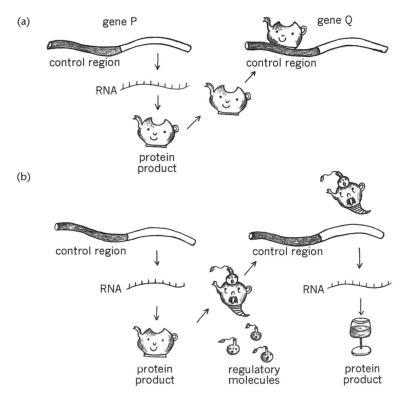

Figure 2.3
The control of gene activity. In (a), the product of gene P (a teapot) binds to the control region of gene Q and prevents transcription. In (b), a regulatory molecule associates with the product of gene P, which changes its shape (a nonfunctional teapot) and makes it unable to bind to gene Q. Gene Q is now transcribed and the mRNA is translated into a protein (a wineglass).

mation, what does it mean? The most obvious answer is that it means that the stretch of DNA embodies in an encoded form the sequence of amino acids for a particular polypeptide chain. But it could also mean that the particular DNA sequence provides the specificity required for a certain type of regulatory molecule to attach itself. These two types of information are very different, so what do we mean when we use this word "information"?

It is surprisingly difficult to find a general definition for "information," yet it is important for us to try to do so, because we are going to be describing and comparing the ways in which different heredity systems transmit information, and how these information transmission systems affect evolution. In everyday language the term *information* is used for a multitude of things. We say that a cloud, something physical, conveys information about the weather; a clock provides information about the time; the smell in a restaurant carries information about the food; newspapers contain information about world events. Biologists would say that a DNA sequence carries information about the sequence of amino acids in a protein, or where a regulator binds; they would also say that a bird's song carries information about its species, and a mother's playful behavior carries information about the world for her child. So what do all these very different "sources of information" have in common? In what sense do all "carry information?"

As we see it (which is from an evolutionary point of view), for something (a source) to contain or carry information, there must first be some kind of receiver that reacts to this source and interprets it. The receiver can be an organism, a cell, or a man-made machine. Through its reaction and interpretation, the receiver's functional state is changed in a way that is related to the form and organization of the source. There is usually nothing intentional about the receiver's reaction and interpretation, although commonly it benefits from it.

This may sound like a rather complicated way of explaining a word whose meaning is obvious anyway, but it is an explanation that is appropriate for all the examples we have mentioned so far and for many others. For example, a train timetable is something that can affect the potential actions of the person who reads it; a recipe for an apple pie can affect the baking activities of a cook; the length of daylight can affect the flowering time of a plant; an alarm call can affect the behavior of the animal that hears it; a DNA sequence can affect the phenotype of the organism. In all cases a receiver can react to the source in a functional way that corresponds to the source's particular form. When the receivers react in such a way, they are interpreting the source's organization. So although they are very different, a train timetable, a recipe, an environmental cue, an alarm call, and a DNA sequence (whether coding or regulatory) are all sources of information.

One of the interesting and important things about information sources is that when a receiver reacts to them and acquires information from them, they usually do not change. For example, the reaction of a human to a recipe, or a cell to a sequence of DNA, or a computer to a piece of software, does not change the recipe, the DNA, or the software. The sources remain exactly the same after the reactions as they were before. A source of information is not like a source of food or materials, which is destroyed as it is used.

Defining information in the general way that we have means that when looking at the heredity systems that transmit information, we can ask questions that help to pinpoint their similarities and differences. How is information in the source organized? How are variations in the source's organization generated? How many variations are possible? In what way does the receiver react to the source? How does the receiver interpret the information? How is information duplicated? By asking such questions, we can detect the special properties of a particular type of information and its transmission. For example, one of the most significant properties of DNA is its linear, modular organization. We can think of a DNA strand as a linear sequence of units or modules (the nucleotides A, T, C, and G) in which each site in the sequence can be occupied by any one of the set of four nucleotides. A nucleotide at a given site can be replaced by any other without it affecting other nucleotides in the strand. This means that a huge number of sequences are possible. How large this number is depends on the length of the sequence, but even when it is not particularly long, the number of different possibilities is awesome. For example, with a sequence of only 100 units, made up of four different modules, 4^{100} different sequences are possible. This is a number that we cannot even imagine—it's more than the number of the atoms in the whole galaxy! And a stretch of 100 nucleotides would be just a tiny fragment of a DNA molecule. When the entire DNA complement of a genome is considered, the number of possible combinations is limitless. Of course, the same can be said about any system of modular units: even with a run of a few hundred, we can construct a vast number of different sequences using just two units, as with the 1 and 0 of computers, and even more with the twenty-six letters that are the basic units of written English.

Another very important, yet very peculiar property of DNA is one that we tend to take for granted. It is that replication is not sensitive to the base sequence that is being replicated. It is much the same as copying with a photocopier, which will reproduce a sonnet of Shakespeare, a page of *Mein Kampf*, and the composition of a chimp at a typewriter with exactly the same fidelity. Other types of copying, such as that which occurs during learning, are different. When we learn something new and try to teach it to someone else, our success in both receiving and transmitting the information depends on what it is about. It is very much easier to teach a child a nursery rhyme of five lines than five lines from the telephone directory. Learning and teaching, which is an obvious way of transmitting information, is sensitive to the form and function of the information, and this sometimes limits what can be learned and transmitted. There is no such limit with DNA replication. The DNA reproduction system is indifferent to the content or function of what is copied and transmitted.

These two characteristics of DNA—the vast number of variations possible because of its modular organization, and the indifference of the replication process to the "content" of the transmitted sequence—mean that potentially it can provide a lot of raw material for natural selection. But these very same characteristics have a downside: they also mean that a lot of nonsensical DNA variations can be generated and transmitted. One of the questions that immediately arises, therefore, is how organisms cope with a potentially vast number of frequently useless or detrimental variations. If the quality of information can be tested only through its functional effect in the next generation, when it is exposed to selection, then it seems to be a terribly wasteful system. In fact, as we shall see later, some of the most ingenious mechanisms in living organisms are direct or indirect solutions to the problem of DNA's potential to vary. In addition, the way in which DNA is actually "interpreted" in the context of a developing organism makes the problem far less formidable than it seems at first glance. To see why, we need to return to the main theme of this chapter, which is the relation between genes and characters.

Genes, Characters, and Genetic Astrology

In terms of what we know today, the outline that we gave earlier of the way in which DNA is transcribed and translated into proteins and how these processes are regulated is much too simple, and we shall have to fill in some of the details later. But for the time being we can use this simple version to start looking at the relation between genes and characters. If we ask how changes in genes alter the way an individual develops, our outline points to one fairly simple answer: a change in a gene's DNA sequence (a mutation) leads to a change in mRNA; this leads to a change in the polypeptide chain of a protein, which in turn causes a change in the visible phenotype.

This wonderfully simple answer has been found to be correct for several "monogenic" diseases—diseases for which the presence or absence of symptoms depends on which alleles of one particular gene are present. Some of these diseases are dominant (only one allele has to be "defective" for the disorder to show itself), others are recessive (both alleles have to be defective for the person to be sick). The classic example of a recessive disorder that results from a simple difference in DNA is sickle cell anemia. This debilitating disease gets its name from the distorted, sickle shape of some of the sufferer's red blood cells. The cause of both the sickling and the sickness is one "wrong" nucleotide in the sequence of the gene coding for one of the two types of polypeptide in hemoglobin, the protein that enables the red blood cells to carry oxygen around the body. This tiny DNA difference—an A instead of a T—leads to just one of the 146 amino acids in the polypeptide chain being different. Nevertheless, the change is enough to make the hemoglobin molecule less capable of carrying oxygen. If both of a person's genes have the mutation, all of their hemoglobin is defective, and that person develops severe anemia, with all its consequences.

We've shown the chain of events leading from the changed DNA sequence to the anemia in figure 2.4. Sickle cell anemia is not a unique example: Tay-Sachs disease and cystic fibrosis are two other monogenic diseases that are beginning to be understood at the molecular level.

In the case of simple monogenic disorders like sickle cell anemia, people with the defective genes always have the symptoms, whatever their

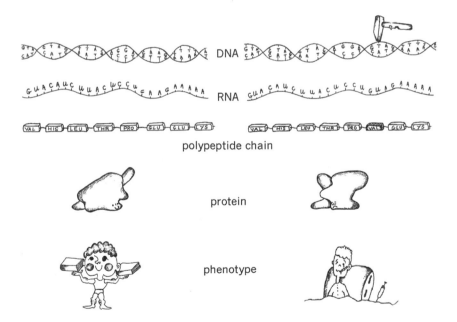

polypeptide chain

protein

phenotype

Figure 2.4
From DNA to phenotype. On the left, normal DNA and the chain of events leading to the phenotype. On the right, sickle cell anemia, in which a changed nucleotide (indicated by the hammer) leads to an abnormal protein which causes sickness.

conditions of life and whatever other genes they have. However, such simple monogenic diseases are not common: they make up less than 2 percent of all the diseases that are known to have a genetic component. For the remaining 98 percent of "genetic" disorders, the presence or absence of the disease and its severity are influenced by many genes and by the conditions in which a person develops and lives. Unfortunately, many people's understanding of the relation between genes and characters is based on the tiny minority of monogenic diseases. The popular view is that genes discretely and directly determine what a person looks like and how they behave. We have genes for this and that (our eye color, our nose shape, how shy we are, our intelligence, our sexual orientation, etc.), and the person you see is largely the sum of the effects of his or her genes plus a little social-educational gloss. The individual is seen as little more than a sophisticated robot, driven by his or her genes.

Obviously, such a conception of how genes act fuels the belief that biotechnology will give geneticists enormous powers. People believe (and are encouraged by some scientists to believe) that in the not too distant future geneticists will be able to find out all about them simply by sequencing their DNA. Not only will the geneticists be able to read and translate the person's "book of life," they will even be able to edit out the mistakes if necessary. Harvey F. Lodish, a leading cell biologist, professor of biology, and member of the Whitehead Institute for Biomedical Research, presented the following scenario when the journal *Science* asked him what he saw in the future for science:

By using techniques involving in vitro fertilization, it is already possible to remove one cell from the developing embryo and characterize any desired region of DNA. Genetic screening of embryos, before implantation, may soon become routine. It will be possible, by sequencing important regions of the mother's DNA, to infer important properties of the egg from which the person develops. This assumes that predictions of protein structure and function will be accurate enough so that one can deduce, automatically, the relevant properties of many important proteins, as well as the regulation of their expression (for example, how much will be made at a particular stage in development in a particular tissue or cell type) from the sequence of genomic DNA alone. All of this information will be transferred to a supercomputer, together with information about the environment—including likely nutrition, environmental toxins, sunlight, and so forth. The output will be a color movie in which the embryo develops into a fetus, is born, and then grows into an adult, explicitly depicting body size and shape and hair, skin, and eye color. Eventually the DNA sequence base will be expanded to cover genes important for traits such as speech and musical ability; the mother will be able to hear the embryo—as an adult—speak or sing. (Lodish, 1995, *Science* 267: 1609)

This scenario, offered in 1995, fits the general public's conception of the genetics of the future. There was a surge of predictions similar to this at the turn of the millennium, when people were thinking about the scientific advances likely in the twenty-first century.

The belief that a person's character is "written in the genes" was one of the reasons for the hysterical public reaction when cloning produced Dolly the sheep. That little lamb conjured up a strange mixture of feelings, because on the one hand it seemed to offer the hope of personal immortality, yet on the other it looked as if our unique individual identity might be at risk. Both notions stem from the belief that the causal relation between genes and traits is simple and predictable—that identical sets of genes will inevitably produce identical phenotypes. Such beliefs are very mistaken, however, and potentially harmful.

We cannot guarantee that in the future there will not be institutes of genetics that pretend to read an embryo's future (and hear it sing) largely from its DNA. If the demand is there, people willing to establish such institutes will certainly be found. However, few professional geneticists (at least in their more lucid moments) believe in such genetic astrology. This is so in spite of the incessant media claims that the gene for homosexuality, adventurousness, shyness, religiosity, or some other mental or spiritual trait has been isolated. Geneticists are usually much more cautious about their work. If you look at the actual scientific papers rather than the newspaper stories about these wonderful genes, you find that what has been discovered is a correlation between the presence of a particular DNA sequence and the presence of the character. Usually it is not at all clear that the DNA sequence is *causally* related to the character, and it is almost always very clear that "the gene" is neither a sufficient nor a necessary condition for the character's development.

Let's take a closer look at one of these traits. Not long ago, an amazed public was informed by the media that the gene for "adventurousness" or, as the scientists preferred to call it, "novelty seeking," had been isolated. A person's decision to do something exciting like becoming a fighter pilot or a revolutionary, or alternatively to be an orderly and conscientious librarian or accountant, is, the journalists told us, determined to a large extent by which alleles of one particular gene they have. However, if we turn to the original scientific papers, we find that the power of this gene is rather less than was proclaimed by the popular media. We discover that some people who have the allele that is correlated with adventurousness are in fact very cautious and conventional, whereas some of those who lack it are nevertheless impulsive, thrill-seeking risk takers. All that can be said is that those who have the allele have a somewhat greater chance of being adventurous. In fact, only 4 percent of the difference among people with respect to their adventurous behavior can be attributed to the particular gene that was investigated; 96 percent of the difference is unexplained by the purported "novelty-seeking" allele. Even the 4 percent that got so much media attention is somewhat problematical, because it is not always easy to classify a person as adventurous or not adventurous. People can be adventurous in some aspects of life, but very conventional in others. Moreover, in their analysis, the researchers apparently did not take into account birth order, a factor which others have found to be a major influence on the development of adventurousness. Children who are born second, third, or later in the family are more adventurous than first-born or only children are. Clearly, this has nothing to do with inheriting a particular allele—it would be a gross violation of Mendelian laws if an allele was more common in first-borns than in other children.

Studying human genetics is not easy, because investigators cannot direct who people should marry and how they should live. There are always many uncontrollable factors that could be influencing what they discover. Even when a study shows that there is a correlation between the presence of a particular allele and some aspect of human behavior, we have to be very cautious about accepting that the relationship is *causal*. For example, we need to know whether the observed association is found in all conditions and in all populations, or only in the particular sample studied by the scientist. One of the reasons why many of the much heralded discoveries of "genes for" various things have ended in embarrassed silence is that when people started following up the original discovery, they found that the correlation did not exist in other populations. It is very rare for the association between a gene and a trait to be simple.

The Tangled Web of Interactions

Earlier we said that for the overwhelming majority of diseases with a hereditary component, more than one gene is involved. Obviously, with these diseases, the causal relationship between differences in genes and differences in traits is more complicated than with monogenic disorders. Yet it is still tempting to assume that they can be explained in essentially the same way. This temptation should be resisted, however, because a very different type of explanation is required. To see why, we are going to make use of an example that the American geneticist Alan Templeton has used to illustrate the complexity of the problem. The disease that Templeton considered is a common one, coronary artery disease, and the gene is *APOE*. Templeton based his discussion on a large survey of a particular American population, so it may not be typical for all groups, but it illustrates the complexity of the relation between genes and traits very nicely.

The *APOE* gene codes for a protein (apoprotein E, or apoE) that helps to carry fats around in the blood. It has three common alleles, which we can call allele 2, allele 3, and allele 4. There is one amino acid difference between the protein variants encoded by alleles 2 and 3, and one difference between alleles 3 and 4. Although these protein differences are small, population surveys have shown that the three alleles are associated with differences in the incidence of coronary artery disease. Comparing people of the three most common genotypes, 2/3, 3/3, and 4/3 (notice that all have one copy of allele 3) shows that people with genotype 3/3 have a below-average chance of developing the disease, those with genotype 2/3 are average, but people with genotype 3/4 are twice as likely as the average person is to suffer from coronary heart disease. So, it looks as if allele 4 is the defective "bad" allele.

Now let us look at something else—at cholesterol, which, as we have all been told many times, can have dramatic effects on our chances of having heart problems. People with a high level of blood cholesterol are much more likely to get coronary artery disease than those with low levels. You might guess from this and the fact that the apoE molecule helps to transport things like cholesterol around in the blood, that the *APOE* gene probably affects cholesterol levels. Sure enough, the population surveys have shown that people with allele 4 do have, on average, high cholesterol. From this it is tempting to conclude that allele 4 causes high serum cholesterol, which in turn leads to an above-average chance of getting coronary artery disease. But it is not that simple!

When both the *APOE* genotypes and the serum cholesterol levels are considered together, the picture becomes very complicated. The first thing you realize is that not all individuals with high cholesterol have an above-average chance of getting coronary artery disease. People who have high cholesterol but are homozygous for allele 3 are no more at risk than the average member of the population; they are at less risk than people with medium or low cholesterol levels who carry allele 4. So high cholesterol is neither a sufficient nor a necessary cause of coronary artery disease. The second thing you realize is that the combination of allele 4 (the "bad" allele) and high cholesterol (which is also bad) is not the worst thing possible! It is people with allele 2 and high cholesterol who have the greatest chance of developing the disease. It is only when you have average or low levels of cholesterol that being a carrier of "bad" allele 4 puts you at a greater risk than others with the same cholesterol level. Allele 4 cannot therefore be thought of simply as a "defective" allele that increases the risk of coronary artery disease.

This probably all sounds very confusing, but it is still only the beginning of the story. As we all know, cholesterol levels are altered by diet, by taking exercise, and by drugs, so these factors also have to be taken into account when we are thinking about the *APOE* gene and coronary artery disease. Moreover, the *APOE* gene is one of more than a hundred that can affect the development of the disease, and several of these genes have alleles whose influence depends on the lifestyle and environment in which the individual develops. For each gene and each allele, the other genes and alleles present (what in the jargon of genetics is known as the "genetic background") and the conditions of life determine the way in which it will affect the development of a trait.

It should by now be obvious why understanding the genetics of this type of disease is so difficult, and why, even if it was thought to be desirable, for most conditions genetically engineering "good" genes into people is not a realistic possibility. Coronary artery disease also illustrates one reason why genetic astrology is unrealistic. Genome sequencing may tell us about our DNA, and it may even tell us about our genes, but the interrelationships among those genes and the environment are so complex that we cannot just add their average effects together and from this predict what a person's strengths and weaknesses will be. The same gene does not always lead to the same phenotype. As biologists have known for a long time, all multicellular organisms, including human beings, have a lot of developmental plasticity: their phenotype depends on a multitude of environmental factors, as well as their DNA.

There is something else that complicates the ambition to predict our phenotype from our DNA. It is that you can sometimes have two or more networks of interactions, with different components, that end up producing identical phenotypes. Certain aspects of the phenotype seem to be remarkably invariant, in spite of genetic and environmental differences. So on the one hand you can have identical genes leading to very different phenotypes, and on the other you can have dissimilar genes producing exactly the same phenotype.

These are not new discoveries. Many years ago, before anything much was known about the intricate ways in which genes are regulated and interact, and long before the concept of genetic networks became fashionable, geneticists realized that the development of any character depends on a web of interactions between genes, their products, and the environment. A visual representation of this idea, which is still relevant and helpful, was developed by the British embryologist and geneticist Conrad Waddington in the 1940s and 1950s. Waddington depicted developmental processes as a complex landscape of hills and branching valleys, descending from a high plateau. In this "epigenetic landscape" (as he called it), the plateau represents the initial state of the fertilized egg, and the valleys are developmental pathways leading to particular end states, such as a functioning eye, a brain, or a heart. A small part of an epigenetic landscape is shown in the upper part of figure 2.5.

The lower part of the figure is Waddington's depiction of the processes and interactions that, quite literally, underlie his landscape. It is a sort of x-ray view through the scene, which shows how the landscape is shaped by the tensions on a network of guy ropes attached to its undersurface. The guy ropes represent the products of genes, and the genes are the pegs in the ground. So if you take a valley such as that on the extreme left, and think of it as representing the development of a normal, functioning heart, it is clear that development depends on the interaction of many genes (the pegs) and their products (the guy ropes). Some valleys are deep and steepsided, so the characters vary little, whereas other valleys are broader and flat bottomed, so the end products are more variable. Changes in genes (pegs), or conditions that alter the interactions of the gene products (guy ropes and their connections), can affect the shape of the epigenetic landscape and the final phenotype. Waddington's visualization of the networks underlying development suggests, however, that often the effects are very indirect.

A vivid recent illustration of the intricacy and sophistication of genetic networks became apparent when geneticists started using genetic engineering techniques to "knock out" (disable) a particular gene and follow the consequences of this knockout on development. Much to their surprise, the scientists found that knocking out genes that were known to participate in important developmental pathways often made no difference whatsoever—the final phenotype remained the same. Somehow, the genome can compensate for the absence of a gene. There are several reasons why it may often be able to do this: first, many genes have duplicate copies, so when both alleles of one copy are knocked out, the reserve copy compensates; second, genes that normally have other functions can take the place of a gene that has been knocked out; and third, the dynamic regulatory structure of the network is such that knocking out single components is

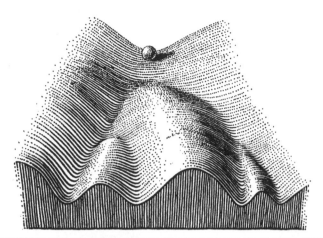

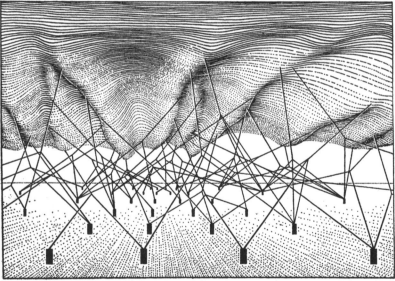

Figure 2.5
Waddington's pictures of epigenetic landscapes. His legend for the upper picture began "Part of an epigenetic landscape. The path followed by the ball, as it rolls down towards the spectator, corresponds to the developmental history of a particular part of the egg." The lower picture had the legend "The complex system of interactions underlying the epigenetic landscape. The pegs in the ground represent genes; the strings leading from them the chemical tendencies which the genes produce. The modeling of the epigenetic landscape, which slopes down from above one's head towards the distance, is controlled by the pull of these numerous guy-ropes which are ultimately anchored to the genes." (Reproduced with permission from C.H. Waddington, *The Strategy of the Genes*, London, Allen and Unwin, 1957, pp. 29, 36.)

not felt. The developmental end product remains intact, at least in most environments.

Knockout experiments show that there is a lot of structural and functional redundancy in the genome, and that the pathways of development are so strongly channeled ("canalized," in the geneticists' jargon) that many differences in genes make very little difference to the phenotype. It was Waddington who coined the term "canalization" to describe this type of dynamic developmental buffering, and it is not difficult to understand it in terms of his epigenetic landscape. If one peg (gene) is knocked out, processes that adjust the tension on the guy ropes from other pegs could leave the landscape essentially unchanged, and the character quite normal. We mentioned in the previous chapter that during the 1960s people were surprised to discover that some genes have many different alleles, most of which make little difference to the organisms. They were selectively neutral, having no detectable effect on average survival or reproductive success, at least in the conditions in which they were studied. Now that we know more about the complexity of the molecular events that go on between the gene and phenotype, we are less surprised to find neutrality of this type. If knocking out a gene completely often has no detectable effect, there is no reason why changing a nucleotide here and there should necessarily make a difference. The evolved network of interactions that underlies the development and maintenance of every character is able to accommodate or compensate for many genetic variations. That is why so many of the potentially deleterious effects of the huge number of variations in the information in DNA are masked and neutralized.

Genes in Pieces

By this time it may seem that the relationship between genes and traits cannot get much more complicated, but there is yet another complicating factor. So far we have assumed that each gene has one polypeptide product, so that at least at this level things are more or less predictable. This is what classical Mendelian genetics led us to expect, and indeed this is how it seemed to be in the early days of molecular biology, when most of the results were coming from research with bacteria. However, in the late 1970s, much to everyone's surprise, it was discovered that the relation between genes and proteins is usually not so simple. In eukaryotes—those organisms whose cells have a nucleus, which includes all plants, animals, fungi, and many single-cell organisms—it is not a straightforward matter of a continuous sequence of DNA nucleotides coding for the sequence of amino acids in a polypeptide. Rather, the DNA sequence coding for a polypeptide is often a mosaic of translated and nontranslated regions. The translated regions, known as "exons," are interrupted by nontranslated ones, the "introns." What happens is that the whole DNA sequence is transcribed into RNA, but before this RNA arrives at the ribosomes, it undergoes a process called splicing. Large protein-RNA complexes— "spliceosomes"—excise the introns from the primary RNA transcript, and join together the remaining exons. It is this processed mRNA that is translated into polypeptides.

That is not the end of the story, however. Splicing is sometimes even more elaborate, because the exon or intron status of a sequence often is not fixed. In at least 40 percent of the RNA transcripts of human genes, different bits can be spliced together. This means that one DNA sequence can give rise to many mRNAs and protein products. We've shown the general idea of alternative splicing in figure 2.6. The "decision" over which polypeptide will be formed depends on developmental and environmental conditions, as well as other genes in the genome.

A particularly extreme example of alternative splicing is seen with a gene called *cSlo*. (*cSlo* is short for *chicken Slowpoke*, because it is a chicken version of the *Slowpoke* gene found in fruit flies, but this is not important here.) The *cSlo* gene, which is active in the hair cells of the chicken inner ear, has

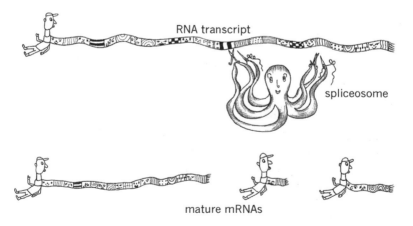

Figure 2.6
Alternative splicing: the spliceosome (represented by an octopus) can splice (cut and sew) together different segments of the original RNA transcript to produce several different mRNAs.

576 alternative splice variants. They code for a protein that has a role in determining the sound frequency to which inner ear cells respond, and the variations in the protein sequence parallel variations in the frequencies to which different cells respond. It seems that having so many versions of the protein enables the chicken to tune its cells and distinguish between the sounds it hears. Geneticists think that homologous genes in the mouse and humans have an even larger repertoire of splice variants. How splicing is regulated—how each cell decides which segments of the primary RNA transcript are to be included in the mRNA that will be translated—is not yet understood.

Although there are still many gaps in our knowledge, molecular biologists have found out an amazing amount about DNA and how the information in it gets used. However, they have also created an unanticipated problem. If we think about regulatory sequences, about alternative splicing, and about other ways (yes, there are even more) in which a single stretch of DNA can produce several different products, we begin to wonder what a "gene" really is. Evelyn Fox Keller, an American philosopher and sociologist of biology, has discussed the problem in her book *The Century of the Gene*. She wrote:

> . . . the gene has lost a good deal of both its specificity and its agency. Which protein should a gene make, and under what circumstances? And how does it choose? In fact, it doesn't. Responsibility for this decision lies elsewhere, in the complex regulatory dynamics of the cell as a whole. It is from these regulatory dynamics, and not from the gene itself, that the signal (or signals) determining the specific pattern in which the final transcript is to be formed actually comes. (Keller, 2000, p. 63)

Clearly, the relationship between genes and visible traits is very different from the way in which it is usually presented to the public. The idea that a gene is a sequence of DNA that codes for a product, and variations in the DNA sequence can cause a difference in the product and hence in the phenotype, is just too simplistic. Coding sequences are only a small part of DNA, and DNA is just a part of the cellular network that determines which products are produced. When and where these products are produced depends on what goes on in other cells and what the environmental conditions are like. Cellular and developmental networks are so complicated that there is really no chance of predicting what a person will be like merely by looking at their DNA. Although it has considerable rhetorical and marketing power, the dream of genetic astrology is just that—a dream.

Changing DNA during Development

We must now introduce one final complication into the story of how information in DNA is expressed in the characters of the individual. It comes as a surprise to many people to discover that not only do cells have a powerful kit of enzymes that can chop and change the RNA that is transcribed from DNA, they also have enzymes that can cut, splice, and generally mess around with DNA itself. Changes in DNA are a part of the normal development of many animals, and through this natural genetic engineering their cells can come to have nonidentical genetic information.

One of the most spectacular examples of developmental changes in DNA is found in our immune system. During the maturation of lymphocytes (the white blood cells that produce the antibodies needed to fight infection and destroy foreign cells), DNA sequences in the antibody genes are moved from one place to another, and are cut, joined, and altered in various ways to produce new DNA sequences. Because there are so many different ways of joining and altering the bits of DNA, vast numbers of different sequences, each coding for a different antibody, are generated. Consequently, the DNA of one lymphocyte is different from that of most other lymphocytes, as well as from that of other cells in the body.

The way that DNA is reorganized in the cells of the vertebrate immune system is remarkable, but it should not be thought of as an isolated peculiarity. Developmental changes in DNA have been found in many other organisms, although not all are quite as spectacular. Some were discovered a long time ago. In the late nineteenth century, when August Weismann and other biologists were trying to unravel what happens during cell division, one of their favorite sources of material was *Ascaris*, a parasitic worm inhabiting the intestine of horses. It is wonderful material to study, because the chromosomes are very few and very large. Unfortunately, although this was not realized at the time, *Ascaris* is also rather unusual, because during early development something known as "chromatin diminution" occurs. We have depicted this in figure 2.7e. Large pieces of chromosome are eliminated from the cell lineages that are going to form the body cells; only the germ line retains unchanged chromosomes. This strange phenomenon is relevant to something that we mentioned in chapter 1, where we described Weismann's ideas about heredity and development. Weismann thought (incorrectly) that during development the hereditary material in the nucleus gets simpler and simpler, both because nuclear division is unequal and because material passes from the nucleus to the cytoplasm in order to direct cellular activities. So when chromatin diminution was discovered in *Ascaris*, he was cautiously delighted, because it was just the kind of evidence that he needed to support his views about both the continuity of the germ plasm and the way in which chromosomes control development.

Ascaris is undoubtedly odd—most animals do not go in for massive chromatin diminution during the formation of their body cells. Yet, when you start looking closely at the chromosomes and DNA of other animals, you discover that the behavior of a surprising number of them is "odd," although their oddity takes different forms. Take the genetic workhorse, the fruit fly *Drosophila*. Do all its cells have the same DNA? Not at all. Some cells are polyploid—during development, the chromosomes replicate, but the cells do not divide, so they contains four, eight, or even sixteen copies of each chromosome, instead of the usual two. This is shown in figure 2.7a. In 2.7b we have shown another type of chromosomal oddity found in *Drosophila*. One of the things that make it such a useful animal for geneticists is that it has several cell types with polytene chromosomes. Polytene chromosomes are formed when DNA replicates many times, and the replicated DNA remains associated so that the chromosomes become

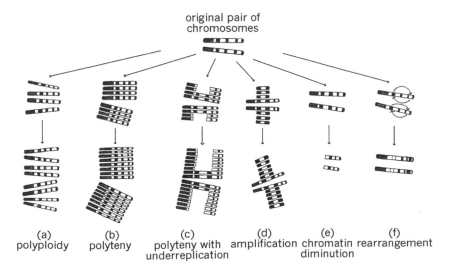

Figure 2.7
Regulated DNA changes: the whole or part of a normal chromosome can be deleted, amplified, or rearranged. (The banding represents that seen when chromosomes are stained with dyes that distinguish between regions with different types of DNA composition.)

multistranded. In the figure we have shown an eight-stranded chromosome. Some polytene chromosomes in *Drosophila* are enormous, with over a thousand strands lined up in parallel, and they can be used to find out where specific genes are located. What is interesting is that not the whole of each chromosome is replicated to the same extent, as shown in figure 2.7c and d. In the larval salivary gland cells, where polyteny is most extreme, heterochromatin is replicated very little. Heterochromatin means "differently colored" chromosome material, because that's how it appears under the microscope, but we now know that heterochromatic regions often contain large blocks of simple noncoding DNA. They do in *Drosophila*, so it's not surprising that in the very active salivary gland cells it is heterochromatin that is replicated least. Other *Drosophila* cells have chromosome regions that are overreplicated, rather than underreplicated. In the follicle cells surrounding the developing eggs in the female, the sites containing the eggshell genes are present in more copies than other regions. The selective amplification of these very active genes obviously makes sense.

It would be easy to go on and describe many other odd types of chromosomal or DNA changes, such as the elimination of half of the chromosomes that occurs in some insects, or the amplification of ribosomal genes that occurs during the development of some amphibian eggs, but we have probably already said enough to make the point that cells are able to make controlled changes in their DNA. These changes are part of normal development, and, like any other developmental process, they are regulated by the cellular environment. Their existence has implications for how we think about the relationship between DNA and the phenotype. If we go back to the Dawkins metaphor that we discussed in chapter 1, which views DNA as "the recipe" and the body as "the cake," we see that the "recipe" can be altered by changes in the body: development (baking) can change the recipe! However, in all of our examples, the regulated changes in DNA occur in somatic cells, not in the germ line, so they do not affect the next generation. The "recipe" in the germ line is not altered. Yet, as the American geneticist James Shapiro has been arguing for some years, the very existence of a cellular machinery that changes DNA in a regulated way should alert us to the possibility that similar processes may occur in germ-line cells and play a part in evolutionary change. He wrote:

These molecular insights [Shapiro is referring to the controlled regulation of DNA alterations] lead to new concepts of how genomes are organized and reorganized, opening a range of possibilities for thinking about evolution. Rather than being restricted to contemplating a slow process depending on random (i.e., blind) genetic variation and gradual phenotypic change, we are now free to think in realistic molecular ways about rapid genome restructuring guided by biological feedback networks. (Shapiro, 1999, p. 32)

In the next chapter we shall look more closely at the claim that the cells' genetic engineering enzymes are involved in the formation of the genetic variation that is passed to the next generation, and therefore have a role in producing the variation on which natural selection acts.

Dialogue

I.M.: You described the supposedly special properties of the genetic inheritance system: that it is modular, that copying is indifferent to content, that information is encoded, and that it is a system that allows unlimited heritable variation. I must say that to me these properties do not seem particularly special. They fit the information transmission system that we are using right now—the linguistic system, which is certainly very different. If two such different systems are similar in these respects, why are you making such a fuss about these properties? How do they help us understand the uniqueness of the DNA system?

M.E.: The properties you mentioned are indeed common to the genetic system and the linguistic system, but this doesn't mean they are trivial or self-evident. Not all transmission systems have these properties. For example, take the case of a mother who transfers molecules of food substances to her offspring through her milk, thereby making their future food preferences similar to hers. The food molecules are not part of an arbitrary combinatorial code, and the information is not organized in a linear sequence of modules. In later chapters we will be dealing with a lot of other cases of information transmission that do not share the properties you listed. You will find that characterizing the units of variation in the way that we did really helps one to think about heredity in different systems. The fact that there are some striking similarities between the genetic and the linguistic systems is important, and we will argue later that it tells us something about the special role that these two systems have had in evolution.

I.M.: OK, I will wait and see. I also have to say that I was not very impressed with your scorn for "genetic astrology." You convinced me that we are further away from the scenario with the singing embryo than I thought, but certainly not that this scenario is impossible. Take your example of coronary artery disease. The population survey *did* show that on average those having allele 4 are significantly more likely to develop the disease than those with other alleles. This means that people with allele 4 can be advised to be more careful. Surely that is an important first step?

M.E.: This type of advice could be dangerous. In general, if you are advising or treating individuals according to the *average* effects of a gene, you may be doing the wrong thing. Remember, that for people with high cholesterol, allele 2 was associated with the disease more than the "bad" allele 4. In this case, people with "good" allele 2 are at greater risk. By looking at the interactions, we can often predict risks much better than by using just the *APOE* genotype or cholesterol levels alone. We can give people better advice. The problem is that in the overwhelming majority of cases where genes are involved in a disease there are probably a lot of interactions, about which we know very little, and there is probably no absolute "good" or "bad" allele. Because of the vast number of interactions, an allele that is good in one genetic or environmental context can often be bad in another. The *average* effect of an allele does not help you to predict what the effect of this allele will be in any individual case. Don't forget that the boy drowned in the pond that the statistician told him was on average only 20 cm deep. Averages mask individual variation—that's the whole point of an average. We have already said that a lot of alleles are *on average* selectively neutral—*on average* it makes no difference which allele an individual has. But it can make a very big difference in some conditions, with positive effects in one environment and negative in another. Yet, *on average*, the effects will cancel out. This does not mean that the allele has the same effects in all conditions.

I.M.: OK, so it will be a long time before scientists understand the complex interactions among the different genes and the environment, and their effect on heart disease in an individual. But eventually this will happen. The computer in your futuristic genetics institute will not base its predictions on the average effects of genes and some crude environmental factors, but on an analysis of the interaction of genes in several well-defined environments. Genetic counseling in the future will be based on a lot more knowledge, including knowledge of the complex genetic networks that you described. Eventually, on the basis of genetic information, it will be possible to make fairly precise predictions about how various traits, including mental ones, are likely to develop. It seems to me intellectual cowardice not to recognize that all traits, including the mental ones, have a firm genetic basis.

M.E.: There are no traits that do not have a genetic basis. The question is whether variation in genes causes variation in traits, and, if it does, how and in what conditions. For some traits scientists will certainly be able to make gene-based predictions, especially in cases where variations in a small number of genes have a large effect on the development of the trait, and the conditions of development are not too diverse. This will be possible for some diseases, and it will be wonderful. If it is possible to predict the risks a person faces on the basis of their genes and the environments they are likely to encounter, it will be a real blessing. But for traits that are influenced by many genes, each of which has small effects, and where development occurs in a complex environment (as it does with respect to behavioral traits), this is very unlikely. The number of possibilities is simply too great. The interpretation of genetic information depends on too many factors. If you think about a trait in which a very modest number of genes are involved, just twenty genes each having two alleles, then we are talking about more than a million genetic combinations, more than a million genotypes. And this is the number of possibilities that have to be considered even without combining the genotypes with different environments! With complex traits like mental ones, tens or hundreds of genes, some with many alleles, are involved in the construction of the trait, and we do not even know how to define "the environment." There are so many social and psychological environments that may be relevant to development—almost as many as the number of people. And these environments are partially constructed by the behavior of the individuals themselves!

I.M.: I am not sure that one cannot develop methods to deal with the complex cases. And you just told me that, although there is a lot of variation, even when you knock out genes you often get a normal phenotype, because there is all this redundancy and compensation in the system. You can't have your cake and eat it! You can't claim on the one hand that there are an infinite number of possibilities, each different from the other, and on the other hand tell me that they are all equivalent and lead to essentially the same phenotype. If what you say is true, it means that the vast majority of genetic and environmental combinations end up the same. Maybe when we understand the developmental networks better, we shall be able to develop methods to predict when we are going to have a deviation from the norm, and in what direction.

M.E.:   You are right. Maybe we shall be able to develop such methods. But you see, in suggesting these methods you are already recognizing that they have to be based on a different way of thinking about the causal relation between genes and traits. The thinking has to be in term of a network, its dynamics, and its built-in compensatory mechanisms. You have to think about the system that interprets information. But even if from the million or so combinations of genes we were just talking about only 1 percent has visible effects on the phenotype, you are still talking about ten thousand combinations. And worse than that, this ten thousand is not a constant—it will be a different ten thousand in a different environment. A lot of geno-types that result in a normal phenotype in normal conditions will show variation under abnormal, stressful conditions.

I.M.: But there must be situations where a small change in an important regulatory gene will affect the expression of many genes. In such a case we could "read" and predict the change in the trait from the small genetic change.

M.E.: This can happen when the genetic system is constructed in a rigid hierarchical manner: when there are regulatory genes with a large effect on the activity of other genes, which then have somewhat smaller effects on yet other genes, and so on. In some cases the system is indeed built in such a way, but this is certainly not the general case. In most cases you find networks that are very much more flexible and fuzzy. Functions are more distributed. In the nonhierarchical cases you cannot read the change in the trait from the change in the DNA sequence of a single regulatory gene.

I.M.: So are you claiming that we shall remain ignorant forever? History shows that such claims have always been proved wrong.

M.E.: No, we shall not remain ignorant, but we believe that the road to understanding does not lie exclusively in the provinces of the analysis of DNA or even proteins. The idea that all we need to do now that we have the DNA sequence is to work out the protein products—the translated bits of the genetic puzzle—and from them predict and remedy illnesses is plainly naive. The suggestion that a knowledge of DNA sequences can solve our health problems and our social problems is, at the global level, ridicu-lous, if not cynical and unscrupulous. As the American geneticist Richard Lewontin keeps reminding us, the DNA doctrine is a politically charged doctrine. If we want to solve 95 percent of the health problems in the world, what we need to do is give people enough to eat, and make sure they can drink clean water and breathe clean air. But even if we are con-cerned with just the well-fed few who may benefit from treatments based on sequenced DNA, we are still being misled. The much-hyped vision of everyone in the future carrying a small magnetic card containing their own DNA sequence, and of scientists using this information to analyze their individual qualities, find out their genetic weaknesses, and solve many of their health problems is unrealistic.

I.M.: But the card envisioned will become real plastic, and even if its use-fulness is limited, it is not something you can ignore. But tell me, how useful has molecular genetics been in medicine so far? We hear a lot about it, but it's difficult to know how much has really been achieved.

M.E.: When in 1998 David Weatherall, director of the Institute of Molecular Medicine at the University of Oxford, posed himself the question, "How much has genetics helped?," his answer was, in effect, "not much yet." The same is still true. However, like Weatherall, we must qualify this answer a little, because molecular genetics has been crucial for detecting some genetic defects before birth. For many of the 2 percent of diseases that are monogenic, you can now test an embryo's cells or tissues and tell the parents whether their child will later develop the disease. This is enormously important. For example, at one time nearly 1 percent of the newborn in some populations had thalassemia, a type of severe anemia. It affected thousands of children around the world. But as a result of genetic counseling, and of prenatal diagnosis and selective abortion, in many populations the number of hopelessly sick children has declined dramatically. So there have been successes. However, for the other 98 percent of genetically influenced diseases, neither prediction nor treatment is as yet possible. The claims we hear so often about the discovery of "the gene" for this or that condition are either naive, confused, or simply misleading. The promise that diseases will be cured by implanting normal genes into genetically sick individuals—what is called "gene therapy"—has also not yet been fulfilled, although there are glimmers of hope for one or two monogenic diseases. Of course, we must be fair—it takes a long time for a new technology to yield real fruits. When the developmental system is better understood, there may be more successes. Nevertheless, it seems to us that the present, almost exclusive, focus on genes is unfortunate and misguided. We should also consider nongenetic transmissible factors. It is known already that heritable nongenetic variations in physiology and morphology can influence these complex diseases, and we shall say more about this in chapter 4. Genes are just part of the system that we need to understand in order to fight disease. Sequencing genes is not a magic key to health.

I.M.: That's a pity. All the same, molecular genetics is providing important investigative tools—you do accept that. I am still wondering what identifying all these genes will tell us. You said that the current estimate is that there are about 25,000 genes in humans, and that's not a vast number. Geneticists should be happy that there are not that many genes to study and manipulate.

M.E.: Well, the number of genes keeps being revised, but the relatively low number was a bit of a shock for the genetics community. People were uneasy. They expected about four times as many.

I.M.: Why? The number of possible interactions among 25,000 genes, many with several protein products, is astronomical. So why be uneasy? If there are fewer genes than estimated, but lots of proteins and lots of interactions, what is the problem?

M.E.: One of the problems is that there is a nematode, a small worm with less than a thousand cells, that has about nineteen to twenty thousand genes, and the fruit fly *Drosophila* has about thirteen to fourteen thousand. That is not so many less than we have. Phenotypic complexity is obviously not related to the number of coding genes, or even to the number of combinations.

I.M.: So what? Why should it be? I would have thought that the idea that humans were more complex simply because they have more genes was acknowledged to be very naive. After all, I keep hearing that we are 99 percent genetically identical to the chimpanzee, so clearly it is not the quantity of genes that matters. I would guess that a lot of evolution does not proceed by increase in gene quantity. But I am afraid that I have lost your thread with respect to evolution. It's very interesting to learn about the wonders of modern genetics and how terribly complicated everything is, but exactly how does it all relate to evolution?

M.E.: The complex relation between genes and the development of traits is relevant to evolution in two ways. First, it channels and limits how we should think about evolutionary mechanisms. If an adaptation involves many genes, we have to think about how the activity of the whole network of interactions can evolve through the selection of mutations, which are usually assumed to be chance events happening in single genes. Inevitably this means that we have to think in terms of development and regulation, and about selection for the developmental, physiological, and behavioral stability and flexibility of genetic and cellular networks.

I.M.: Before you make your second point, I need to get this one straight. If I understood you correctly, you said that no serious geneticist really believes in genetic determinism, because the phenotype is always the result of very complex interactions between genetic information and the conditions of life. So the popular talk about genes "for" complex traits and about animals as "genetically driven robots" is the result of people not understanding or ignoring what individual development really entails. That's fine. But it seems to me that you are now saying something more than this. You seem to be saying that since the genetic networks that affect development are so constructed that it takes changes in many genes to make a selective difference, and that a change in a single gene often has little effect, the single gene should not be regarded as the unit of evolutionary change. I don't see why this should be the case. Why should-n't a single mutation make a small difference to the network's activity, and have either a beneficial or detrimental effect? What is wrong with the idea that beneficial mutations will slowly accumulate and affect the way organ-isms work and behave?

M.E.: A change in a single gene can occasionally have consistent effects on the organism's ability to survive and reproduce. But often the average effect of a mutation is likely to be selectively neutral: with certain other genes in certain environments it improves an organism's chances of leaving offspring, whereas in other circumstances it has the opposite effect. So, yes, your diagnosis of our position is correct. The claim that one can think about evolutionary change in terms of individual gene mutations that have, on average, small, additive, beneficial effects and accumulate over a long time to form a phenotypic adaptation is problematical and often untenable. The complex interactions between genes, and between genes and environments, mean that the effects of genes on the reproduc-tive success of individuals are often nonadditive. What we know about development tells us that we should be thinking about networks, not single genes, as the unit of evolutionary variation.

I.M.: The way you think about development obviously influences your views on the adequacy of the gene-centered approach to evolution. Do those who think in terms of selfish genes or gene selection necessarily have a very deterministic view of development?

M.E.: No, certainly not. It works the other way around. Genetic deter-minists—those who see humans and animals as genetically manipulated puppets—always think about evolution in terms of gene selection, but people who see evolution in terms of gene selection are usually not genetic determinists. They know that development involves a complicated inter-play between both genetic and nongenetic factors. However, gene selec-tionists do think that evolution is driven by the stepwise accumulation of single gene changes.

I.M.: What difference does it make? It is genetic differences that enable adaptation, after all. Thinking about canalized networks rather than indi-vidual genes just makes things more complicated.

M.E.: When the network is your unit of evolutionary variation, then what you concentrate on is the evolution of the phenotypic trait that it affects. You focus on variations in traits, not genes, and follow the transmission of variations in traits. Moreover, when the unit of evolutionary change is a network of interactions, then the way the network is constructed, the constraints on its structure, its robustness and flexibility, become very important targets of selection. The processes leading to developmental plasticity (the capacity to adjust in response to conditions) and canalization (stability in the face of environmental or genetic perturbations)—the type of things that Waddington wrote about—become of major importance in evolutionary change if you take this view. Plasticity and canalization are usually network properties, not properties of single genes.

I.M.: I'm sure that those gene-centered evolutionists who are not genetic determinists would agree that developmental plasticity and canalization are important. Yet, from my reading of popular literature in biology, I find that people who have a gene-selectionist approach to evolution often tend to have a somewhat deterministic view of development. Even Richard Dawkins, who is clearly not a genetic determinist, sometimes talks about lumbering robots, and uses other such telling metaphors.

M.E.: Yes, ideas about evolution sometimes do seem to influence how people think about development, rather than the other way around. In particular, in the hands (or pens) of nongeneticists, the gene selectionists' assumptions commonly lead to a view of individual development in which plasticity is rather limited. They have a problem, because if they grant too much plasticity (or too much canalization) to a trait, it would mean that single gene differences would not be coupled with alternative selectable phenotypes. And as you pointed out, the rhetoric does suggest that this has sometimes put them on a slippery intellectual slope. When gene selectionists talk about lumbering robots, about genes controlling organisms and so on, it is usually not the clever, emotional, and sometimes saintly robots of Isaac Asimov that their words conjure up, but the string puppets of an old-fashioned puppet show. This is particularly evident when some of them talk about human sociobiology. However we will leave that issue until chapter 6.

I.M.: So what is the second message that we have to take from what you wrote about the ways that genes affect development?

M.E.: Our second point is really a reiteration of the point made by Shapiro: since so many organisms have cellular systems that alter DNA during development, we have to recognize that DNA can change in response to environmental cues. We therefore need to study the systems of nonrandom DNA change to see how much and what kind of heritable genetic variation is formed through them, and whether they influence genetic variation in the next generation. If they do, it changes the way we have to think about the role of the environment in evolution.

Genetic Variation: Blind, Directed, Interpretive?

In 1988, the American microbiologist John Cairns and his colleagues dropped a small bombshell on the biological community. For more than fifty years, right from the early days of the Modern Synthesis, biologists had accepted almost without question the dogma that all new heritable variation is the result of accidental and random genetic changes. The idea that new genetic variants—mutations—might be produced particularly when and where they were needed had been dismissed as a heretical Lamarckian notion. In reality, however, there was little evidence against it. The rate at which new mutations are produced is very low, so detecting them at all required a lot of searching among large numbers of animals or plants; deciding whether or not mutations were produced randomly was effectively impossible. Only for bacteria were there techniques that allowed vast numbers of organisms to be screened relatively easily, and it was these organisms that had provided the main evidence that mutation is random. Experiments carried out in the 1940s and 1950s seemed to show that for bacteria the conditions of life had no effect on the production of new mutations.

It was this conclusion that John Cairns and his associates challenged in 1988. They argued that the earlier experiments had been overinterpreted. Their own experiments suggested that some mutations in bacteria *are* produced in response to the conditions of life and the needs of the organism. The generation of mutations is therefore not an entirely random process. This was not the first time that experimental evidence suggesting nonrandom mutation had been reported, but the scientific stature of John Cairns, and the publication of his group's findings in *Nature*, the leading British science journal, meant it could no longer be ignored. The *Nature* article led to a spate of responses and comments in both the scientific and popular press. The idea of nonrandom mutation was seen by many as a challenge to the well-established neo-Darwinian theory of evolution, and although some people's reaction was to suggest mechanisms that might underlie the production of induced mutations, others were extremely reluctant to accept that they could occur at all. They offered alternative interpretations of the experimental results—interpretations that did not require that mutations were formed in response to the environmental conditions. The upshot of all the arguments was that it was quickly realized that there was really no good evidence of any kind to show that all mutations are random accidents. It was equally clear, however, that a lot more experimental work was needed before it could be firmly concluded that some mutations are formed in response to environmental challenges.

We don't want to go into the details of all the claims and counterclaims that resulted from the work following up the 1988 *Nature* paper. On balance we think that the experimental evidence that is now available suggests that Cairns and his colleagues were probably wrong; they were not dealing with mutations that were produced in direct response to the environmental challenge they imposed. However, what emerged from the work their paper stimulated and subsequent molecular studies is important, because it has resulted in a far less simplistic view of the nature of mutations and mutational processes. There is now good experimental evidence, as well as theoretical reasons, for thinking that the generation of mutations and other types of genetic variation is not a totally unregulated process.

In this chapter we want to look at the whole question of where the variation that underlies the genetic dimension of evolution comes from. Essentially, it has two sources: one is mutation, which creates new variations in genes; the other is sex, through which preexisting gene variations are shuffled to produce new combinations. We shall focus mainly on mutation, particularly on nonrandom mutation, but first we want to say something about the variation generated through sexual reproduction, and how this process has been shaped by natural selection.

Genetic Variation through Sex

Sexual reproduction is the most obvious source of genetic variation. In animals like ourselves, it creates enormous diversity by producing new combinations of the genes existing in parents. From personal experience we know how very different the children in a human family can be, and how the kittens in a litter are often totally dissimilar, even on the rare occasions when we are quite sure only one father was involved. This variation, which is the outcome of sexual reproduction, is not adaptively linked to the particular environment in which the parents live. Neither is it linked to the environment that the offspring are likely to inhabit in the future. Even though our children may think otherwise, they are not automatically superior and better fitted for this world than we are. The variation generated by sexual processes is blind to function, and to the present and future needs of the lineage.

The diversity produced through sexual reproduction has three sources. We described two of them in chapter 1. The first is the mixing of genes from two nonidentical parents, which leads to an offspring that is different from both. The second, which makes offspring different from each other, stems from the way chromosomes are distributed into sperm and eggs. In most animals and plants, almost all chromosomes come in pairs, with one copy of each chromosome being inherited from each parent. During meiosis, the cell division that leads to gamete production, the number of chromosomes is halved, so each sperm or egg ends up with only a single copy of each chromosome. Which particular set it gets is a matter of chance. If an organism has four chromosomes, two copies of chromosome A and two copies of B, which we can write as $A^m A^p B^m B^p$ (where m means a chromosome was inherited from the mother, and p means it was a paternal chromosome), there are four possible combinations in the gametes; $A^m B^m$, $A^m B^p$, $A^p B^m$, and $A^p B^p$. Obviously, if there are more pairs of chromosomes, there are more possibilities. For a human being, who will normally have twenty-three pairs of chromosomes, over 8 million different combinations are possible, so a lot of variation is generated by the random distribution of chromosomes when sperm and eggs are formed.

The third source of variation is something we haven't mentioned before. It is the recombination of genes brought about by a process known as crossing-over. During meiosis, the members of each chromosome pair come together and the pairing partners exchange segments. So, if the sequence of genes on chromosome A^m is $l^m m^m n^m o^m p^m q^m r^m s^m$, and on A^p it is $l^p m^p n^p o^p p^p q^p r^p s^p$, after crossing-over you might have chromosomes $l^m m^m n^m o^p p^p q^p r^p s^p$ and $l^p m^p n^p o^m p^m q^m r^m s^m$. Old combinations of alleles that were linked and inherited together are broken up, and new combinations are formed. Since crossing-over can occur at different sites in different germ cells, this process of recombination generates an almost limitless amount of variation in the gametes.

From what we have said so far, it may sound as if the shuffling of chromosomes and genes during sexual reproduction is a haphazard and unregulated process. It would be quite wrong to think of it in this way, however. Sexual reproduction is an intricate evolved system, which has been modulated by natural selection in various ways. It is an expensive process, because it means organisms have to spend time and energy on the complicated processes of meiosis and gamete production, and sometimes also on the production of males and finding a mate. And at the end of all this investment the offspring produced might receive such a poor combination of genes that they are less able to survive and reproduce than their parents. So why not keep the parent's genome intact by reproducing asexually? Why don't all organisms clone themselves and dispense with the need to produce costly males? What are the advantages of sexual reproduction?

There is no short and simple answer to this question. The problem of the origin and maintenance of sex is one of the most perplexing in evolutionary biology, and there have been endless debates about it. Its origins may be tied up with selection for systems that repair damaged DNA, and DNA repair may still be one of its major functions, but there is no consensus about the adaptive significance of sex in present-day organisms. Fortunately, we need not go into all the arguments here, because the point we want to make is a simple one: it is that for both the individual and the population, there are potential advantages as well as disadvantages to sexual reproduction. Most evolutionists would agree that in the short term, in a stable environment, asexual reproduction, which preserves the parent's well-adapted combination of genes, is best. The snag is that parental genomes cannot be preserved for ever. Even totally asexual lineages change, because mutations are inevitable. Some harm their carriers, and will be weeded out by natural selection, but many may remain and accumulate. Consequently, in the long run, asexual lineages may deteriorate and go extinct. In contrast, if organisms reproduce sexually, the shuffling and recombination of parental genes means that some offspring may be lucky and get dealt a set of genes with fewer damaging mutations than either of their parents. Sexual reproduction can therefore preserve lineages by preventing the accumulation of deleterious mutations. Another advantage is that if competition for resources is intense, then at least some sexually produced offspring may have genotypes that make them good competitors. In the medium to long term, in changing environments, by bringing together beneficial mutations arising in different individuals, sex will lead to faster evolution than would be possible in asexual lineages.

Since there are both potential advantages and disadvantages to sexual reproduction, it is not surprising that it is used to different extents and in different ways in different species. What we see today is a whole spectrum of modes of reproduction and modifications of the sexual process. Some lineages manage without sex at all, although most (such as the whiptail lizard, *Cnemidophorus uniparens*) are probably of recent origin and have a rather limited evolutionary future. Other species and groups (for example, aphids, yeasts, water fleas, and many plants) reproduce both sexually and asexually. Even when reproduction is always sexual, species differ in how sex comes into the life cycle. In some there are two separate sexes, whereas in others (such as earthworms and garden snails) a single individual is capable of producing both sperm and eggs. With species of the latter type, some are capable of self-fertilization, whereas others need a sexual partner. When we look at the details of gamete production, we find that the number, structure, and behavior of chromosomes differ from species to species. There is one species of *Ascaris*, the parasitic worm that we mentioned in the previous chapter, that has only a single pair of chromosomes, but most plants and animals have a number in double figures. Chimpanzees, for example, have twenty-four pairs, which is one more than we have but many fewer than the thirty-nine pairs of the dog. The structure of chromosomes and how they behave is not uniform either. They come in many sizes and shapes, and the rate of crossing-over between homologous chromosomes differs between species, between the sexes (there is none at all in female butterflies and male *Drosophila*), between chromosomes, and even between regions of chromosomes.

What do all these differences in sexual reproduction mean? The answer is that for many of them we don't know, and they may have no adaptive significance at all. However at least some of them are thought to be adaptations that determine how much genetic variability there is in the next generation. Take the species that have both sexual and asexual generations: generally, they reproduce asexually when conditions are constant and good, but sexually when things change or life becomes stressful. Aphids, for example, commonly reproduce asexually throughout the summer, but before they overwinter, they have a sexual generation. Similarly, *Daphnia*, the water flea, reproduces asexually when environmental conditions are good, but when life gets tough it switches to sexual reproduction and produces resistant eggs that can survive poor conditions. This makes evolutionary sense. If an individual is doing well and its environment is not changing, asexual offspring, who have the same set of genes, will probably do very well too. So why change? If it ain't broke, don't fix it! Avoiding sexual reproduction will not only preserve a good set of genes, it will also double the rate of reproduction, because there is no need to produce males. As anyone who has battled with aphids on their roses knows, the asexual strategy can be very successful. But if conditions change, so that offspring are likely to experience a different environment (as the aphids may after summer is over), investing in sexual reproduction is a better bet. Although costly males have to be produced, at least a few of the varied sexual offspring may survive new conditions. We have shown the advantage of variation generated through sexual reproduction in figure 3.1.

There is some evidence that another aspect of sexual reproduction, the amount of crossing-over between chromosomes, has also evolved to fit the conditions of life. It tends to be lower for species living in uniform, stable environments, and higher for those living where conditions are less predictable. The suggested explanation is that natural selection has led to low recombination rates in constant conditions, because offspring do best if they have much the same genotype as their parents. But when lineages have repeatedly encountered varied or varying conditions, high recombination rates have been selected, because variety among the offspring has increased the chances that some will survive. We know from laboratory experiments that the average rate of recombination differs between populations of the same species, and that selection can change recombination rates. We even know some of the genes and alleles that affect recombina-

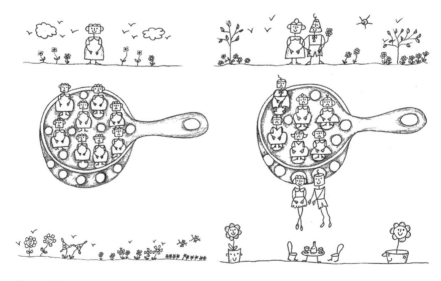

Figure 3.1
An advantage of sexual reproduction. On the left, the identical progeny of a well-adapted asexual individual cannot pass through the selective sieve of changed conditions, so the line becomes extinct; on the right, a few of the varied offspring of sexual individuals pass through the sieve and continue the lineage.

tion. So, although the evidence that average recombination rates are related to ecological conditions and the lifestyle of the species is not extensive, it would be surprising if the rate of recombination had not been adjusted by natural selection.

Even though there are many gaps in our knowledge, biologists take it for granted that most differences in sexual reproduction are of adaptive significance. They have no difficulty in suggesting how some plants benefit from having elaborate mechanisms that prevent self-fertilization, or what advantages there are in the ability of some animals to change sex, or why some organisms produce many offspring and some only a few. They can also suggest how structural changes in chromosomes, which rearrange the genes and alter the likelihood of recombination between them, can be of selective importance. It is true that they cannot think of adaptive explanations for every variation in sexual processes, and indeed acknowledge that for some there may be none. Nevertheless, no one doubts that natural selection has influenced whether, when, and how sexual reproduction is used, and how much variation it generates.

One final point about sex: it is not always tied to reproduction. In bacteria and viruses, there are no paired chromosomes, no meiosis, no gametes, and no sexual reproduction. Nevertheless, they do have various sexual processes in which, through recombination mechanisms similar to those of other organisms, genetic information is exchanged between chromosomes that originate in different individuals. Fortunately, we need not go into the details of these. At the molecular level, all recombination is extremely complex, with bits of DNA unwinding, breaking and rejoining, and complementary base-pairing taking place between nucleotide chains from different chromosomes. Needless to say, a lot of enzymes and other proteins are involved. Many of these are parts of what in the previous chapter we called the cell's natural genetic engineering system—the kit of enzymes and other molecules that enable specific DNA changes to be made in certain cells during development. However, the most important role and evolutionary origins of the components of this kit are almost certainly in something much more fundamental—in the maintenance of DNA. This will become clearer after we have said something about the ultimate source of all genetic variation—mutation.

Variation through Mutation

Changes in DNA sequences are inevitable. They result from imperfections in the DNA copying process, from changes introduced when mobile elements (jumping genes) move from one place to another, from spontaneous chemical changes, and from the effects of damage caused by the chemicals produced in normal cellular activities. In addition, external physical agents such as x-rays and ultraviolet radiation, or chemical agents such as mustard gas and LSD (lysergic acid diethylamide), cause lesions in DNA. Many of these agents are known to increase the risk of cancer, which is often associated with genetic changes in somatic cells. When DNA changes occur in germ-line cells, they can affect the next generation.

Long-term Darwinian evolution through the genetic system depends on these DNA changes. But there is a paradox here, because if DNA is changeable, its effectiveness as a carrier of hereditary information is reduced. If only very imperfect copies of the information that has enabled survival and reproduction are transmitted, evolution by natural selection would be very slow, if not impossible. Information needs to be durable, as well as somewhat changeable. So how can DNA, which is not an intrinsically stable molecule, function so effectively as a carrier and transmitter of information?

The answer is that DNA can do its job because organisms have a whole battery of mechanisms that protect and repair it, ensuring that existing nucleotide sequences are well maintained and are copied accurately. Cells have proteins that scavenge for and degrade molecules that would damage DNA; if damage does occur, there is another set of proteins that can repair it, sometimes using a recombination process that substitutes a similar undamaged sequence from elsewhere. When DNA is replicated, there are systems that check that each nucleotide added to the growing daughter strand is the correct (complementary) one, and remove it if it is not. After the new daughter strand is synthesized, it is proofread, and if mismatched nucleotides are found, they are corrected. Thanks to these and other proofreading and correction systems, the error rate during the replication of human DNA is only about one in every ten thousand million nucleotides. Without them, it has been estimated it would be nearer one in a hundred.

This amazing system for maintaining the integrity of DNA has presumably evolved through natural selection for DNA-caretaker genes. Lineages with poor DNA maintenance and sloppy replication failed to survive, because they kept changing, producing all sorts of new mutations, most of which were detrimental. Such lineages had a lot of variation, but less heredity; good sets of genes were not transmitted accurately. Lineages with better mechanisms for looking after their DNA continued, because they transmitted accurate copies of the genes that had allowed them to survive and reproduce. In this way, natural selection has ensured that there is a good genetic engineering kit for DNA maintenance, and that mutation rates are generally low. Mutations do occur, but not too often.

Randomness Questioned

We now have to return to the problem with which we opened this chapter. The question we have to ask is whether the few mutations that still occur are all rare and random mistakes, the consequences of the remaining imperfections in the genomic surveillance, repair, and maintenance systems, or are they something more than this? Is there any specificity about where and when mutations occur?

It is really rather surprising that although biologists have always accepted that environmental factors can influence when and where variation is generated through sexual processes, until recently they have been very reluctant to concede that the same is true for variation coming from mutation. They accepted that the average rate of mutation has been adjusted by natural selection, and that because genes differ in size and composition they mutate at different rates. But the idea that mutations might be formed specifically when and where needed was rarely even considered. It was simply assumed that all mutations are blind mistakes, the outcome of faults in the system. The only specificity that was recognized was the sequence specificity of some mutagenic agents. Ultraviolet radiation, for example, tends to cause lesions in regions of DNA where there are two or more thymines, one after the other. However, such T-T sequences are scattered throughout the genome, being present in all sorts of genes coding for all sorts for proteins with all sorts of roles, so the lesions caused by UV are not function specific. Many other mutagens also have some sequence specificity, but they, too, are not specific to particular genes or functions. In general, it has been assumed that mutations are not adaptive, and they are not developmentally controlled. They are certainly not the cell's response to a need. They are mistakes, and if they make any difference at all at the phenotypic level, they are almost always sorry mistakes. Only very rarely will a chance lucky mistake increase the likelihood that a cell or organism will leave descendants.

Today, many geneticists will agree that the view of mutation that we have just outlined is inadequate, and like them, we are going to argue that not all mutations are haphazard mistakes; rather, some mutations are "directed." "Directed mutation" is part of the jargon of genetics, and it does not mean that we or other biologists believe there is some guiding intelligence or "hand of God" directing changes in DNA according to the organism's needs. Such ideas have no part in scientific reasoning (and through their absurdity ridicule religion as well). Our argument is simply that evolution by natural selection has led to the construction of mechanisms that alter DNA in response to the signals that cells receive from other cells or from the environment.

No one can deny that directed DNA changes are possible, because they occur in development. We described some of them in the previous chapter. The most familiar example is the chopping and changing of DNA that goes on in the cells of the immune system, but there are many others. These regulated DNA changes are part of normal development, just like regulated changes in the transcriptional activity of genes, or in RNA splicing and translation. They are guided not by higher powers, but by a cellular system that is the product of genetic evolution. Even the most conservative neo-Darwinian evolutionists do not worry about this type of directed DNA change. They see the changes as a part of development, and regard the mechanisms behind them as relevant only to the evolution of development.

What does disturb many evolutionary biologists is the idea that some of the mutations that are the raw material of evolution are not the result of blind accidents. That is why the claim by John Cairns and his colleagues, that they had found directed mutations in bacteria, set the cat among the pigeons in 1988, although in fact the idea that genomic changes are not blind to function was not a new one. It had been central to the thinking of a few geneticists for some time. In 1983, in a lecture given when she received the Nobel Prize, Barbara McClintock said:

In the future, attention undoubtedly will be centered on the genome, with greater appreciation of its significance as a highly sensitive organ of the cell that monitors genomic activities and corrects common errors, senses unusual and unexpected events, and responds to them, often by restructuring the genome. (McClintock, 1984, p. 800)

At the time, McClintock's views were still regarded as rather unorthodox, and they were probably almost unknown to most evolutionary biologists. What her experimental work had suggested to her was that when cells cannot respond to stresses effectively by turning genes on and off or by modifying existing proteins, they mobilize systems that alter their DNA. The new genetic variation that is produced in stressful conditions (e.g., after a sharp temperature change or prolonged starvation) is *semi*directed in the sense that it is a response to environmental signals, but it does not lead to a unique and necessarily adaptive response. It falls somewhere between totally blind variations, which are specific neither in their nature nor in the time and site in the genome where they occur, and totally directed variations, which are reproducible adaptive changes that occur at specific sites in response to specific stimuli.

Although the whole issue of stress-induced mutation is still very controversial, McClintock's views have subsequently been at least partially vindicated. The origin of new genetic variation is certainly a lot more complex than was previously assumed. No longer can we think about mutation solely in terms of random failures in DNA maintenance and repair. We now know that stress conditions can affect the operation of the enzyme systems that are responsible for maintaining and repairing DNA, and parts of these systems sometimes seem to be coupled with regulatory elements that control how, how much, and where DNA is altered. So what terms are appropriate to describe the types of mutations that are now being revealed?

To help explain the situation, which is surrounded by controversy and bedeviled with a lot of awkward terminology, we will use a thought experiment. Imagine three human tribes, which have three different lifestyles and ways of coping with problems. The members of the first tribe are called the "Conservatives." They have a long written history, and the young members of the tribe are obliged to memorize this history and learn the lessons of the past. The past is sacred, and past actions form a binding example for the members of the Conservative tribe to follow. Many of the tribe's men and women are busy maintaining their vast archives, memorizing and transmitting the wisdom written in their books. These books have the answer to the changing patterns of life. When things change, the Conservatives rely on their knowledge of the way their ancestors handled the situation, and do the same. This often solves the problem, but when facing the totally unknown or unpredictable, the Conservatives are rather helpless. Only very rarely and by accident do some members of the tribe stumble on a solution and manage to survive such an emergency. The survivors then become sanctioned as "ancestors," and their deeds are added to the sacred books.

The people of the second tribe, the "Explorers," have the very opposite philosophy. They see that the world is in constant flux and often changes, and believe that, for many things, past experience is limiting and misleading, so should be forgotten. The present and future require a constant process of reevaluation. The Explorers stress the central role of individual discovery, and the lack of any preconceived ideas. When they encounter an emergency, whether similar to past emergencies or completely new, they respond by encouraging everyone to find new creative solutions to the problem. Usually someone does find a solution, which is then adopted by all the members of the tribe, but often many perish before this solution is found.

The members of the third tribe, the "Interpreters," respect the past, just like the Conservatives, but they are not bound by it in the same way, and encourage exploration and controlled deviation from tradition. According to their philosophy, their ancestors had divine wisdom, and their sacred words must be the inspiration for all times, but the deep wisdom of the ancestors is written in metaphorical language. This means it requires a new interpretation in each generation, according to the changing needs of the society, but still in accordance with certain rules. When they face a state of emergency that is similar to one described in their books, interpretation is easy and they act more or less like their ancestors, slightly updating their behavior according to the needs of the present. They give freedom to their interpretive imagination, and try to find new solutions that do not contradict the old wisdom.

Figure 3.2 shows how the three tribes would respond to an outbreak of an unknown sickness in their community. The Conservative tribe's doctors find nothing in their sacred books that matches this particular illness, so all give a tonic and continue as before. Every patient dies unless a doctor makes a mistake, forgets what he is supposed to do, and by chance hits on an effective treatment. The members of the Explorer tribe approach sickness very differently. They ignore past experience and the precise symptoms, and simply try every cure they can think of, using the same range of treatments as they would apply to any type of sick person, sick cow, or sick rhubarb crop. Patients die quickly, although sometimes, through luck, a treatment that works is found before everyone dies. The members of the third tribe, the Interpreters, look for the illness in their books, and although they may find nothing there that exactly fits this particular illness, they see that there have been illnesses with similar symptoms in the past. They then mobilize their interpretive talents and improvise on the basis of the remedies that their ancestors used. Although some are unsuccessful and the patients die, they eventually find a good treatment.

We believe that the behavior of the three tribes is similar to three possible biological strategies for dealing with adverse conditions. The behavior of the first tribe, the Conservatives, is similar to a strategy of responding to every situation with well-established physiological responses or precise directed mutations. These evolved responses "solve" problems that are similar to those faced by the lineage in the past, so are adequate for normal development and day-to-day living. If circumstances remain similar to

Figure 3.2
The response of the Conservative (top), Explorer (middle), and Interpreter (bottom) tribes to a novel situation, an unknown sickness. The Conservative doctors offer nothing other than a tonic; the Explorers try every conceivable type of conventional and alternative medicine; the Interpreters try treatments that are similar to those that were effective for the same type of illness in the past.

those of the past, this strategy is successful, but in new and unpredictable conditions, it fails. Only a rare and lucky mistake—a beneficial chance mutation—allows the lineage to survive.

The behavior shown by the Explorer tribe when dealing with unknown or changing situations is equivalent to a biological strategy of enhancing the rate of random mutation. This strategy is always costly, because there is always a considerable lag between encountering a problem and producing a successful mutation that solves it, and before this happens many individuals may die from the nonbeneficial mutations that are induced. But there is a good chance that in a large population a favorable mutation will occur, and the lineage will survive. If the group is small, the chances of its extinction are high.

The third type of behavior, that of the Interpreters, is analogous to a biological situation in which the response to adverse conditions is to produce mutations that are not entirely random, but are also not precisely directed. They are "interpretive" in that where or when they happen is based on the evolutionary past, although there is a random element in exactly what happens. We are going to focus mainly on this type of mutational process in the rest of this chapter, because we believe that it has had an important role in evolutionary adaptation.

Acquired, Required, Interpretive Mutations?

In order to flesh out what we mean by interpretive mutations, we are going to describe four different situations in which the mutations that occur sit somewhere between random and directed. Most of the examples will be based on what we know from microorganisms, but some of the phenomena we describe have also been found in other groups, notably in plants. Although we are calling them all "interpretive" mutations, the processes that occur in the four situations fall at different points on the spectrum between totally random and totally directed mutation.

The first situation involves what we will call *induced global* mutation. Imagine that some organisms find themselves in an environment in which they can no longer survive or reproduce. Their only hope of salvation is that a lucky mutation will crop up and enable them to deal with their adverse circumstances. If mutation rates are low, which they usually are, the chances that any will survive are slim. But if they have mechanisms that kick in in stressful conditions and increase the rate of mutation throughout the genome, things may be better. Many individuals will perish quickly (they get mutations that make matters worse), but the chances that one or two will have a liberating mutation are enhanced. It's a bit like desperately poor people buying lottery tickets. By buying tickets they have a chance of becoming rich, although most will certainly become penniless even more quickly. Notice that with the type of strategy we are describing here there is no increase in the relative frequency of specifically *beneficial* mutations. The tactic is close to the way the Explorer tribe deals with its problems: try everything in the hope that something will work. We've shown it in figure 3.3.

Thanks to studies made in the last twenty years, we now know that mutation rates in bacteria are indeed enhanced when they encounter an environment that is so hostile that they completely stop growing and reproducing. In such conditions, a spate of new mutations is generated throughout the genome. Every single mutation is random, in the sense that it is not function-specific, but the general genomic response—the

Figure 3.3
Induced global mutation: on the left, in normal conditions, the mutation rate is low (few umbrellas); on the right, in conditions of acute stress (cloud and storm), mutation rates throughout the genome are high (more umbrellas) and some mutations happen to be adaptive (open umbrella).

increased mutation rate—may be adaptive. The phenomenon has been studied most intensely in bacteria, but something similar happens in plants. One of the things that Barbara McClintock discovered many years ago was that stress conditions lead to a massive movement of mobile elements in the genomes of plants. She regarded this as an adaptive response, which provided an important source of new variation.

Increasing the mutation rate in drastic emergency conditions is obviously a tactic that can sometimes work. There is a chance that it will provide a beneficial mutation. Natural selection could therefore favor genetic changes that result in this response to stress. Lineages with mechanisms that enable individuals to enhance their mutation rate (buy a lot of lottery tickets) when life gets really tough have a better chance of not going extinct. Although most individuals perish, at the level of the lineage stress-induced global mutation is an adaptive response.

Not everyone accepts that stress-induced mutation is an evolved adaptation, however. Some people argue that the spate of mutations that occurs in adverse conditions is simply a by-product of stress-induced failure. When cells are stressed, especially when starved, one of the things that may happen is that they are no longer able to produce the proteins needed for DNA maintenance and repair. It may even be that starved cells are obliged to turn off their DNA-caretaker genes to save energy. If so, faults will occur and remain uncorrected. In other words, there will be a lot of mutations. In this case, the generation of mutations is just a pathological symptom of the problems cells are experiencing, not an evolved adaptive response to adverse conditions.

People can and do argue about whether *induced global* mutation is an evolved adaptive response, or something pathological that may incidentally have beneficial effects, but there is no doubt that our second type of nonrandom mutation process—*local hypermutation*—is an adaptation. With induced global mutation, the mutations produced are nonrandom because they occur at a *time* when they are likely to be useful; with local hypermutation, changes are produced at a genomic *place* where they are useful. Certain regions of the genome have a rate of mutation that is hundreds or thousands of times higher than elsewhere (figure 3.4). In the jargon of genetics, they are "mutational hot spots." The genes in these hot spots code for products that are involved in cellular functions requiring a lot of diversity. That is what makes the high local mutation rate adaptive.

The English geneticist Richard Moxon and his colleagues have studied local hypermutation in *Haemophilus influenzae*, a bacterium that causes

Figure 3.4
Local hypermutation: on the left, the low mutation rate (few umbrellas) found in most regions of DNA; on the right, a mutational hot spot, where some of the many mutations happen to be adaptive (open umbrellas).

meningitis. Like other pathogens, this bacterium has a life full of challenges. It encounters several very different microenvironments as it invades and colonizes different parts of the body, and has a nonstop fight with its host's immunological defenses. As we described briefly in the last chapter, mammals have a magnificent immune system in which regulated DNA rearrangements and mutation enable cells to constantly produce the new types of antibody that are needed to do battle with pathogens. Yet *H. influenzae* frequently manages to evade its host's ever-changing defenses, and also copes with the varied environments it encounters in different parts of the host's body. It does so because it possesses what Moxon has called "contingency genes." These are highly mutable genes that code for products that determine the surface structures of the bacteria. Because they are so mutable, subpopulations of bacteria can survive in the different microhabitats within their host by changing their surface structures. Moreover, by constantly presenting the host's immune system with new surface molecules that it has not encountered before and does not recognize, the bacteria may evade the host's defenses.

What, then, is the basis for the enormous mutation rate in these contingency genes? Characteristically, the DNA of these genes contains short nucleotide sequences that are repeated again and again, one after the other. This leads to a lot of mistakes being made as the DNA is maintained and copied. To explain exactly how this happens would involve going into a lot of detail about DNA replication and repair, which we want to avoid, but it is easy to see the general nature of the problem. Suppose that you have a sequence ATATATAT in one strand, which is paired with the complementary sequence TATATATA in the other. It is not difficult to imagine that during replication the two strands might slip out of alignment so that there is an unpaired AT at one end, and an unpaired TA at the other. This misalignment could lead to the unpaired nucleotides being removed by the DNA caretaker system, or alternatively they might be given pairing partners. The result would be mutations—sequences that are either two nucleotides shorter or longer than before. Repeats also enhance the chances of different chromosome regions pairing with each other, breaking and recombining, which provides more potential variation. Because the number of repeats can increase or decrease, this type of mutation is readily reversible, so lineages frequently switch from one phenotype to another.

It is difficult to find an appropriate term for the type of mutational process that occurs in contingency genes. Moxon refers to it as "discriminate" mutation, and the term "targeted" mutation may also be appropriate. Whatever we call it, there is little doubt that it is a product of natural selection: lineages with DNA sequences that lead to a high mutation rate in the relevant genes survive better than those with less changeable sequences. Although the changes that occur in the DNA of the targeted region are random, there is adaptive specificity in targeting the mutations in the first place.

The contingency genes of *H. influenzae* are not an isolated example. Similar highly mutable genes, with DNA sequences that seem to have been selected for mutability, have been found in other pathogens that are constantly at war with their host's immune system. They have also been found in species of snakes and snails that use poison to capture their prey and defend themselves against predators. The high mutability of their venom genes is thought to be an adaptation that enables the animals to keep up with changing predators and prey, and counter the evolved resistance to the venom that predators and prey develop.

It should be clear that the high mutation rates just described are not controlled responses to changed physiological conditions. Mutation goes on all the time. The mutations in contingency genes are "acquired" only in the evolutionary sense, not in any physiological sense. Our third type of mutational process, *induced local* mutation (figure 3.5), is different, however, because it happens in response to changed conditions. It involves a smaller increase in the rate of mutation (five to ten times the average mutation rate elsewhere), but occurs especially in those genes that help the organism to cope with the new situation. Mutations are therefore both induced by the environment and are specific to the gene that can save the day. In no sense is this type of mutation random—the mutations are both required and acquired.

Barbara Wright has found nonrandom mutation of this type in her studies of the gut bacterium *Escherichia coli* (always known as *E. coli*, because *Escherichia* is such a mouthful). To understand her experiments, we have to remember that when bacteria are semistarved, a whole array of mechanisms that protect the cell and allow it to survive for a bit longer are brought into operation. Genes that in good times are active, because their products are needed for reproduction, are turned off; others that are normally kept repressed, because their activities are unnecessary and would

Figure 3.5
Induced local mutation: on the left, the mutation rate in normal conditions (few umbrellas); on the right, a localized increase in the mutation rate in response to stress (a local storm) results in an adaptive mutation (open umbrella).

be wasteful, are selectively mobilized. Among the genes that are turned off in times of plenty are those that are needed to synthesize amino acids, because amino acids are normally readily available in the food. However, whenever a particular amino acid is in short supply, the relevant gene becomes active, and the cell makes the amino acid for itself.

What Barbara Wright did was look for mutations in a defective copy of one of the amino acid genes. Because the gene was faulty and produced a nonfunctional product, just turning it on was not enough to rescue the cell when there was a shortage of the amino acid. It also needed a beneficial mutation that changed the faulty gene. Using various genetic tricks, she compared the rate of mutation when the required amino acid was present and the bacteria could grow vigorously, with the rate when the amount of amino acid available was insufficient. In the latter conditions the bacteria could survive, but only just. She found that in these stressful conditions, the rate of mutation in the defective gene was much higher than normal, and, most importantly, the elevated mutation rate was specific to that particular gene. Increased mutation in this case depended on a combination of two factors: first, the shortage of the amino acid, which activated the gene; and second, the presence of a cellular emergency signal that is formed in times of crisis. The outcome of these two factors was that the gene relevant to the crisis conditions became more mutable, so the chances that a cell would have the lucky mutation that enabled it to survive increased.

The fourth and final type of interpretive mutational processes can be called *induced regional* increased mutation (figure 3.6). Not much is known about it, and it may overlap with some of the previous categories, but it is particularly interesting because it has been found in multicellular organisms. Sometimes a change in conditions, for example a substantial short-term rise in temperature, increases the rate of mutation in a specific set of genes by several orders of magnitude. The mutations produced are not known to be adaptive, but since the process is a unique and very specific response to particular environmental circumstances, no one can call it random. Naturally, we would like to know whether what is seen is an adaptation to stress, and if it is not an adaptation now, whether it could have been one in the past. But we do not know.

This type of transient regional mutation has been found in the mustard plant, *Brassica nigra*. A heat shock leads to the loss of some of the many copies of the DNA sequences that code for rRNAs (the RNAs that are part of the ribosomes). The adaptive significance of this genomic response is not clear—there is no evidence that the loss of these genes increases the

Figure 3.6
Induced regional mutation: on the left, the mutation rate in normal conditions (few umbrellas); on the right, the mutation rate in several specific regions is enhanced by mildly stressful conditions (a gentle storm).

reproductive success of individuals in which it has occurred. However, the decrease in the number of copies is transmitted to the next generation, because although it happens in somatic cells, some of them give rise to reproductive tissue. Similar heritable changes in the number of copies of rRNA and other repeated sequences have also been found when flax plants are moved to different nutritional conditions. At present, the mechanism of mutation is unknown, but the presence of repeated sequences suggests that recombination processes may be involved.

We have summarized in table 3.1 the various categories of mutations that we have described. Looking at the table, you can see that there are many types of DNA alteration that do not fit tidily with descriptions in terms of "random" or "directed." There is no difficulty with the specific and adaptive responses that occur during development, which are clearly directed, or with the haphazard mistakes of various kinds, which are clearly random or blind, but there are many interesting cases that fall in the twilight zone between the two extremes. If we think in terms of

Table 3.1

Type of genetic change	Targeted to a specific gene or region?	Induced or regulated?	Adaptiveness of the type of change	Type of DNA alteration
Classic blind mutation	No	No	None	Changes in bases, mistakes in repair and replication, movements of mobile elements, breakage and rejoining, etc.
Induced global increased mutation	No	Yes, by extreme stress	None, but elevation of the general mutation rate may be adaptive	Elevation of overall blind mutation rate
Local hypermutation	Yes	No	Yes	DNA sequence organization leads to high mutability in specific regions
Induced local increased mutation	Yes	Yes, by nonextreme stress	Yes	Mutation targeted to specific active genes
Induced regional increased mutation	Yes	Yes, by changed environment	None (as far as is known)	Mutations targeted to particular DNA repeated sequences
Developmental	Yes	Yes, regulated by developmental signals	Yes	Precise genomic changes and mutations in well-defined regions

an axis of change with the "extreme blind" and the "extreme developmental" genetic changes at either end, the mutations that we term "interpretive" sit somewhere between the two. Some, like those resulting from a stress-related global increase in mutation, are very close to the blind end of the axis, whereas local and regional mutations that result from specific physiological changes are semidirected, so are closer to the developmental end.

With local mutation, there is a measure of randomness in what is produced, but this randomness is targeted or channeled, because the changes occur at specific genomic sites and sometimes in particular conditions. These mutations are particularly interesting, because they are likely to be adaptive. Instead of evolutionary salvation coming from searching for and finding a needle (the exceedingly rare beneficial mutation) in a huge haystack (a large genome), the search is for a needle in a small corner of the haystack, a corner that is well pinpointed. There is still a need to search, but the search is now informed. The cell's chances of finding a mutational solution are enhanced because its evolutionary past has constructed a system that supplies intelligent hints about where and when to generate mutations.

Evolved Genetic Guesses

Even if we didn't have all the new experimental evidence showing that mutation is sometimes localized and under environmental or developmental control, the evolutionary arguments for expecting it are very powerful. It would be very strange indeed to believe that everything in the living world is the product of evolution except one thing—the process of generating new variation! No one doubts that how, where, and when organisms use sex, which reshuffles existing genetic variation, has been molded by natural selection, so surely similar selection pressures should also influence how, when, and where variation is generated by mutation. In fact it is not difficult to imagine how a mutation-generating system that makes informed guesses about what will be useful would be favored by natural selection. In our judgment, the idea that there has been selection for the ability to make an educated guess is plausible, predictable, and validated by experiments. As the American geneticist Lynn Caporale has said, "chance favors the prepared genome." The preparedness is, of course, evolutionary!

Once it is recognized that not all mutations are random mistakes, the way one sees the relationship between physiological or developmental adaptation and evolutionary adaptation begins to change. We are used to thinking of them as very different: physiological and developmental changes involve *instruction*—what happens in cells or organisms is controlled by internal or external regulatory signals; evolutionary changes involve *selection*—some heritable variants are preferred to others. In the jargon of philosophers of biology, the physiological and developmental processes that underlie a phenotype are "proximate causes," while evolutionary processes—natural selection and whatever else has constructed the phenotype during evolutionary history—are "ultimate causes."

Yet, if the generation of some heritable variation is under physiological or developmental control, how distinct are the two types of causes? Seeing evolution purely in terms of selection acting on randomly generated variation is wrong, because it involves instructive processes too. As we see it, the dichotomy between physiology/development and evolution, and between proximate and ultimate causes, is not as absolute as we have been led to believe. They grade into one another. At one extreme there are purely selective processes, acting on chance variation, while at the other there are purely instructive processes, which are totally physiological or developmental and do not involve any selection. Between these extremes we find the majority of the processes in the real world, which are to varying degrees both instructive and selective. Some developmental changes, such as those occurring during the development of the immune system, also involve selection, whereas some evolutionary changes, particularly in bacteria and plants, may have instructive components. In other words, Darwinian evolution can include Lamarckian processes, because the heritable variation on which selection acts is not entirely blind to function; some of it is induced or "acquired" in response to the conditions of life.

This view of the origins of heritable variation affects something that we discussed in chapter 1—the distinction that Dawkins makes between replicators (genes) and vehicles (bodies). According to Dawkins, the gene is the unit of heredity, variation, and evolution, whereas the body is the unit that develops. The gene, the replicator, controls the body-vehicle that carries it, but is unaffected by developmental changes in that body. However, if, as the evidence suggests, what happens in the body can affect the processes generating changes in genes, the distinction between replicators and vehicles becomes blurred. Development, heredity, and evolution are too interdependent to separate them.

Dialogue

I.M.: Let me attempt to summarize your argument. Your main point is that not all mutations are the random, chance changes in genes that they were once thought to be. Whether, when, and where mutations happen, and how many of them there are, sometimes depends on the conditions the organism is experiencing. The reason for this is that there are evolved systems that change the genome in response to environmental challenges. Did I get it right?

M.E.: Yes. And in chapter 9 we'll have more to say about how these systems may have evolved.

I.M.: Good. I want to start, then, by looking at the first argument you used to support this view. You said that it is not difficult to think of evolutionary reasons why some animals and plants always reproduce sexually, whereas others use sex only rarely; or why some plants usually self-fertilize, whereas others never do; or why in some chromosome regions there is almost no recombination, in others much more. You then suggested that since evolutionary biologists are happy with the idea that past natural selection has influenced when and how much variation is generated through sexual processes, there can be no theoretical objections to accepting that the same is true for variation generated through mutation. If the production of one type of variation has been modulated by natural selection, why not the other? I agree that this makes sense, but arguments from plausibility are neither proofs nor evidence. It seems to me that your argument would gain some circumstantial support if the two systems that generate variation were linked in some way. Are they? Are sex and mutation mechanistically related?

M.E.: At the cellular level, yes, to some extent they are. The mechanisms that lead to crossing-over—the recombination of genes through an exchange of chromosome segments during meiosis—are related to mutation. It's an enormously complicated subject, which is tied up with the way cells repair DNA. We shall be saying more about the relationship in chapter 9. A careful answer would be that there is overlap between the enzymatic systems that control recombination and those that produce interpretive mutations. The full extent of this overlap is not yet known.

I.M.: Does that mean that environmental conditions can have effects on recombination as well as on mutation? You haven't said much about it, other than that those interesting creatures that have both sexual and asexual options switch to the sexual one when life begins to get tough. This is the equivalent of induced global mutation, I think, because it generates a lot of variation at a time when it is likely to be useful. Do you have anything more specific? Do you ever get a stress-induced increase in recombination in particular regions of chromosomes—the equivalent of induced local or regional mutation?

M.E.: In the fruit fly *Drosophila*, heat stress increases general recombination rates. What is particularly interesting is that some regions become more recombination-prone than others. For example, regions that *never* normally recombine, like the tiny chromosome 4, suddenly start recombining. And some regions that are usually very reluctant to undergo recombination show a thirty-fold increase in recombination rate. So it is not an indiscriminate process. But what, if anything, these induced localized increases in recombination mean in terms of adaptive advantage is not clear.

I.M.: How does all this induced genomic change fit with the central dogma of molecular biology? You have been arguing that what happens during the lifetime of an organism can affect the amount and type of genetic variation found in the next generation. Yet if the central dogma is valid, there is no transfer of information from proteins back to RNA and DNA. So how can something that happens at the level of the whole organism, which surely means proteins, affect the genome of the next generation? Don't we need to assume some backtranslation—information being transferred from proteins to DNA? Surely this cannot happen!

M.E.: The argument that the central dogma means that developmental adaptations to the conditions of life cannot affect what is transmitted to the next generation is an old one. It was used especially during the 1960s, when the unidirectional nature of information transmission (from DNA to proteins and not vice versa) was recognized. For example, in 1966 John Maynard Smith wrote, "The greatest virtue of the central dogma is that it makes it clear what a Lamarckist must do—he must disprove the dogma." Recently Ernst Mayr, one of the founders of the Modern Synthesis, has echoed this view, describing the central dogma as "the final nail in the coffin of the inheritance of acquired characteristics." But in the light of what we know today, they are both wrong. Backtranslation is not necessary for acquired characters to be inherited, for the very good reason that most "acquired characters" do not involve a change in the amino acid sequences of proteins at all. Think about what happens when there is a cellular response to changed conditions. What changes in the cell? Is it the amino acid sequence of a protein? Usually it is not. What changes is which genes are switched on and which are switched off. It is the *amounts* of the various proteins, not their sequences, that are altered. Backtranslation is irrelevant to transmitting such alterations. A genetic change that simulates the acquired change would have to be in a regulatory region of DNA, not in a protein coding sequence. Even if a cellular response does involve an altered amino acid sequence, the chances are it is a consequence of altered splicing or translation, not a change in the DNA coding region. So a genetic change simulating the acquired change is again likely to affect regulatory sequences, not coding sequences. The types of genetic change that affect the regulation of gene activity are those that alter either the number of copies of genes, or the nucleotide sequences in control regions, or a gene's location on the chromosome. These are what we often see with interpretive mutations.

I.M.: This leads me to two questions, which in a sense are the opposite of each other. The first is: If so good, why so little? What I mean is, why is it so difficult to find examples of these directed or semidirected mutations, when they could provide so much potential benefit to organisms?

M.E.: The thought experiment gives a partial answer. The Conservative tribe's strategy of always doing the same thing is effective only if situations repeat themselves exactly. It doesn't work if something a bit different crops up. Similarly, a *precise* directed mutational response to change is unlikely to be a good solution to a cell's problems, because usually environmental conditions that are exactly the same do not reoccur repeatedly. We would therefore not expect finely tuned directed mutation systems to evolve very often. The most effective genomic response to most changed conditions is through an educated guess, and an improvisation on the basis of the guess—through what we called interpretive mutation systems. More and more of these are being discovered, particularly in bacteria, although we still know very little about them.

I.M.: This leads me to my other, opposite, question. In chapter 2 you emphasized how complex the relation between genes and traits is. You said that usually a change in a gene, if it has any effect at all, will have many effects, particularly in multicellular organisms. If so, a new mutation might be beneficial in one cell type, say a liver cell, but have detrimental effects in another, say a nerve cell. Surely it is likely that the overall effect of a mutation in all its many contexts will be bad. Even most of what you called interpretive mutations seem a bit problematical. The chances that any kind of directed mutation could serve the organism well in all the different environments and in all cell types seem rather small to me, maybe as small as with a random mutation. Why should we expect to find directed or semi-directed mutations at all?

M.E.: You are touching on a very fundamental problem here. For characters to be adaptively modified through any kind of induced mutation, a change at the level of the organism has to feed back to produce a corresponding change at the level of the gene. This is difficult to envisage in complex multicellular organisms. In bacteria or other single-cell organisms, it is not so difficult to imagine how a change in the cell's state could affect the genome in a way that will be adaptive. We gave an example of this type of genomic response (the "local" type of induced mutation) when we described how the mutation rate of a defective gene in an amino acid biosynthesis pathway in *E. coli* increases when the amino acid is in short supply. However, even here, although the mutations are highly targeted, there is randomness in the changes produced within the targeted region. But you are right: when a system is complex, and the many interactions between genes and the environment make the phenotypic effects of genes very indirect, transferring information from the organism to the DNA becomes less likely. This is another reason why we would not expect back-translation. Even if information could be transferred from an altered protein to the DNA sequence that codes for this protein, it would lead to an adaptive change only in the relatively rare cases when the gene-protein-trait relation is very simple. Usually it is not.

I.M.: So the more complex the organism, the less likely it is that they will have systems that enable directed genetic change to occur?

M.E.: Yes and no. Don't forget that directed genetic changes are found *within* complex organisms—we depend on them for our immune responses. They are adaptive because they are restricted to only one type of cell. So the basic machinery for making controlled genomic changes is certainly present, even in organisms like us. Yet, as far as we know, it is not used to produce directed changes in the genes that are passed from one generation to the next. One reason for this may be that, however localized in the genome the mutations are, in complex organisms "directed" mutations would have "random" effects on the organism as a whole, simply because of all the cellular interactions.

I.M.: You are implying that although some microorganisms have evolved systems that enable them to do a bit of Lamarckian evolution by transmitting genetic information that has been modified in response to their conditions of life, more complex organisms cannot do this. Am I right?

M.E.: We think it is unlikely that complex organisms have systems that enable adaptive changes to be induced in the genes transmitted to their offspring, although we wouldn't rule it out. If you think about it from an evolutionary point of view, multicellular organisms are in an odd situation. On the one hand there are many circumstances in which passing on some induced, "acquired" characters would be beneficial. Yet, on the other hand, their option to transfer adaptive information through induced changes in DNA becomes less as biological complexity increases.

I.M.: You agree then that complex organisms do not go in for Lamarckian evolution?

M.E.: Not at all. As we have said elsewhere, not everything that is inherited is genetic. There are systems that transmit information between generations at a *supragenetic* level. With these, adaptations that occur during life are coupled far more directly with the information that the organism transmits to the next generation. Consequently, through the supragenetic inheritance systems, complex organisms can pass on some acquired characters. So Lamarckian evolution is certainly possible for them. In the next three chapters we describe these additional inheritance systems—the epigenetic, the behavioral, and the symbolic—and show how they can have both direct and indirect influences on evolutionary change.

II   Three More Dimensions

The idea that DNA alone is responsible for all the hereditary differences between individuals is now so firmly fixed in people's minds that it is difficult to get rid of it. When it is suggested that information transmitted through nongenetic inheritance systems is of real importance for understanding heredity and evolution, two problems arise. The first is that for most people the genetic system seems quite sufficient to explain everything. They invoke Occam's razor: if one system can explain everything, why do we need to look for others? The second problem is that even when people agree that there is no escape from the mass of experimental data showing clearly that there are other, nongenetic, inheritance systems, they find it difficult to know how to think about them and their significance in evolution. We are all deeply conditioned by what we know about the genetic system, and tend to attribute its properties to other types of inheritance and to evaluate them in terms defined by the genetic system. We should not do this, of course, but it is difficult to change our habits of thought.

Some years ago, after several very frustrating and largely unsuccessful attempts to get our point of view across to colleagues and students, we found an analogy for how different heredity systems could work alongside the genetic system. Since it seemed to satisfy many people, we will repeat it here. Think about a piece of music that is represented by a system of notes written on paper, a score. The score is copied repeatedly as it is passed on from one generation to the next. Very rarely, uncorrected mistakes occur during copying, and perhaps an impertinent copier sometimes makes a tiny deliberate alteration, but with the exception of such small and rare changes, the piece of music is transmitted faithfully from generation to generation in the form of the written score. The relationship between the score and the music is analogous to the genotype/phenotype distinction. Only the genotype (the score) is transmitted from one generation to the next; the phenotype (the particular performance, the actual interpretation of the piece) is not. Changes in the genotype (mutations) are passed on; changes in the phenotype (acquired characters) are not.

This was the situation until new ways of transmitting music were invented. The technologies of recording and broadcasting made it possible to transmit performances by recording them, editing them, copying them onto tapes or disks, and broadcasting them. Now, through these new technologies, the actual interpretations of the music can be transmitted as well as the written musical score. In terms of the genotype/phenotype analogy, the recording and broadcasting systems transmit the "phenotypes" of the pieces, rather than the "genotypic" instructions in the score. A phenotype, one particular performance, is affected by the notes in the written score, the skill of the musicians, the nature of the musical instruments, the general musical culture, and so on. Importantly, it is also affected by the interpretations of the score that the conductor and musicians have heard in the past—by earlier phenotypes. The relationship between the two systems of transmission is usually unidirectional—a change in the score alters performances, whereas the performance of the music usually does not change the score. However, occasionally a performance may alter the score: a particularly popular interpretation of the music may lead to a version of the score that includes notational changes that make it easier for the interpretation to be reconstructed. In this case, a phenotype affects a genotype. In all cases, by opening up a new channel of information transmission, the new technologies can affect the way the music is played.

The recording-broadcasting transmission system is based on a completely different technology from that of copying of the score, and likewise the heredity systems that we are going to discuss in the next three chapters are completely different from the DNA system. They do not come *instead* of the DNA transmission system (the written score); they are additional to it. The genetic system is the basis of all biological organization, including the organization of the supragenetic heredity systems we are going to consider, but these additional systems allow variations in a different type of information to be transmitted. The variations occur at higher levels of organization—at the cell, organism, or group level. They may be quite independent of variations at the genetic level, in just the same way that variations in recorded performances may be independent of variations in the score. The genetic system, like a score, defines the range of possibilities, and when this range is wide and many heritable phenotypes are possible, a lot of interesting evolution can occur through natural selection acting on these variant phenotypes.

In the next three chapters we shall describe some very disparate types of heredity systems, all of which allow phenotypic variations to be transmitted from one generation to the next. In chapter 4, we look at the evolutionary implications of cellular inheritance systems. In chapter 5 we focus on behavioral transmission in nonhuman animals, and see what it means for evolution. Chapter 6 will be about human symbolic systems and cultural evolution. For the time being, as far as it is possible, we will ignore variations in the genetic system, and also the interactions of the different systems of inheritance with each other and with the genetic system. We come to all of these topics in part III.

4 | The Epigenetic Inheritance Systems

A person's liver cells, skin cells, and kidney cells, look different, behave differently, and function differently, yet they all contain the same genetic information. With very few exceptions, the differences between specialized cells are epigenetic, not genetic. They are the consequences of events that occurred during the developmental history of each type of cell and determined which genes are turned on, and how their products act and interact. The remarkable thing about many specialized cells is that not only can they maintain their own particular phenotype for long periods, they can also transmit it to daughter cells. When liver cells divide their daughters are liver cells, and the daughters of kidney cells are kidney cells. Although their DNA sequences remain unchanged during development, cells nevertheless acquire information that they can pass to their progeny. This information is transmitted through what are known as *epigenetic inheritance systems* (or EISs for short). It is these systems that provide the second dimension of heredity and evolution.

Until the mid-1970s, the existence of epigenetic inheritance was barely recognized. Developmental biologists devoted most of their efforts to trying to find out how cells became differentiated. They were concerned with the signals that switched genes on and off, and with the cascade of events that led to cells in one place becoming specialized for one particular function, while those somewhere else were induced to have a different function. The emphasis was on how cells acquired their specialized roles rather than on the complementary problem of how, once the appropriate genes had been turned on and off, cells remembered their new epigenetic state and transmitted it to their progeny. In 1975, two rather speculative articles drew attention to the problem by suggesting a possible solution to it. Robin Holliday and John Pugh, two British biologists, and Arthur Riggs in America independently suggested a mechanism that would enable states of gene activity and inactivity to be maintained and transmitted to future cell generations. Their ideas generated a lot of interest, and after a rather slow start the study of cell memory and epigenetic inheritance began to take off. It was given even more impetus when it was realized that understanding epigenetic inheritance was going to be crucial to the success of cloning and genetic engineering projects.

Today, epigenetics is quite a buzzword, and biologists are well aware of the existence of EISs and their importance in development and medicine. However, there is still a reluctance to recognize that they may also have a significant role in evolution. So, to illustrate how EISs can affect evolution, we are going to resort to another thought experiment. The scenario that we are going to describe will show that evolution is possible on the basis of heritable epigenetic variation even when there is no genetic variation at all. To avoid any misunderstanding, we need to stress at the outset that we do not underestimate the importance of genetic variation in evolution. We are using the thought experiment merely to show that it is possible to think about evolutionary change based solely on variations transmitted by nongenetic cellular inheritance systems.

Evolution on Jaynus

Imagine that on Jaynus, a planet not too far or too different from our own, there is life. The organisms found there are very diverse, having all sorts of amazing shapes and behaviors, although their complexity does not exceed that of a jellyfish (figure 4.1). All Jaynus creatures multiply solely by asexual processes: there is nothing like the meiotic cell division that leads to the production of gametes in Earth's animals and plants, and there is no sexual reproduction of any other type. But, just as here on Earth, there are several types of asexual reproduction. Some creatures multiply by shedding buds from the adult body; in most others multiplication is through single cells that become detached, start dividing, and develop into adults; and in a few it occurs through the assembly of cells from several different individuals to form a kind of "embryo," which then begins the developmental process.

Jaynus organisms have a genetic system that is based on DNA, and replication, transcription, and translation are much the same as on Earth. However, there is one very extraordinary thing about the DNA of Jaynus creatures—*every organism has exactly the same DNA sequences*. From the simplest organism, a tiny unicellular creature, to the enormous fanlike colonial worms, the DNA is identical. Their genomes are large and complex,

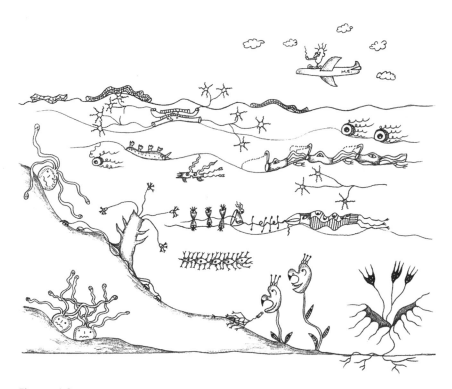

Figure 4.1
Life forms on Jaynus.

but no organism deviates from the universal standard sequences because there are cellular systems that check DNA and destroy any cell suspected of carrying a mutation.

The evolutionary history of Jaynus organisms began about 2 billion years ago, when a large chunk of our own planet became detached and disintegrated into meteorites. These meteorites contained the simple life forms that had evolved on Earth, and one of them, carrying its living cargo (in a state of suspended animation, of course), reached Jaynus. Some organisms survived, and since conditions on Jaynus allowed life to flourish, they evolved into various unicellular and multicellular forms. Present-day organisms are all descendants of a common ancestor—a floating, colonial, mattress-like form, with the same genome as that now found in all of them. Through natural selection, this ancestor's descendants evolved the "suicide" system for genetically deviant cells, but they still diverged to form all of the many types of organisms found today. Adaptation to different habitats led to structural and functional modifications of the original "mattress" that were hereditary and cumulative. In the shallow parts of the stormy sea, some individuals adhered to flat rocks, evolving a "stem" and flat, leaflike structures that absorb light, energy, and the organic material that is constantly and spontaneously formed on Jaynus. These individuals had an advantage over their free-floating sisters, since they did not break up as easily and could absorb nutrients and other materials more readily, so natural selection led to the accumulation of adaptations in this direction. In other more open habitats, the original mattress fragmented into small balls whose outer cells produced rapidly moving flagella; from this state one line evolved by further fragmentation into single cells, which divided very rapidly and parasitized other species.

Now imagine what happened when Earth scientists arrived on Jaynus and started studying the living things they found there. From the similarity to life on Earth, particularly the presence of the DNA inheritance system, they were quickly able to guess at the early evolutionary history of the organisms, but the lack of genetic differences between them amazed and at first mystified them. How did the wormlike and plantlike creatures evolve from a common, simpler ancestor if the genome was fixed? There was no shortage of heritable phenotypic variation within populations of Jaynus creatures, and the heritable differences among morphological types ("species") were enormous, but what could be the basis of these hereditary differences?

After a short initial period of confusion and disbelief, the scientists focused on the cellular heredity systems, the EISs, which they were familiar with through work with organisms on Earth. When they looked at these systems in the Jaynusites, they discovered that all hereditary variation and evolution in these organisms is based on very sophisticated cellular heredity systems. Variations in the functional states of cells, in cell architecture, and in cellular processes can all be transmitted from generation to generation. Sometimes, depending on their mode of multiplication, variations in the organization of whole tissues and organs are transmitted. Because EISs play a double role, being both response systems *and* systems of transmission, the scientists concluded that the role of directed or interpretive variation had been much larger on Jaynus than on Earth.

We will explain the dual role of EISs in more detail later. Meanwhile, let's continue with the thought experiment and imagine how, in order to clarify the unfamiliar mode of heredity and evolution on Jaynus to its readers, who were conditioned to think about DNA-based inheritance and evolution, the *Daily Earth* attempted to explain the bizarre phenomena. The headline read

Not So Different After All

At last scientists are beginning to make sense of what went on as life evolved on Jaynus. And the remarkable thing is that what shaped those strange creatures is something we know quite a lot about from studies of how our own bodies are formed. The differences between the "epibeasts," as scientists affectionately call them, are much the same as the differences between your lungs and your liver, your kidneys and your skin, your blood and your brain.

Professor Paxine Mandela, head of the Epigenetics Institute in Burkly, put it this way: "All tissues and organs are made up of cells, and almost every type of cell in your body has exactly the same DNA. What makes your liver, lung, kidney, skin, blood, and brain cells different is not different genes or DNA, but different use of the information encoded in DNA. To keep it simple, you can think about genes being 'on' or 'off'—active and involved in producing some product, or inactive with no product produced. The whole genome—all of the genes in a cell—can then be likened to a huge switchboard, with genes in the switched-on condition showing a red light, and switched-off genes showing green. If you compare the switchboard in different types of cell, you'll find the patterns of red and green lights are different. They have different combinations of genes switched on.

"Cell switches are turned on and off at certain critical stages during development, as tissues and organs form. Once established, much of the pattern of red and green lights on the switchboard is locked in, and the same pattern is inherited by daughter cells. So, the various cell types breed in their own image—skin cells don't produce kidney cells, they produce more skin cells. And liver cells produce liver cells, and kidney cells produce kidney cells. We call the cellular systems responsible for the maintenance and trans- mission of patterns of gene activity and other cell states 'EISs,' short for 'epigenetic inheritance systems.'"

You may be asking by now what this has got to do with Jaynus life. Well, according to Professor Mandela, EISs underlie a lot of what goes on in Jaynus organisms. The sur- prising discovery that all those weird creatures have exactly the same DNA forced sci- entists to look for other ways in which they could pass on their features, and this is how Professor Mandela got involved. She and her colleagues have now discovered that what we see on Jaynus is much the same as what we would see here on Earth if each of your organs was an independent creature that could breed. Imagine kidney creatures that could bud off tiny kidney buds that developed into mature kidney creatures, and heart creatures that could shed buds that developed into heart creatures; think of sheetlike skin creatures multiplying by fragmenting, or blood cell creatures multiplying by simple cell division. That's what goes on on Jaynus. There's no sex when Jaynus's epibeasts multiply—it's all asexual splitting and budding and aggregation. And just as there's no difference between the DNA in our various organs, there's no difference between the DNA of the various Jaynus creatures. They all have the same genes, but how they use those genes is very different, and these differences are transmitted from one generation to the next. Each species of epibeast, just like each organ, has a characteristic set of epigenetic patterns which it passes on.

By now you have probably guessed how evolution by Darwinian natural selection is possible for Jaynus's epibeasts, even though they all have the same genes. Epibeasts have wonderful systems for keeping their DNA perfect and unchanging, but the patterns of lights on their switchboards and other bits of their non-DNA inheritance systems do get changed, so new variants appear. Sometimes they arise through errors, and sometimes because an environmental feature forces a switch to be thrown, but however they are produced, if the variant helps the epibeast to survive and multiply better, the lineage will change.

So, although they are unlike the animals and plants on our own planet, whose evolution is thought to be based on the selection of changes in DNA, evolution in Jaynus creatures still depends on the selection of heritable variants. But the variations in epibeasts are passed on by their elaborate EISs, and new variants can arise when conditions alter the way their genetic information is used. The interesting question now, according to Professor Mandela, is to find out how important EISs have been in the evolution of life here on Earth. "After all," she stresses, "EISs are not unique to Jaynus organisms. Epibeasts are not so different, after all."

Let us leave life on Jaynus and return to biological reality on our own planet. We used the imaginary scenario of life on Jaynus to focus atten-

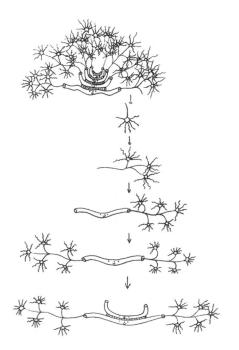

Figure 4.2
Asexual reproduction of *Jaynusi zeligowska*.

tion on EISs and their role in evolution. Biologists now know quite a lot about these systems, but they tend to associate them with ontogeny—with the processes through which the fertilized egg develops into an adult organism with specialized cells, organs, and organ systems. They stress the role of EISs in the determination and regulation of cellular activities, but usually overlook their evolutionary potential. What we are now going to do is describe four broad categories of EISs, which we will characterize first by looking at their role in cell heredity, and then by looking at the wider, evolutionary issues. It may be helpful to keep in mind the idea that EISs are additional "transmission technologies." In the same way that recording and broadcasting are technologies that transmit interpretations of the information contained in a musical score, EISs transmit interpretations of information in DNA. They transmit phenotypes rather than genotypes.

Self-Sustaining Loops: Memories of Gene Activity

Through the first type of EIS, daughter cells can inherit patterns of gene activity present in the parent cell. They do so when the control of gene activity involves self-sustaining feedback loops. This type of system was first described theoretically by the American geneticist Sewell Wright in 1945, and by the late 1950s examples had been found in bacteria. Subsequently it has been found in every living organism that has been studied, and its significance as a cell memory system has become clear.

The essence of a self-sustaining system is that A causes B, and B causes A. The simplest example is one in which a temporary cue turns a gene on, and that gene's product then ensures the continued activity of the gene. Figure 4.3 shows how the system works. When gene A is active, a protein is produced which, among other things, acts as a regulator, attaching itself to the control region of gene A, keeping it active long after the original inducing cue has disappeared. Following cell division, if the level of A's protein remains high enough in each daughter cell, it will continue to act as a positive regulator and the gene will remain active in both cells.

This feedback system means that in the same noninducing conditions there can be two genetically identical types of cell, in one of which the gene is active, and in the other it is inactive. The difference between the two cell types stems from the different histories of their ancestors—on whether or not they received the initial cue that switched the gene on. This cue might have been an external environmental change, or an internal developmental or regulatory factor. Or, occasionally, an ancestral cell's state may have changed as a result of "noise"—through random

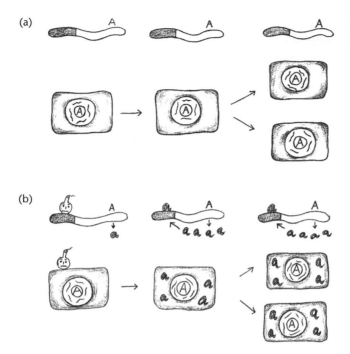

Figure 4.3

Self-sustaining loops: in (a), gene A is inactive and daughter cells inherit the inactive state; in (b), a temporary stimulus (bomb) induces gene A to be active and produce its product. Because daughter cells inherit the product, they also inherit the active state.

fluctuations in the cellular environment that turned the gene on or off. Whatever the cause of the gene being switched on, for as long as the amount of its protein product does not fall too much, it will remain active after cell division. The inheritance of the active or inactive state is simply an automatic consequence of more or less symmetrical cell division.

Most self-sustaining loops are a lot more complex than the simple one we have just described. They contain several genes, several regulatory regions, and several proteins. The principle is much the same, however. The most important difference between simple and complex loops is that the latter can be very stable, whereas a simple loop can easily be perturbed by changed conditions.

If we think about a self-sustaining loop as an information system, what can we say about the organization of the transmissible information? Even a simple loop has component parts (the coding sequence of the gene, its regulatory region, and its protein product), but its functional state (whether active or inactive) depends on the interactions between them. The state of the loop is therefore transmitted from generation to generation as a whole, and it varies as a whole. Thus, it is the loop that is the unit of heritable variation. Following Maynard Smith and Szathmáry, we will call information that is organized in such a nondecomposable manner *holistic*. It is very different from the information in modular systems such as the DNA system, where the components (the nucleotides A, C, T, and G) can be changed without destroying the whole.

The functional state of a self-sustaining loop is heritable, but how much evolution can there be with such a system? The holistic nature of the loop means that it can have very few functional states. A simple loop usually has two—active and inactive—so there are only two variants. With only two variants, natural selection cannot accomplish very much, except switch between them according to circumstances. This is not very interesting from an evolutionary point of view. However, every cell has many different self-sustaining loops, which may be independent of each other. If a cell has just twenty autonomous self-sustaining loops, each of which can have two states, there are over a million possible functional variants of the cell. There is definitely evolutionary potential here, and natural selection could lead to interesting adaptations. But in order to have this large amount of variability, we have to treat each feedback loop as a component in a collection of loops, and look at the combination of different active/inactive loops in the whole cell. What is transferred between generations is part of the cell phenotype—it's a set of patterns of gene activity.

Structural Inheritance: Architectural Memories

The second type of epigenetic inheritance is very different from the previous one, since it is concerned with cell structures, not gene activities. Alternative versions of some cellular structures can be inherited because existing structures guide the formation of similar ones in daughter cells.

The most remarkable examples of structural inheritance have been found in ciliates, a group of unicellular organisms that have orderly rows of cilia— short, hairlike appendages—on their outer surface, the cortex. Like many other aspects of their morphology, the organization of the ciliary rows is inherited. For example, the average number of ciliary rows differs in different strains. There is nothing very remarkable about that, of course, but what is remarkable is the peculiar type of cortical inheritance revealed by some experiments carried out by the American geneticist Tracy Sonneborn and his colleagues in the 1960s. Because of their relatively large size, it is possible to perform microsurgery on ciliates such as *Paramecium*. So what Sonneborn and his coworkers did was cut out a piece of the cortex and rotate it through 180 degrees before reinserting it. They then looked at the descendants of the operated-upon organisms. Amazingly, the offspring inherited the change: they, too, had an inverted row of cilia. It was as if the descendants of a person whose leg had been amputated inherited the same handicap.

Similar experiments with other ciliates as well as *Paramecium* have shown that various altered cortical structures can be inherited for many generations, but the mechanisms behind this are far from being understood. Those currently working in this field believe that some kind of three-dimensional templating is involved. Somehow, a structure in the mother cell acts as a template that directs the assembly of the protein units that form a similar structure in a daughter cell. Although we do not know exactly how this happens, the key point for present purposes is that it is the *organization* of the cortex that is changed and inherited, not the constituent components. The same building blocks are used to build several different self-templating heritable structures.

The idea that preformed structures play a crucial role in cell heredity has been taken up and extended by the British biologist Tom Cavalier-Smith. He has considered how the many types of membrane in an ordinary cell are formed. Cell membranes, such as the plasma membrane that surrounds the cell, or those of the internal membrane system known as the endoplasmic reticulum, or those around the mitochondria (the small energy-generating organelles in present-day cells) differ from each other in composition as well as in location. For example, the nature and organization of their proteins is different. Such membranes cannot assemble without guidance. Their persistence and continuity depend on preexisting membranes, which template the formation of more membranes with the same structure. Through this templating, the membrane grows and is eventually divided between daughter cells. Cavalier-Smith calls the set of self-perpetuating membranes the "membranome" of the cell because like the genome it carries hereditary information in its structure. He believes that some of the most dramatic events in early evolution, including the formation of the first true cells, the origin of various bacterial groups, and the emergence of the first eukaryotic cell, were associated with and dependent on changes in the membranome. As he sees it, the evolution of life cannot be understood without recognizing the importance of structural inheritance. "The popular notion that the genome contains 'all the information needed to make a worm' is simply false," says Cavalier-Smith.

Recently, interest in structural inheritance has grown for a practical although rather unfortunate reason. Certain disease agents, which have potentially devastating effects, seem to have self-templating properties. These agents, called prions, do not contain DNA or RNA; they are made of proteins. Prions are associated with diseases of the nervous system, such as BSE (bovine spongiform encephalopathy, commonly called mad cow disease), scrapie in sheep, and CJD (Creutzfeldt-Jakob disease) in humans. The history of the research that led to the discovery of prions is an interesting one, which has a lot to teach us about the politics and sociology of science, but this is not the place to go into details. A good starting point for our story is with the Fore people of New Guinea, who in the early part of the twentieth century were still relatively isolated and practicing a Stone Age type of culture. These people were found to have a high incidence of a debilitating condition they called *kuru*, meaning "shivering" or "trembling." As well as having tremors, people with kuru became increasingly unsteady on their feet, developed blurred speech, and showed various behavioral changes. Inevitably they died within a year or two of the symptoms first appearing. The Fore people attributed the disease to sorcery, and early Western visitors also assumed that it had a psychosomatic cause, but in the 1950s it was recognized as a degenerative disease of the nervous system. At that time it affected about 1 percent of the population. But what caused it?

For some time investigators thought that kuru might be a genetic disease, because it was unique to the Fore people, and tended to run in families. It affected mainly women and children, which genetically is a little odd, but with the help of some creative reasoning, the pattern of transmission could be explained in terms of the inheritance of a defective gene. There were certain things that didn't quite fit the Mendelian model, however. For example, women who married into an affected family would often develop kuru, although there was no reason to think they carried the purported kuru gene. All alternative explanations of the cause of the disease seemed equally unsatisfactory, however. Dietary deficiencies did not fit the facts, and it could not be a normal infection, because neighboring groups with whom the affected villagers traded and interacted did not catch the disease.

Nevertheless, eventually it was shown that kuru is caused by an infectious agent. When Carleton Gajdusek (an American virologist who was to receive a Nobel Prize in 1976) and his coworkers injected samples of brain tissue from people who had died of kuru into the brains of chimps, these animals developed a comparable disease about a year and a half later. Their brain tissue was then capable of infecting other chimps, whose tissues could infect other chimps, and so on. Kuru was clearly transmissible. And the sad fact is that something not unlike Gajdusek's experiments had been happening among the Fore people. Kuru was a result not of their genetic inheritance, but of their mourning rituals, in which women and children dismembered, cooked, and ate the body of their dead relatives, including the brain. Men and older boys were affected far less than women and children were, because they lived separately from them, and only rarely took part in the mourning rites. Thankfully, as the Fore people abandoned cannibalism in the late 1950s, the number of deaths from kuru declined dramatically, although for several decades it continued to appear in previously infected people.

The studies of Gajdusek and others showed that the causative agent of kuru, like diseases such as scrapie and CJD, was an infectious agent, but it was one with some very unusual properties. It was very resistant to heat, chemicals, and radiation, and did not cause an inflammatory response. The incubation periods for the diseases were very long. In fact, many of the usual properties of viruses and viral infections were missing, including the presence of infectious nucleic acids. So what was the nature of these very unconventional "slow viruses," as they were called?

In the 1980s, Stanley Prusiner, a Nobel Prize winner in 1997, started advocating the then very unfashionable idea that the infectious agent causing the degenerative brain diseases was made solely of protein. He suggested that "*pro*teinaceous *in*fectious particles," or "prions," are proteins with an abnormal conformation and, importantly, they are able to convert the normal form of the protein into their own aberrant form. Figure 4.4 shows the essence of the idea (we've left out other molecules that may be

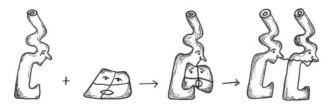

Figure 4.4
Structural inheritance: an aberrant prion protein (the long form) contacts the normal protein (square form) and induces it to change its conformation into its own aberrant shape.

needed for the conformational change). Once prions are present, there is therefore a chain reaction, with more and more abnormal protein accumulating. The differing chemical and physical properties of the aberrant form of the protein affect the structure and functioning of brain cells, thus causing the symptoms of the diseases.

When BSE erupted in England in the mid-1980s, damaging the livelihood of farmers, interest in prions began to grow even more. BSE was soon linked to the practice of feeding cattle with sheep- and cattle-derived protein supplements (thereby making cattle cannibals too!). Aberrant proteins in the feed initiated the transformation of proteins in the cattle that had eaten it, and when the bones and flesh of these animals became cattle feed, more cattle were infected, and so on. Eventually it was reluctantly acknowledged that a disease that was occurring in humans, "new variant CJD," was also likely to have been caused by eating the flesh of infected cattle. The public panic that followed this admission caused even more problems for farmers, because many people gave up eating British beef, even though the feeding and slaughtering methods were changed. Unfortunately, because prion diseases have a long latent period, the problems created by BSE remain with us. It is still too early to know the full effect that eating meat from cattle carrying the BSE agent will have on human health, or how effective the measures taken to eradicate it will have been.

With hindsight, the agricultural practices and attitudes that generated the BSE crisis in England are puzzling. When Gajdusek gave his Nobel lecture in 1976, he summarized what was then known about the spongiform encephalopathies, including the way that the causal agent can be transferred from one species to another. It was not known what it was, but the infectious agent's ability to cross the species barrier was well established long before the BSE crisis. We now know that prions can sometimes infect other species because they are able to convert the corresponding protein into an abnormal form, even if it has a slightly different amino acid sequence.

One good side effect of the BSE crisis was that it gave prion research a boost and made biologists far more aware of this type of heredity. Very different types of prion have been identified in yeast and in the fungus *Podospora*. These prions can be transferred from one cell generation to the next, and template the formation of similar prions in the daughter cells. Their discovery has provided an explanation for some cases of non-Mendelian inheritance in yeast that had puzzled geneticists for a long time. Moreover, unlike the prions of mammals, which damage the cells containing them, yeast and fungal prions seem to do the cells no harm. In fact there is evidence, which we will come to in chapter 7, that some of them may have adaptive roles.

Prions may even have adaptive roles in multicellular organisms. A protein with prion-like properties has recently been found to underlie the ability of the sea slug *Aplysia* to remember its past experiences. Remarkably, cell memory and the organism's memory seem to be related! The scientists who discovered this believe that it may be just the beginning, and that there are probably many more proteins whose functional importance is related to their prionlike properties.

What can we say about the way information is organized and transmitted in structural inheritance systems? Information is holistic, of course, because the properties of prions and other self-templating cellular entities reside in their 3D structure. The information affects cell phenotypes and is transferred when the parental conformation is reconstructed. In contrast to the DNA system, there is no specialized replication machinery that is able to copy any structure, regardless of the way the component units are organized. The ability of a structure to be reconstructed in daughter cells is inherent in its organization. Most variations in the conformation or organization of a prion or other structural unit are probably not self-perpetuating at all, although work with mammalian prions has shown that a single protein can produce several prion "strains," characterized phenotypically by different incubation times and differences in the nature and distribution of the brain lesions. Even so, it is likely that the number of self-templating organizations that a structural complex can assume is small, so evolution at the level of the single structure is severely limited. But, as we argued for self-sustaining loops, if every cell has many independent heritable structural complexes, the amount of variation at the cell level can be enormous, and interesting evolution is therefore possible.

Chromatin-Marking Systems: Chromosomal Memories

The third type of EIS is known as the chromatin-marking system. Chromatin is the stuff of chromosomes—it is the DNA plus all the RNA, proteins, and other molecules associated with it. In eukaryotes, small proteins called "histones" are a necessary part of chromosomes. They play a major structural role in compacting DNA. Slightly less than two turns of DNA, a length of about 146 nucleotide pairs, is wound around a core of eight histones (two molecules of each of four types) to form a beadlike structure known as a nucleosome, from which the tails of the histone molecules protrude. With the help of another type of histone, which links each

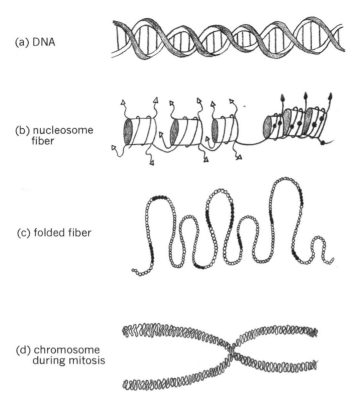

(a) DNA

(b) nucleosome fiber

(c) folded fiber

(d) chromosome during mitosis

Figure 4.5

Stages of DNA packaging. The DNA molecule shown in (a) is wound around histone cores to form the nucleosome fiber shown in (b). The region on the left in (b) represents active chromatin with a loose structure, whereas in the inactive region on the right the chromatin fiber is condensed and has differently modified histone tails. The nucleosome fiber is further folded (c), and before cell division chromatin is compacted even more; (d) shows its very compact form during the stage in mitosis when the chromosomes have replicated but daughter chromosomes have not yet separated.

nucleosome core to the DNA between them, the string of nucleosomes is twisted into a chromatin fiber, which is further compacted into an assembly of loops. Figure 4.5 shows the stages of chromatin packaging.

Although its structure is so complicated, chromatin is not something fixed and unchanging. The same DNA sequences can be packaged differently in different cell types, and at different times during a cell's life span. Not surprisingly, how a region of DNA is packaged, and the nature and density of the proteins and other molecular entities associated with it, determine how accessible it is to the factors necessary for the genes in it to be transcribed. The structure of chromatin therefore affects the likelihood that genes will be active. We touched on this earlier when we mentioned the regulatory molecules that bind to DNA, thereby enabling or preventing transcription.

The non-DNA features of chromatin that are of interest to us here are those that are transmitted from generation to generation and enable states of gene activity or inactivity to be perpetuated in cell lineages. Such alternative heritable differences in chromatin have come to be known as "chromatin marks." For more than a quarter of a century now, it has been realized that finding out how these marks are established, how they function, and how they are transmitted to daughter cells is one of the keys to understanding development. There are several different types of chromatin mark, but the first to be recognized, and the one about which we now know most, is DNA methylation. It was speculation about the epigenetic role of DNA methylation by Holliday and Pugh in England and Riggs in America that launched the modern study of EISs in 1975.

Methylated DNA, which is found in all vertebrates, plants, and many (although not all) invertebrates, fungi, and bacteria, has a small methyl group (the chemists write it CH_3) attached to some of its bases. The quantity and distribution of methylated bases varies widely between groups, but in many the methyl group is attached to the base cytosine (C). Cytosines can therefore exist in either a methylated (C^m) or unmethylated (C) state. Adding this methyl group does not alter the role of cytosine in the genetic code. If a DNA sequence codes for a protein (and remember that a lot of DNA does not), the protein that is produced will have exactly the same amino acid sequence whether or not some or all of the cytosines are methylated. What the methylated bases in and around the gene do is not alter the protein, but influence the likelihood that it will be transcribed. Usually (but not always) genes in densely methylated regions are not transcribed, although exactly how methylation has this effect is not understood. Sometimes it may affect transcription directly, by interfering with the binding of regulatory factors to a gene's control region. Alternatively, it may act more indirectly, through a set of proteins that bind specifically to methylated DNA and prevent the transcription machinery from getting to work on it. In whatever way it works, the different patterns of methylation that characterize different cell types are part of the system that determines which genes are permanently silent and which can be transcribed.

Methylation patterns do more than influence how easily genes can be turned on and off. They are also a part of the heredity system that transfers epigenetic information from mother cells to daughter cells, and we have a fairly good idea of how this works (see figure 4.6). Methylation patterns can be reproduced (at least in vertebrates and plants) because they hitchhike on the semiconservative replication of DNA. Commonly, methylation occurs in the cytosines of CG doublets or CNG triplets (N can be any one of the four nucleotides). Because the nucleotides in the DNA molecule are paired, and C is always paired with G, a CG doublet on one DNA strand is always paired with GC on the other. When the cytosines are

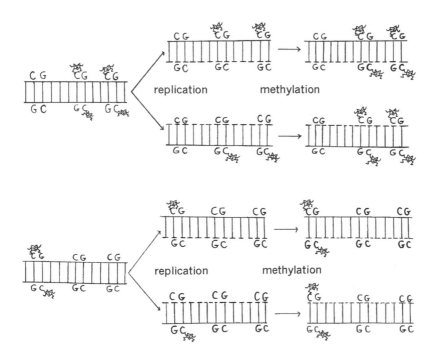

Figure 4.6
Replication of two methylation patterns. Solid lines are parental DNA strands; dashed lines are daughter strands; the icons on some of the Cs represent methyl groups.

methylated, the symmetry is the same—C^{m}G is always paired with GC^m. However, when DNA replicates, the newly formed strand is unmethylated, so a C^{m}G doublet in the old strand is partnered by a nonmethylated GC doublet in the new strand. This asymmetrical situation doesn't last long, because it is recognized by an enzyme, methyltransferase, which promptly attaches a methyl group to the cytosine of the new strand. Since the enzyme ignores sequences that are nonmethylated on both strands, the old pattern of methylation is reconstructed in the daughter DNA molecules. Like other replication processes, this one isn't perfect and errors are made, but there seem to be backup systems that maintain the overall pattern of methylation. Figure 4.6 shows both how a DNA sequence can have different sites methylated, and how the different patterns of methylation—the marks—are reproduced when DNA is replicated.

During development, methylation marks change, although it is far from clear how the new marks are generated. It could happen through enzymatically adding or removing the methyl groups, or by a change in chromatin structure that affects the access of the enzymes that maintain methylation patterns. Whatever the mechanisms that establish them, these regulated changes and the subsequent maintenance of methylation patterns are essential for normal development. The clearest evidence of this comes from studies of mice whose methyltransferase genes have been knocked out. They develop abnormally, and die before birth. We also know that the methylation patterns of tumor cells often differ from those of normal cells. With some colon cancers, for example, the first obvious sign of the transformation of a normal cell into a tumor cell is a change in DNA methylation. We do not know the cause of such transformations, but one possibility is that chemical agents directly or indirectly alter methylation patterns, and hence affect the normal activity of the genes that regulate cell growth and division. Several people have suggested that one of the causes of our increasing health problems as we get older is the accumulation of accidental changes in methylation marks that make our cells work less and less efficiently. There is a little experimental evidence that supports this conjecture, although it is probably only one of many causes of aging changes.

Methylation is the chromatin-marking EIS about which we know most, so what can we say about the way in which information is organized and transmitted through this system? The most obvious thing is probably that it is very similar to DNA. As with the copying of DNA sequences, copying methylation patterns depends on the activity of enzymes that will reproduce any pattern, irrespective of the information it carries. Another similarity is in the way information is organized—it is modular. Often (though not always) it is possible to change the methylation state of one cytosine without affecting any others. Potentially, therefore, the number of different methylation patterns a gene can have is very large, even if the DNA sequence is unchanged.

Methylation is not the only chromatin-marking system. This is obvious if we think about invertebrates, which often have little or no methylated DNA. The fruit fly *Drosophila*, for example, has so little that until recently it was thought that it probably had none at all, yet cell phenotypes are undoubtedly transmitted to daughter cells. There are good experimental reasons to think that they are able to do so because the protein complexes that bind to DNA and influence gene activity can also act as heritable chromatin marks. Several models for the "copying" and transmission of these protein marks have been proposed, and we show one of these in figure 4.7. We have to admit, however, that although it is clear that protein-based chromatin marks exist, scientists are still a long way from understanding how they are inherited.

Chromatin structure is so extraordinarily complicated that it is inevitable that ideas about how it is organized and the way that it transmits information are surfacing all the time. Recently a lot of attention has been focused on modifications of the nucleosomal histones. Some of the amino acids in the tails of histones can be modified by enzymes that add or

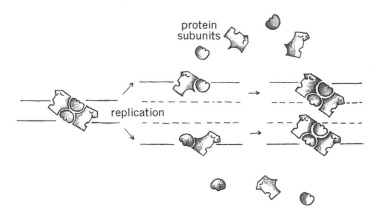

Figure 4.7
Inheritance of protein marks. Prior to replication, protein subunits are bound symmetrically to the two DNA strands; following replication, the subunits bound to the parental strands (solid lines) guide the assembly of similar units on the daughter strands (dashed lines).

remove small chemical groups such as acetyls or methyls. The presence or absence of these groups alters the interactions of histones with each other, with DNA, and with other proteins, so they affect how tightly DNA is packaged and how accessible it is to the transcription machinery. For example, the addition of acetyl groups usually leads to a looser chromatin structure, so transcription is more likely, whereas removing acetyl groups and adding methyl groups usually results in a very condensed chromatin fiber, which does not allow the DNA to be transcribed.

People are now beginning to work out how the modifications of various sites in histones occur, and how they have their effects. There is talk of a "histone code," because it looks as if combinations of differently modified histones form marks that affect the binding of specific regulatory factors. Whether it is a true code remains to be seen, but there is little doubt that histone modifications are a crucial part of the chromatin-marking system that determines gene activity. At present little is known about how histone marks are duplicated, although there are some clues. Nucleosomes are disrupted when DNA replicates, and the histones segregate randomly to daughter molecules. However, they remain associated with the same region of DNA, and somehow seed the reconstruction of a chromatin structure similar to that of the parent molecule. The information associated with histone marks is therefore inherited by the daughter cells.

By describing methylation marks, protein marks, and histone marks separately, we have probably made it look as if they are independent aspects of chromatin structure. They are not, of course. There are, for example, some close correlations between histone modifications and DNA methylation, which suggests that they are causally related. We still have a long way to go before we understand the details of how the various chromatin marks are established and interpreted, but we do know that they are often highly specific and localized. They are induced by the signals cells receive during embryological development or in response to changed environmental conditions. Once induced, the information about cellular activities that is carried in a chromatin mark can often be transmitted in the cell lineage long after the inducing stimulus has disappeared. The chromatin-marking systems are therefore part of a cell's physiological response system, but they are also part of its heredity system.

RNA Interference: The Silencing of the Genes

Our fourth type of EIS, RNA interference (RNAi, for short), is in some ways very different from the others. It wasn't recognized until the late 1990s, and we still know little about it, but what we do know is exciting. It demands a new way of thinking about information transmission among cells, and opens up fantastic opportunities for manipulating cells, combating disease, and engineering new qualities into organisms.

The discovery of RNAi was to a large extent a result of scientists' failures, rather than their successes. What happened was that people who tried to engineer new or altered functions into plants and animals by using experimental tricks to add DNA or RNA to them were constantly being frustrated. The genes they were interested in unexpectedly became silent. For example, what would you expect if you added an extra copy of the gene that helps make the purple pigment in petunia flowers? You would surely expect this extra copy to make the flowers a darker purple, or at worst to have no effect. Instead, the scientists found the flowers were often white (colorless) or variegated (white and purple). Both the new gene and the old genes had somehow been turned off. Equally surprising cases of gene silencing were found in very different experiments in the nematode worm *Caenorhabditis elegans* (usually spoken of as *C. elegans*) and in some fungi. For some time these anomalous findings were described and discussed independently as "cosuppression" (in plants), "quelling" (in fungi), and "RNA interference" (in nematodes), but eventually it was realized that they had things in common. They are now collectively spoken of as RNA interference, and comparable silencing has been found in several other animals.

RNA interference, which leads to the stable and cell heritable silencing of specific genes, has very peculiar features, some of which are shown in figure 4.8. First, it depends on small RNA molecules known as siRNAs (small interfering RNAs), which originate from much larger mRNA molecules with unusual sequences and structures. Such abnormal RNA molecules, which are probably either double-stranded to begin with or become double-stranded when they are recognized as abnormal, are detected by an enzyme, appropriately called "Dicer," which chops them into small pieces, twenty-one to twenty-three nucleotides long. These pieces are the siRNAs. Amazingly, these tiny bits of RNA can cause the destruction of copies of the abnormal mRNA from which they were derived. They probably do so by base-pairing with the complementary sequence in the aberrant mRNA, thereby guiding another enzyme to degrade the molecule. Any normal mRNAs with complementary sequences are also degraded.

A second odd feature of the RNAi system is that, at least in some organisms, the siRNAs are amplified, so numerous copies are present. A third strange property is that siRNAs (or something else derived from or associ-

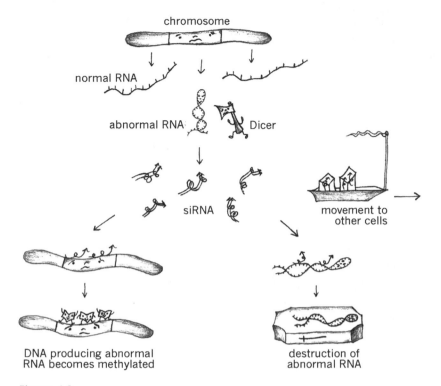

Figure 4.8

RNA-mediated silencing. At the top, a structurally abnormal RNA is produced. This RNA is recognized and chopped into fragments by the enzyme Dicer. The resulting fragments, siRNAs (small interfering RNAs), associate with protein complexes with whose help they destroy copies of the abnormal RNA (right-hand side); they may also interact with the stretch of DNA that produced the abnormal RNA and inactivate it with a methylation or protein mark (left-hand side). In some organisms, siRNA also somehow moves to other cells (middle right).

ated with the RNAi system) can migrate in the body, moving from cell to cell and even reaching different types of cell. For example, the silencing caused by RNAi can move from the rootstock of a tobacco plant through the vascular system to a graft 30 cm away. A final feature of RNAi is that in some cases the association of siRNA with the gene from which the original mRNA was produced creates a stable methylation or protein-binding chromatin mark which is transmitted to subsequent cell generations. Silencing through the RNAi system is then doubly effective, because it not only destroys the existing abnormal RNA, it also inactivates the gene that produced it.

The RNAi system that was uncovered through attempts to manipulate new functions into plants and animals obviously didn't evolve to thwart genetic engineers, so what is its natural function? Its strange properties make it seem likely that its role is to defend cells against invading viruses and the activities of genomic parasites—the transposons or jumping genes, which can replicate and move around the genome. Both viruses and transposons tend to generate double-stranded RNA. Many viruses do so because they have RNA as their genetic material, and when they replicate, this RNA temporarily becomes double-stranded. Transposons generate double-stranded RNA for a variety of reasons, one of which is that their DNA sequences are sometimes repeated side by side but in reverse order, so the RNA transcribed from them can form hairpin structures. Since healthy cells do not produce double-stranded RNA, such molecules can serve as a trigger for the RNAi system, which degrades both the double-stranded RNA and any similar single-stranded sequences. When experimenters introduce extra copies of genes into genomes, often more than one copy becomes inserted, and some insert back-to-front, so they, too, are liable to form double-stranded RNA. This triggers the RNAi system, which silences not only mRNA from the introduced genes but also that transcribed from the organism's own copies of the gene.

The idea that RNAi is a cellular immune system makes sense of a lot of its odd properties, such as the amplification of siRNAs and the ability of silencing to spread around the organism. If parasites are about, the more protection, and the wider the protection, the better! The observation that plants with mutations making RNAi defective are more virus sensitive is consistent with an antiparasite role for this system. So is the finding that defective RNAi genes in *C. elegans* cause a mobilization of their jumping genes.

Defects in RNAi genes such as Dicer may do more than upset antiparasite defenses. They can also have dramatic effects on development, and this is the clue to their significance for epigenetic inheritance. Recently it has become clear that another class of small RNAs is cut from double-stranded precursor RNAs by Dicer. They were first identified in studies of mutations that upset the timing of events in *C. elegans* development. The mutated genes turned out to be rather unusual, because they do not code for proteins. Rather, their products are untranslatable RNAs that can form double-stranded hairpin structures. These molecules are recognized and processed by Dicer and other RNAi components to form RNAs twenty-one to twenty-three nucleotides long that are very similar to siRNAs, although single-stranded. These small RNAs then recognize and pair with complementary sequences in target mRNAs, and by so doing prevent their translation. Mutations in their precursor DNA prevent the proper small RNAs from being formed, so their target mRNAs are not silenced, and normal development is disrupted.

There are differences in the production and effects of these small developmental RNAs and siRNAs (e.g., they do not cause their target RNA to be degraded), but they have a lot in common, and are now being referred to collectively as microRNAs, or miRNAs. Almost every organism that has been studied has been found to have many different miRNAs, some of which are specific to particular tissues. A few are already known to be involved in development, and the belief is that this class of small RNAs will be found to play a crucial role in regulating cellular activities and developmental decisions.

As an EIS, the RNAi system is probably very powerful. Through the miRNAs, information that silences specific mRNAs can be transmitted not only to daughter cells but also to more distant cells and tissues. Unfortunately, at present we know so little about the system that we can only guess at how this information and its transmission are affected by environmental and developmental changes. What determines if there is amplification and how much there is in the system? Can the formation of double-stranded structures be influenced by internal or external conditions? Why do some miRNAs silence their targets by destroying them, while others just prevent their translation? How many different mRNAs can a single type of miRNA silence?

We probably shall not have to wait long to get answers to most of these questions, because the discovery of RNAi has provided scientists with an incredibly powerful tool and there are many incentives to understand it fully. Through RNAi it is possible to selectively inactivate virtually any gene, simply by introducing an appropriate synthetic siRNA into the cell. The potential benefits from applying this technique are staggering. It is already being used to find out what the many genes that have been discovered through genome sequencing actually do. By making siRNA that is homologous to a part of a gene and engineering it into a cell, that gene can be silenced, and from the effects of this silencing its function can often be deduced. Commercially, RNAi is now being used to silence unwanted genes, such as those that speed the deterioration of ripe tomatoes or cut flowers. Perhaps the most exciting prospect is the use of the RNAi in medicine, where it may well produce the greatest revolution in treatment since the introduction of antibiotics. Preliminary work has already shown that when applied to human cells in culture, siRNAs designed to match sequences in the poliovirus genome inhibit the virus's replication by degrading its RNA. Similar promising results have been obtained with siRNA targeted at the genome of the AIDS virus, HIV. There is still a long way to go, and snags have already been encountered, but the prospects for using RNAi to produce new treatments look good.

RNAi is the last of our four EISs, and we now want to move on to some of the wider implications of epigenetic inheritance. Before doing so, however, we need to stress that although we described the four categories of EISs as if they were independent of each other, in reality they are not. The siRNAs of the RNAi system, for example, can prevent further transcription by mediating the formation of chromatin marks on the DNA sequence from which their precursors were transcribed. There is therefore overlap between the chromatin-marking and RNAi systems. Other systems are probably also linked: a protein that maintains a self-sustaining loop could be part of a chromatin mark or even a prion. Exactly how the different components of the cellular memory systems are interrelated is likely to be extremely complex, but we can be fairly confident that the transmission of cell phenotypes depends on a mixture of heritable structural elements, biochemical loops, replicated RNA molecules, and chromatin marks. All are potentially variable. This means that an awful lot of heritable variation is possible through EISs alone.

Transmitting Epigenetic Variations to Offspring: Monstrous Flowers and Yellow Mice

No one doubts that EISs have been important in evolution. They were obviously prerequisites of the evolution of complex organisms, where developmental decisions have to be transmitted to daughter cells, and where the long-term maintenance of tissue functions depends on cell phenotypes being stable and transmissible. What is far more controversial is the idea that epigenetic variations can be transmitted not only in cell lineages but also between generations of organisms, and that these variations play a significant role in adaptive evolution. This idea was the basis of the thought experiment with which we began this chapter. Although genetically identical, Jaynus creatures could evolve because they passed on some of their epigenetic variations, and through natural selection the accumulation of these variations led to adaptive changes. The question we now have to ask is whether the same kind of thing happens with creatures here on Earth. Do plants and animals transmit epigenetic information to their offspring?

Think first about epigenetic inheritance in unicellular organisms. We have already said that both yeast and *Paramecium* can transmit structural variants, and since all cells have self-sustaining loops, this mode of information transmission must also be present in unicellular organisms. Chromatin-marking systems are certainly present, and RNAi has been found in protozoa. So all four types of EISs are present in single-cell eukaryotes. Even bacteria have epigenetic inheritance. Some transmit methylation marks, and the French microbiologist Luisa Hirschbein and her colleagues found epigenetic inheritance in *Bacillus*. When this bacterium is made to have two chromosomes instead of one, the genes on the extra chromosome become inactive, presumably through proteins binding to their DNA. Daughter cells inherit and transmit this chromosome's inactive state for many generations. These are preliminary findings, but some microbiologists believe that they are only the tip of a very large iceberg. What we know already shows that bacteria and other unicellular organisms certainly transmit epigenetic information, so interesting evolution must occur along the epigenetic axis in these groups.

There are no theoretical problems in accepting that the evolution of unicellular organisms can occur through natural selection of epigenetic variants, although it is remarkable how little notice is taken of this possibility, which has implications for understanding and treating diseases. With multicellular organisms, the situation is different. If reproduction is by asexual fragmentation or budding, then again there are no theoretical problems, and it is easy to see how inherited epigenetic variants can be material for natural selection. Take a plant that propagates itself vegetatively through shoots that have the power of taking root. Shoots may acquire different epigenetic modifications in response to differing conditions when they were formed, or even because conditions are not the same in all parts of the plant. When these shoots take root and become independent plants, they may compete with each other, and their epigenetic heritages will influence their chances of surviving. Through classical Darwinian selection over many generations of asexual reproduction, epigenetic variants could become more stable and cause long-lasting changes. So, for the many plants and animals that can reproduce by some type of fragmentation, variations transmitted by EISs could play an important role in their evolution.

It is when we think about the transmission of epigenetic variants through sexual generations that theoretical difficulties arise. The main problem is that the fertilized egg has to be in a state that allows descendant cells to differentiate into all the various cell types. It must therefore start from a kind of neutral or unbiased epigenetic state, and for many years it was taken for granted that all memories of the "epigenetic past" had to be completely erased before cells became germ cells. This assumption ruled out any possibility that induced epigenetic variations could be inherited. The discovery in the 1980s that the epigenetic slate is not wiped clean—that some epigenetic information does pass from one generation to the next—was therefore totally unexpected. Yet it really should not have come as such a surprise, because there were some very telling hints around that should have alerted geneticists to the possibility.

For over three thousand years people had known that when a female horse is crossed with a male donkey, the offspring is a mule, whereas a cross between a female donkey and a male horse produces a very different-looking beast, a hinny, with a thicker mane and shorter ears. Both mules and hinnies are sterile. They are genetically identical, yet phenotypically different. For a long time people tended to assume that the differences between mules and hinnies were due to "maternal effects"—the result, perhaps, of differences in the wombs of horses and donkeys. But there were many other indications that the maternal and paternal chromosomal contributions to the next generation are not always the same. In the 1960s, Helen Crouse studied chromosomal behaviour in the fly *Sciara*. This is one of the insects that modifies its genome during development—it eliminates chromosomes from both somatic and germ-line cells. What Crouse found was that the eliminated chromosomes were always those from the father. In fact males transmit to their offspring only those genes that they inherited from their mother. This is a totally bizarre system, and we still have little idea what it is all about, but Crouse highlighted something important. For this elimination to happen, maternal and paternal chromosomes must be tagged in a way that makes them recognizably different to the developing fly—they must therefore be "imprinted" by the parents.

Parental genomic imprinting was subsequently recognized in many other groups, most notably mammals, and became the object of molecular studies in the 1980s. The main impetus for this was a practical problem that geneticists encountered when they engineered foreign genes into mammalian genomes. They found that crosses involving the introduced genes (known as transgenes) often didn't obey Mendel's laws properly. Just like crosses between horses and donkeys, it mattered which way around the cross was made. Some transgenes were expressed only when they were inherited from the father, and were silent when inherited from the mother. With other transgenes the pattern was reversed, so they were active only if they were inherited from the mother. Comparable parent-of-origin differences in gene expression were found for ordinary genes, and today more than seventy differentially imprinted normal genes have been identified in mice. Often the different activities of maternally and paternally transmitted genes are associated with differences in their methylation marks. It seems that during egg production, chromosomes acquire a "maternal" set of chromatin marks, whereas the same chromosomes acquire a different, "paternal" set during sperm production. Both types of parental mark are necessary for normal embryonic development, but exactly how they are established and how they affect development are still being worked out. It is important to find out, because several human diseases are associated with faulty imprinting.

Imprints are intrinsically transient. When a chromosome passes from one sex to the other, the marks that it originally carried are erased, and new sex-specific marks are established. Such constantly changing epigenetic marks are not likely to be raw material for adaptive evolution. So although imprinting proves that epigenetic modifications can be transmitted to the next generation, evidence of something more stable and enduring is needed to support the claim that epigenetic variations can be the basis of evolutionary change. Such evidence exists. The discovery that epigenetic marks can persist for many generations was another outcome of the problems encountered when biologists tried to engineer foreign genes into plants, hoping to endow them with new and useful qualities. Often they successfully inserted the transgene into the host's genome (usually in many copies), and much to everyone's satisfaction it was expressed. But then, after one or two generations, the desirable gene product was formed no more. At first it was thought that the foreign DNA must have been lost from the host's genome, but in many cases it was found that the transgene was still present, but it had been permanently inactivated by heavy methylation. The transgene's methylation mark and its associated inactivity were inherited for many generations.

A few years later it was discovered that it is not only marks that silence experimentally introduced transgenes that can be transmitted to future generations. Modified patterns of methylation and associated changes in the activity of ordinary genes can also be inherited. We want to describe just two examples of this, one botanical and the other zoological. The two we have chosen are particularly interesting, because they both show how easy it is to mistake heritable epigenetic differences for genetic differences.

The first example involves a morphological variant of toadflax. Just over two hundred fifty years ago, Carl Linnaeus, the famous botanist who laid the foundations of our present system for naming and classifying plants, described a newly generated species. This was no small thing for Linnaeus to do, since for the greater part of his very long and fertile scientific life he believed that all species had been formed by God during the Creation, and had remained the same ever since. To accept the idea that a new species had been generated recently and naturally was difficult for him. But since Linnaeus based his system of classification on the reproductive parts of plants, he had to classify the newly discovered variant as a new species because its floral structure was clearly very different from that of the normal toadflax, *Linaria vulgaris*. The five petals of the normal form are organized in such a way that the upper and lower parts of the flower are distinctly different, whereas the new variant, *Peloria* (the name came from the Greek for "monster"), was radially symmetrical, with five spurs instead of the single spur of the normal form. Figure 4.9 shows the difference. For Linnaeus the peloric variant was so extraordinary that he said (in Latin, of course), "This is certainly no less remarkable than if a cow were to give birth to a calf with a wolf's head." He thought the plant might be a stable hybrid, produced through *L. vulgaris* being pollinated by some other species. But as a religious man, Linnaeus was always uncomfortable about his new peloric species.

Peloric variants are found in other species, including the snapdragon *Antirrhinum*, and they have fascinated many of the great figures in biology, including Goethe, Darwin, and de Vries. Darwin made crosses between the peloric and normal varieties of snapdragon and, although he didn't realize it, his numerical results showed quite good Mendelian ratios, with the peloric form being recessive to the normal. If only he had met up with his

Figure 4.9
The peloric (left) and normal (right) forms of *Linaria vulgaris*.

contemporary Gregor Mendel he would have known how to interpret his results, and maybe the history of evolutionary biology would have been quite different! Hugo de Vries, who was one of the rediscoverers of Mendel's laws in 1900, also studied peloric variants. He believed that the peloric form of *Linaria* was a mutation, and found a rate of change from normal to peloric of about 1 percent. Today we would regard this as a very high mutation rate.

During the last two decades, when a lot of genetic research has been focused on developmental processes, botanists have been studying the molecular basis of mutations that alter flower shape, including the famous peloric variant of *Linaria*. What Enrico Coen and his colleagues at the John Innes Institute in England found when they compared the normal and peloric forms was very surprising. There was certainly a difference between them, but it was not a difference in DNA sequence. The morphological change was due not to a mutation, but to an epimutation: the pattern of methylation of a particular gene in the normal and peloric plants differed. So this variant, which has played such a significant role in the history of botany, turned out to be neither a new species (as Linnaeus thought), nor a mutation (as de Vries and others thought), but a fairly stable epimutation. There is a certain irony in that! It is not clear what caused the methylation change in the first place, but once formed it seems to have been transmitted, more or less steadily (although there is residual instability), for many generations. Over two hundred years after Linnaeus's specimen was collected, the peloric form of *Linaria* was still growing in the same region.

Our second example of an inherited epigenetic variant is one that was found in the laboratory mouse. It involves coat color, which has been a favorite trait for genetic analysis ever since the very early days of Mendelian genetics. The usual "mousy" brownish color of the fur is known as "agouti," but there are many genes and alleles that alter this color. The Australian geneticist Emma Whitelaw and her colleagues worked with a mutant mouse strain in which there was a small extra bit of DNA (originating from a transposon) in the regulatory region of a coat color gene. The presence of the extra bit of DNA interferes with the normal formation of pigment, but the extent of the interference is not the same in all mice: in some the coat is yellow, in others it is mottled with blotches of agouti, and in yet others it is completely agouti-colored, so they are described as "pseudoagouti."

There is nothing very exciting about this, since we know that environmental and developmental factors can affect the expression of a gene and lead to variant phenotypes. However, what was surprising and caused a flurry of excitement is the way in which the phenotypic variations are inherited. Yellow mothers tend to have yellow offspring, mottled ones mottled offspring, and pseudoagouti ones tend to have more pseudoagouti offspring than the other two types of mother (see figure 4.10). Since there are no differences in their DNA sequences, something else must be responsible for the inheritance of the variation. In Whitelaw's experiments it was not possible to lay the blame for variation on the usual scapegoat—different unidentified modifier genes—because the mice were genetically identical. Another favorite excuse for unexplainable inherited variation, "maternal effects," was ruled out by transferring embryos between mothers; the investigators showed that the uterine environment did not influence coat color. It turned out that the variation was correlated with the methylation pattern on the extra bit of DNA from the transposon, and this pattern was passed on to the next generation through the egg. In other words, the inherited differences in coat color occurred because the epigenetic slate was not wiped clean before each new generation.

There are several other interesting aspects of this work with yellow mice. The first is that the heritable phenotype is affected by environmental

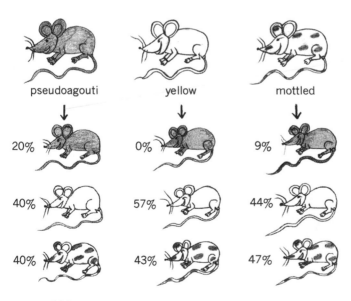

Figure 4.10
Inheritance of variable yellow coat color in mice; the proportions of the different types of offspring from each type of mother are shown.

conditions, in this case the mother's diet. Pregnant females fed with methyl-supplemented food have proportionally fewer yellow offspring and more pseudoagouti ones. The second is that the mutation affects more than fur color: yellow mice are obese, diabetic, more susceptible to cancer, and shorter-lived than their nonyellow littermates. The third is that there are many transposon sequences in mammalian genomes—at least 40 percent of the genome is derived from them. Most are silenced by methylation, but many probably retain the ability to influence the activity of neighboring genes. Taken together, these findings suggest that there may be a large number of genes whose phenotypic effects are influenced by epigenetic marks, and the nature of these marks may sometimes depend on environmental conditions. In the absence of detailed molecular studies, most epigenetic differences would be identified as genetic differences, so at present we simply do not know how much heritable variation is caused by stable epigenetic marks.

There is another big unknown in transgenerational epigenetic inheritance—the significance of RNAi. Through RNAi, gene-specific silencing signals can pass from parent to offspring, as well as from cell to cell. For example, if double-stranded RNA with a sequence that matches part of a specific *C. elegans* gene is injected into this nematode's gut, that gene is silenced throughout the worm because its mRNA is destroyed by the RNAi system. Silencing can also be transmitted through several generations of offspring. There is therefore some kind of RNA-mediated transfer of information through the germ line. How much information is transmitted in this way we do not know. However, if small RNAs capable of affecting development are as common as recent work suggests they may be, it could be quite a lot.

We now want to try to bring things together and to a conclusion by looking at what all this evidence that epigenetic variations can be transmitted to the next generation means for evolutionary theory. In a nutshell, it means exactly what the thought experiment about Jaynus creatures was suggesting: it means that because it provides an additional source of variation, evolution can occur through the epigenetic dimension of heredity even if nothing is happening in the genetic dimension. But it means more than this. Epigenetic variations are generated at a higher rate than genetic ones, especially in changed environmental conditions, and several epigenetic variations may occur at the same time. Furthermore, they may not be blind to function, because changes in epigenetic marks probably occur preferentially on genes that are induced to be active by new conditions. This does not mean that all induced changes are adaptive, but it does increase the chances that a variation will be beneficial. The combination of these two properties—a high rate of generation and a good chance of being appropriate—means that adaptation through the selection of epigenetic variants may be quite rapid compared with adaptation through genetic change.

Heritable epigenetic differences may also play an important part in what Darwin called "the mystery of mysteries"—the origin of species. Speciation is a topic evolutionary biologists argue about a lot, but most will agree that usually new species are initiated when populations become isolated from each other by a geographic or ecological barrier. While separate, the populations change, and the changes that occur prevent them interbreeding successfully if they meet up again. It is usually assumed that the changes are genetic, but we believe that they may often be epigenetic. During periods of isolation, two populations will commonly experience different conditions, perhaps because one is colonizing a new island, or is using a new food source, or experiencing a different climate. If so, new epigenetic marks might be induced in both somatic and germ-line cells. These may do more than affect how well the organisms function in their new environment; they might also affect their ability to interbreed with other populations. Studies of imprinting have shown that the marks on chromosomes from the two parents have to be complementary if development is to proceed normally. So, if two previously isolated populations have acquired different epigenetic variations, incompatibility between the marks transmitted by sperm and egg may prevent a hybrid embryo from developing normally. Even if a viable hybrid is produced, modified chromatin marks may mean that gamete formation is upset, and the hybrid is sterile. In this way the formation of the initial reproductive barrier that is the crucial part of speciation can be an outcome of epigenetic, rather than genetic, changes.

To end this chapter we want to mention just one other type of inherited variation. So far, we have been dealing with the transfer of information through cellular inheritance systems. But information transfer also occurs at higher levels of organization. There is a good example of this in Mongolian gerbils, where the mother's uterine environment may have strong heritable effects on her female offspring's development. A female embryo that develops in a uterus in which most embryos are male is inevitably exposed to a high level of the male hormone testosterone. This high level of the hormone is information for the embryo, and it affects her subsequent development. As she grows up, she develops some special characteristics, such as late sexual maturity and aggressive territorial behavior, and, most remarkably, when she reproduces, her litter has more males than females. Since most of her embryos are male, her female offspring develop, just as she did, in a testosterone-rich uterus, so they grow up to have the same behavioral and physiological traits as their mother. They, too, will produce male-biased litters, and so the cycle continues. In this way the developmental legacy of the mother is transferred to her daughters—there is nongenetic inheritance of the mother's phenotype. Consequently, two female lineages that are genetically identical can be very different behaviorally and have different sex ratios, simply because they transmit different nongenetic information.

Dialogue

I.M.: I need to wrap a wet towel around my head! You have described such a mishmash of transmission systems that it is far from clear to me that calling them all "epigenetic" is of any value at all! It seems to me that their roles are very different. But before we tackle that problem, I have a general question about the relation between the two dimensions of inheritance. In your music metaphor you suggested two possibilities. One was that the score (the genetic system) can affect the performance (the epigenetic system), but not vice versa. The alternative was that there could be two-way interactions, with performances leading to changes in the score as well as changes in the score affecting performances. In this case the inheritance systems would affect each other. But there is also a third possibility—the written score might disappear and be totally replaced by the recording system. One system of heredity might eliminate the other. Does this happen?

M.E.: In existing organisms, which all have a nucleic acid–based inheritance system, it is inconceivable that the DNA inheritance system will be eliminated by another one that operates at a higher level. But theoretically it is possible that one heredity system can replace another. It may well have happened at an early stage in the evolution of life, during the murky period between chemical and biological evolution. Many theorists suggest that heredity during these early stages was not based on nucleic acids, and that the nucleic acid systems came later and replaced the primitive heredity systems. Maybe such a replacement will also occur in the distant future—if we create intelligent, reproducing, and evolving robots, they may eventually eliminate us. This would be equivalent to the elimination of one heredity system by another.

I.M.: That makes sense. Now I want to go back to your definitions. Why are you lumping all these EISs together? They clearly have different properties.

M.E.: All the EISs transfer information from cell to cell, and that's their common denominator. You can carve up biological heredity differently, and classify inheritance systems according to the type of mechanism involved, rather than the level at which they operate. We do this anyway, by distinguishing between the different types of EIS. But yes, we are lumping together things that are very different. The structural inheritance category is probably too broad, because the propagation of prions, the self-perpetuation of membranes, and the inheritance of ciliary patterns may have nothing much in common other than that they all involve some kind of 3D templating. But at present we just don't know enough about what is going on. Nevertheless, putting all the different systems into one category—that of cellular heredity—is useful. It alerts us to the limitations of the system and a very basic constraint, which is that the cell is the focus of this type of heredity. The last system we described, the transfer of developmental legacies at the whole-organism level, really is distinct, and we agree that it should be given a category all of its own.

I.M.: There is another problem of definitions or concepts: you call all these cellular systems "inheritance systems" and talk about their evolutionary potential. But what strikes me as important is their role in development, in ontogeny. Why didn't you focus on their developmental role?

M.E.: Because what is so important about these systems is that they are both heredity systems and regulatory systems. There are many types of regulatory loops, but only a subset of them has a structure that allows functional states to be transmitted from cell to cell; there are many proteins and cell structures that can vary, but only a small subset, which includes prions, has properties that allow the transmission of variants. There are many different types of chromatin structures, but only some of them lead to the reproduction of marks in daughter cells. And we simply don't know the extent of cell-to-cell transmission of silencing through RNAi.

I.M.: I realize that you are focusing on regulatory systems that have special hereditary properties, but why do you treat them as being of evolutionary rather than developmental significance?

M.E.: They are both at the same time, and that is what makes them so fascinating. We are interested in the significance of EISs in evolution precisely because their evolutionary effects cannot be separated from their physiological and developmental role. One cannot make a neat distinction between the physiological/developmental and evolutionary aspects of heritable epigenetic variation. It may be that things get confusing because these days the word "evolution" evokes ideas of change through purely selective processes and blind variation. Instructive processes and directed variation are associated only with development. For some time we have felt that a new term, which would describe processes that are concurrently evolutionary and developmental, selective and instructive, is necessary. We thought of "evelopment," but have not used it much.

I.M.: It's not a very beautiful word, but it may do! You certainly need to conceptualize this mixture. I am now coming to one of the main problems I had with this epigenetic dimension of yours. I can understand how epigenetic variations are transmitted from generation to generation in unicellular organisms, and how this can affect evolution. They are not too different from the interpretive mutations in bacteria and plants that you described in the last chapter. But I still fail to understand how evolution on the epigenetic axis will work in multicellular organisms, even asexually reproducing ones. What happens if an epigenetically variant cell arises? It must have an advantage at both the tissue level and the organism level if it is to survive and replace other variants. Aren't you asking too much?

M.E.: No. It is true that replacing other variants at the tissue level is not enough. Cancer cells survive and multiply, and spread throughout tissues, but they often destroy the organism. If an epigenetic variant is to spread through a population of organisms, it must be beneficial (or at least neutral) at the organism level. Even if it is not beneficial at the tissue level, it will spread if it benefits the organism as a whole. If the variant has an advantage at both the cell lineage and whole-organism levels, evolution will be particularly rapid.

I.M.: Are there any examples of natural selection of epigenetic variants at the lineage or tissue level and at the whole-organism level?

M.E.: At the lineage level, yes, there are. We have already mentioned the cancers that seem to be initiated by heritable epigenetic events—by changes in methylation patterns among other things. As for natural selection at *both* the lineage and the whole-organism level—no, there are no examples that we are aware of, but the experimental work has not been done. There may be something of this sort in plants. It is possible that the *Linaria* case belongs to this category: the epigenetic variant was selected first at the lineage level, and then was not selected against (or was even positively selected) at the whole-plant level.

I.M.: Not so fast! There is another problem—many plants reproduce sexually, through pollen and eggs. I do not quite understand how a successful somatic variant can be transferred to the gametes. And even if it was, it would do no good. A wonderful epigenetic variant of an animal's skin cell or of a plant leaf cell cannot develop into a whole organism, which has other types of cells too! The most that could happen is the development of more skin tissue or leaf tissue, not a whole organism.

M.E.: You have really raised two different questions, so we will answer each in turn. The first question is whether epigenetic variations can be transferred from somatic cells to germ cells. You are right—in sexually reproducing organisms, for an epigenetic variant to be transferred to the next generation, it must be present in the gametes. There are three ways in which the germ cells can acquire a variation. First, a new variation may originate in the germ line. We expect to find this type of variation in all groups of organisms. Second, if somatic cells that harbor the new variation can develop into germ cells, an originally somatic variant can become a germ-line variation. This can occur in organisms that have no separation (or late separation) between germ line and soma, where somatic cell lineages can give rise to germ cells. For example, in the African violet, a leaf can develop into a whole plant with full reproductive organs, so an epigenetic variation in leaf cells could be transmitted to the next generation. Third, if there is some process of information exchange between somatic cells and the germ line, a germ cell can acquire a variation that initially occurred in somatic cells. This is what sometimes seems to happen through RNAi mechanisms.

Now to your second point: you asked how a particular variant cell type could be the basis for development when, before a whole organism with its many diverse cell types can develop, all information about specific cell types has to be erased. Of course, you are right—an epigenetically altered cell must be able to give rise to functional gametes with full developmental potential. A fertilized egg that could become only a variant skin cell wouldn't have much future. But a fertilized egg that has chromosomes with marks that can lead to the development of a new variety of skin cells is something different. Obviously such marks must not interfere with other aspects of development. If you think about the peloric variant of *Linaria*, the marks associated with the gene that affected flower structure did not, as far as we know, affect other aspects of development.

I.M.: It seems to me that the type of epigenetic variations it is possible to transmit is rather limited.

M.E.: No more limited than for any other variation, including genetic variations! All have to pass through the sieve of selection. A genetic mutation that causes a fertilized egg to differentiate into a skin cell is a dead end, and will be eliminated. In a multicellular organism there are a lot of constraints on all types of variation—any variant has first to pass through the bottleneck of development to produce a viable organism.

I.M.: So how is cloning possible? As I understand it, what you do is take a somatic cell, maybe a skin cell, and fuse it with a fertilized egg whose nucleus has been removed. The somatic nucleus then undergoes some mysterious changes, and functions like a normal egg nucleus enabling the development of an embryo and eventually a young animal. How can it do all this if it has all the epigenetic marks of a somatic cell? Why doesn't it just develop into a blob of skin or whatever tissue it was that provided the nucleus? How are all the marks erased? And how are the specific parental imprints that you said were necessary for development re-established? Everything you have told me suggests that cloning should be impossible!

M.E.: Cloning really is a remarkable feat. Obviously, the memories of where the nucleus came from have to be erased in the egg. No one yet knows how this happens. As for parent-specific imprints, don't forget that the somatic nucleus that is implanted into the enucleated egg has chromosomes from a male and a female parent, so it is possible that some imprints are preserved and can be reinstalled. But we would expect to find a lot of mistakes in this process, because the somatic cell cannot go through all the many epigenetic changes that sperm and egg undergo during their developmental history. Unlike gametes, a cell used for cloning is not epigenetically prepared for its new and dramatic role as the foundation for the development of a whole new organism. It's therefore not surprising that people have found a lot of abnormalities in cloned animals. Most embryos never even implant in the uterus, and those that do usually die before birth. Many of those that are born have problems. Even famous Dolly, who survived for six years and had several offspring, was the single success out of 277 attempts to clone the mother sheep. She developed premature arthritis, which may have been the result of problems with her epigenetic resetting. Epigenetic inheritance is certainly a big obstacle to fast and easy cloning.

I.M.: That cloning is possible at all seems like a miracle to me! But let's go back to the effects of these marks in evolution. If marks can be erased during the development of gametes and during cloning, how reliable is their transmission? Are they transmitted as faithfully as genes? From what you said it seems that many epigenetic variations are likely to be erased in at least some of the reproductive cells as they are reset in preparation for the next round of development. If so, the transmissibility of an epigenetic mark will be less than that of an allele of a gene. Is this so?

M.E.: There are few data on how faithfully epigenetic variants are transmitted. But yes, in some cases, transmissibility of an epigenetic variant is less than for a genetic variant, which in sexual organisms is inherited by half the parent's offspring. Occasionally an epigenetic variant may be transmitted to more than 50 percent of the offspring, because an epigenetic mark on one chromosome can sometimes convert the allele on the other chromosome to its own image. How common this is we don't know, so we don't want to make too much of it.

I.M.: When transmissibility is less than 50 percent, doesn't it mean that the frequency of the epigenetic variant must decrease in every generation? Surely only quite strong selection would maintain it in a population. If the epigenetic variant gives only a small benefit, it will disappear!

M.E.: It might, but not if the environment continues to induce the variant. This would compensate for its low transmissibility. And since epigenetic variants are more likely than genetic variants to have phenotypic effects, selection for or against a variant can be quite strong. Low transmissibility therefore may not be such a problem. But we really know very little about the fidelity with which epigenetic variants are transmitted, or even whether the transmissibility remains constant. It may fluctuate, being different in different environments, and we would expect it to be altered by selection. It could be made more reliable through, for example, natural selection for less erasable chromatin marks.

I.M.: But you would still lose some adaptive variations that make only small selective differences. Maybe that is why your Jaynus creatures did not progress beyond the complexity of a jellyfish! I have another problem with transmissibility. I can see how variants might be transmitted through the chromatin-marking EIS, but I cannot imagine how in a multicellular organism a self-sustaining loop or structural element can be inherited and be the basis of evolutionary change.

M.E.: You have to assume that what is transmitted are components of an activity or a state that biases the reconstruction of the same activity or state in the next generation. For a prion disease, it could be a prion that is transmitted in the egg that starts the templating process off again in the next generation. The membrane systems of eggs are of course self-perpetuating, and we simply do not know whether and what type of minor variations in them can be inherited; large changes would almost certainly lead to cell death. For a self-sustaining loop the egg might contain molecules of a binding protein that can initiate and maintain the activities of the gene that produced it. We think that part of the problem in envisaging this type of inheritance is that we usually think of transmission in terms of copying, rather than in terms of reconstruction. But we agree with you that the transmission of epigenetic variants from one generation of multicellular organisms to the next is much more likely with chromatin-marking systems or RNAi. Since marks may have no phenotypic effect in most cells, they can be passed on through the germ line without jeopardizing gamete function and early development. Similarly, miRNAs will affect only those tissues in which their target mRNA is expressed.

I.M.: The RNAi system is unusual, isn't it, because it's a way in which information can be transferred from the soma to the germ line. Are there any other routes of communication between soma and germ line?

M.E.: Yes, theoretically at least. Ted Steele's somatic selection hypothesis, which he began developing in the late 1970s, suggests a route. Steele is an Australian immunologist and he based his ideas on some of the things seen in the development of the immune system. You will recall from chapter 2 that during the maturation of the cells that produce antibodies, new DNA sequences are generated through the cutting, moving, joining, and mutation of the original sequence. The result is that an enormous number of cell types, with DNA coding for different antibodies, is produced. Some survive and multiply; others do not. What Steele suggested was that where you have a situation like this, in which diversity among somatic cells is followed by selection, copies of the mRNA in the selectively favored (and therefore common) cells may be picked up by viruses and carried to the germ line. There, through reverse transcription, the mRNA information can be copied back into DNA.

I.M.: Isn't that contrary to the central dogma?

M.E.: No. The critical part of Crick's central dogma was always between nucleic acids and proteins: the amino acid sequence in a protein cannot be reverse-translated into DNA or RNA. There was never any problem with the idea that RNA could be backtranslated into DNA. It's not difficult to imagine how it could happen through complementary base-pairing. And by the time Steele suggested his hypothesis, reverse transcription had been discovered. So Steele's hypothesis is plausible, and indeed there is some experimental evidence supporting it. It's one of the ways in which somatic events could change the germ line. In addition, in mammals, information about somatic changes can bypass the germ line completely, yet still reach the next generation. For example, acquired immunity to pathogens may be transmitted from mother to offspring through the placenta and through the milk. So you see, there are several different ways in which information from the soma might reach the next generation.

I.M.: There are certainly more things in heaven and earth than are dreamt of in my philosophy! But I'm still worried about the central dogma. Is the transfer of information from protein to protein, as happens with prions, compatible with it?

M.E.: You are right to be worried. Crick said in 1970 that there are three types of information transfer that are unknown and the central dogma postulates never occur: protein to protein, protein to DNA, and protein to RNA. At that time, scrapie, which we now know to be a prion disease, was beginning to interest and puzzle biologists, and Crick recognized that it might be a problem for the central dogma. Interestingly, in the last sentence of his article he wrote, "the discovery of just one type of present day cell which could carry out any of the three unknown transfers would shake the whole intellectual basis of molecular biology, and it is for this reason that the central dogma is as important today as when it was first proposed." It seems that, according to Crick himself, the central dogma should now either be abandoned or modified.

I.M.: Biologists certainly shouldn't be so dogmatic about the significance of the central dogma! But let me return to the evolutionary importance of epigenetic inheritance. I understand that the transfer of epigenetic information from one generation to the next has been found, and that in theory it can lead to evolutionary change. But has anyone ever found any heritable epigenetic variation that is adaptive—that gives a selective advantage to those inheriting it? You mentioned prions, cancer, transposons, strange peloric flowers, and so on, but none of these seems very adaptive to me. Is there any evidence for adaptive epigenetic variants?

M.E.: No, there is no direct evidence.

I.M.: Aha!

M.E.: Don't rejoice too soon. When people started to study genetic variation at the beginning of the twentieth century, they too studied abnormal phenotypes—things like white eyes, wrinkled wings, and so on in fruit flies; frizzled feathers in chickens; and all the strange mouse and guinea pig mutants. And a lot of biologists doubted that these mutations could have any evolutionary significance. They thought that all Mendelian mutants were pathological. It was some time before potentially advantageous mutations were found and it was possible to demonstrate their selective advantage in some conditions. But your question is a little odd. If you accept that heritable epigenetic variation is possible, self-evidently some of the variants will have an advantage relative to other variants. Even if all epigenetic variations were blind, this would happen, and it's very much more likely if we accept that a lot of them are induced and directed.

I.M.: Theoretically, yes, but I just wonder about reality. However, I want to go back to the very last part of the chapter. As you confessed, you jumped from the cell level to the organism level. What is the connection, if there is one, between the cellular heredity that you described and the organism-to-organism transfer in the Mongolian gerbils, which occurs through the passage of molecules in the mother's womb? This type of transmission seems to me like a positive self-sustaining feedback loop with the environment, which in this case is the mother.

M.E.: Yes, you can look at it like that. And in fact that is a good way of seeing a lot of the organism-to-organism transfer of information that we discuss in the next chapter. In animals with a nervous system, there is a new option for information transfer—through social learning. This is really a distinct level of information transfer, but as you will see, it has a lot of properties in common with some of the systems we have been dealing with in this chapter.

Nonbiologists will probably sigh with relief when they see the title of this chapter. After discussing genes, biochemistry, and molecular biology, about which nonbiologists do not have ready intuitions, we turn to behavior. Here, the layperson usually feels a lot more at home. We are all sharp-eyed observers of behavior, and feel that our personal experiences qualify us to understand many of the complex processes that are related to behavioral change. We know that there are many ways of learning, and that we learn from each other as well as on our own. As nature lovers and pet owners, we are well aware that animals can learn a great deal too. Mammals and birds, the animals with which we are most familiar, learn from their personal experiences, from their owners, and from each other, often displaying remarkable abilities as they do so. But how important is such learning in evolution? Of course, all biologists agree that in many circumstances learning is enormously beneficial, and that the capacity to learn has evolved genetically, but is learning also *an agent* of evolutionary change? For example, how does the fact that animals learn from each other affect the evolution of their behavior?

The current fashion among evolutionists, seen particularly in the writings of those who study human behavior, is to stress the genetic basis of behavior, and especially that of best-selling, sex-related behavior. These evolutionists maintain that the behavioral strategies for things like finding a mate, or becoming socially dominant, or evading danger, or finding food, or caring for infants are to a large extent genetically determined and evolutionarily independent of each other. Each has been shaped through the natural selection of genes that led to the construction of a specific behavioral module in the brain, which tackles that particular "problem." This is an interesting point of view, and we are going to examine it in some detail in later chapters, but in this chapter we want to look at something very different. As far as we can, we want to deal with the third dimension of heredity and evolution, the behavioral dimension, in isolation from the first, genetic, dimension. This means that we are going to be looking at behavioral evolution that does *not* depend on selection between genetic variants.

Evolution among the Tarbutniks

It is not easy for biologists to think about behavioral evolution without automatically resorting to ideas about selection among variant genes, so to help overcome this we will again use a thought experiment. This one is about tarbutniks, who evolved in the minds of Eytan Avital and Eva Jablonka in 1995, and are described more fully in their book *Animal Traditions*. What follows is an abridged and slightly modified version of what they wrote there.

Tarbutniks are small rodentlike animals, which got their name from the Hebrew word *tarbut*, which means "culture." One of the interesting things about them is that they are all genetically identical. They have perfect DNA maintenance systems, so their genes never change. In this they resemble the Jaynus creatures of the previous chapter, but unlike Jaynus creatures, they also have mechanisms that completely prevent any transgenerational transmission of epigenetic variations. Tarbutniks can therefore inherit neither genetic nor epigenetic variations from their parents. This does not mean that they are all identical, of course. Chance events during their development result in small differences in their size, fur color, the proportions of their body parts, and also in their calls and various aspects of their learned behavior. In fact, there is quite a lot of variation among tarbutniks, but—and this is the important point—there is no correlation in appearance or behavior between parents and offspring, because the differences between individuals are not inherited. And since the variation is not hereditary, these tarbutnik populations cannot evolve.

Tarbutniks live in small family groups consisting of a pair of parents and several different-aged offspring. They begin life as rather helpless creatures, relying on their mother's milk for food, but they grow rapidly, and soon begin to accompany their parents on foraging expeditions. As they do so, they learn about their environment. They discover how to open nuts and get at the seeds inside by trial and error, but it takes a lot of attempts before they hit on the right way of doing it. They also learn the hard way that black-and-red striped beetles have a nasty taste and should be avoided. Being able to learn from experience is obviously very important for their survival, but tarbutniks are very odd, because they do not learn from each other. The fruits of their individual experiences are never shared with their peers, their parents, or their young. Each tarbutnik has to find out about its environment through its own experiences. In every generation, through trial and error, each tarbutnik has to reinvent the wheel for itself.

Now imagine that tarbutniks are suddenly endowed with a capacity for something the behavioral biologists call "social learning" or "socially mediated learning." In other words, tarbutniks can learn from and through the experiences of others. Since young tarbutniks live with older individuals and have intense daily contact with them, they can acquire a lot of information from adults, especially from their parents, as figure 5.1 shows. The learning from parents and peers that takes place before the youngsters become independent is particularly important for them, but tarbutniks continue to learn as adults, both from each other and from their young.

Figure 5.1
Socially mediated learning in tarbutniks: the youngster is introduced to carrots by its mother (top), and consequently devours them enthusiastically when adult (bottom).

The capacity to learn from others may seem like a very small modification of tarbutnik life, but it has profound effects, because it enables patterns of behavior to spread in the population. A tarbutnik that either by accident, or through trial-and-error learning, or by observing the activities of individuals from another species, discovers how to crack open a nut may transmit this useful information to its descendants. Even if it does not have many offspring itself, the nut opener may pass on its skill to someone else's young, just by being a sociable and caring neighbor; even "bachelor" and "spinster" tarbutniks can transmit the useful new behavior to "cultural offspring."

We will come to how animals "transmit" behavior later, but we should say here that when using this word we do not want to suggest that the processes involved are active or automatic or intentional. An animal "transmits" behavioral information only in the sense that through its behavior other animals acquire that information. By acquiring information from or through others, changes in behavior are inherited (not always by blood relatives), and may become established in the population. We will be using "inheritance" for any of the socially mediated transmission and acquisition processes that result in the reconstruction of an ancestor's behaviors or preferences in its descendants.

Adding socially mediated learning to tarbutnik life means that new habits, skills, and preferences can be transferred from generation to generation. This is important, because if some behavioral variations are inherited through socially mediated learning, Darwinian evolution is possible. Imagine that a tarbutnik learns through its own experiences that by squatting in a depression in the ground it is less likely to be spotted by predators. This behavior improves its chances of surviving and reproducing. Its fortunate offspring do not have to rediscover the useful hole-squatting habit for themselves, because they learn it from their experienced parent. Some of the family's neighbors learn it too. Soon, tarbutniks start elaborating on the behavior they have learned, deepening the depression by digging. This protects them not only from predators but also from inclement weather, so they thrive and the digging habit spreads. Digging sometimes produces a burrow with two entrances, which allows them to escape predators even more easily, and this habit also spreads through learning. Eventually, because the tarbutniks who survive longest are those who spend a lot of time in their burrows, females start giving birth there. This protects both mother and young, and the youngsters readily learn the burrow-using habit to which they have been exposed from birth. In this way, as the inventions or chance discoveries of individuals are selected and accumulate, tarbutniks develop a new burrow-living tradition, and their whole lifestyle changes.

The way tarbutniks communicate can also evolve through social learning. Imagine a situation in which tarbutnik youngsters often fail to hear and respond to their parents' alarm call when it is made in thick vegetation. By accident a parent discovers that its youngsters do respond to a call of a different pitch, and it begins to use this more audible call when in thickets. The youngsters now have a better chance of surviving. Use of the new alarm call spreads as the young learn it from their parent, and later use it with their own offspring. Thanks to the benefits of the new call and the ease with which it is learned, a new calling tradition is established.

Communication between mates can also undergo "cultural" change through social learning. Red berries are a favorite tarbutnik food, so imagine that by chance a male discovers that females who manage to snatch his berries from him are also more available for mating. He learns to allow them to steal from him, and in this way gets more partners and sires more offspring than his competitors. His offspring and their observant young friends learn his successful type of behavior, and it spreads. Gradually they discover and learn that actively offering berries rather than just allowing them to be stolen is an even better way of securing a mate. The new berry-offering tradition spreads and becomes established in the population.

Tarbutnik evolution can be taken even further. Imagine that severe floods have made a river change its course, and the initial population is split into two isolated parts. The two subpopulations experience somewhat different conditions and learn different things. In subpopulation A, males continue to woo the females by offering them red berries. However, in the area where subpopulation B is living, there are no red-berry bushes, so males learn to offer their prospective mates already-cracked nuts, which are a local delicacy. Now imagine what happens when, after many generations, members of the two subpopulations meet and begin courtship (figure 5.2). The females from subpopulation A expect red berries, and do not respond to the nut-offering males from B; similarly, nut-requiring females from B are not interested in the red-berry–offering males from A. Because of their different courtship traditions, there are no "mixed matings" between members of the two subpopulations. They are culturally isolated from each other. Each subpopulation has become a "cultural species."

In the scenarios we have just described, tarbutnik populations evolved through the selective retention and transmission of variant patterns of learned behavior. Through natural selection, their culture changed. Now

Figure 5.2
Cultural reproductive isolation in tarbutniks. Berry-loving females respond to the advances of berry-offering males, but reject the advances of nut offerers.

if tarbutniks were real animals, some biologists would be grinding their teeth at seeing the word "culture" in our account. They would insist that for animals to have culture, a lot more than we have suggested is needed. For them, culture is limited almost entirely to humans and perhaps a few primates. Other biologists take a far less restricted view of culture, and readily accept that it is widespread in the animal kingdom. Obviously then, since culture is such a loaded and problematic term, we must try to define what we mean when we use it. There are several definitions in the biological literature, most of which are rather similar in spirit to the one that we are going to use. We see culture as *a system of socially transmitted patterns of behavior, preferences, and products of animal activities that characterize a group of social animals*. The transmitted behaviors can be skills, practices, habits, beliefs, and so on. Once we have defined culture in this way, "cultural evolution" can be defined as *the change, through time, in the nature and frequency of socially transmitted preferences, patterns, or products of behavior in a population*.

The thought experiment with tarbutniks showed that behavioral-cultural evolution is possible without any genetic variation. Of course, biological reality is not so simple. There are no animals that are devoid of genetic variation. We now know that the amount of genetic variation in populations of real organisms is enormous—far more than any geneticist imagined fifty years ago. But as we pointed out in chapter 2, in most cases genetic variants have only very small effects, and these are not the same in every individual. It is therefore reasonable to assume that in a real, genetically variable population in which behavior is transmitted through social learning, most cultural evolution is to a large extent independent of genetic variation. For example, the population-specific dialects of songbirds such as starlings, or of groups of sperm whales, which cannot be explained only in terms of individual adaptations to local conditions, are probably consequences of cultural evolution that is independent of any genetic variation in the populations. Cultural differences between human groups are also likely to be largely independent of their genes. Such independence may not remain indefinitely, because in some circumstances the genetic and cultural systems inevitably interact. We deal with this in chapter 8. For the time being we want to forget genes and look more closely at the behavioral inheritance system—at the ways in which information is transmitted between generations through animals interacting with and learning from or through others.

Transmitting Information through Social Learning

"Learning" can be defined in a very general way as *an adaptive (usually) change in behavior that is the result of experience.* "Social learning" or, more precisely, "socially mediated learning" is therefore *a change in behavior that is the result of social interactions with other individuals, usually of the same species.* It has been classified in many ways, but we are going to distinguish just three major routes of behavior-affecting information transfer—three types of behavioral inheritance systems (BISs). The first is very similar to the transmission mechanism we described for Mongolian gerbils, since it is based on the transfer of behavior-influencing substances. The second is based on socially mediated learning in which individuals observe the conditions in which the behavior of experienced individuals is taking place, as well as the consequence of such behavior; although the inexperienced individuals do not imitate, they use what they observe to reconstruct a similar behavior. The third BIS involves imitation. These different ways of acquiring information from others are certainly not independent of each other, and there are many cases that are intermediate or a mixture and cannot easily be pigeonholed. Any real socially learned behavior (for example, learning from others what is good to eat) may depend concurrently on several different types of learning.

Inheritance through the Transfer of Behavior-Influencing Substances: On Preferring Juniper Berries and Carrot Juice

We want to start looking at the ways in which knowledge, habits, preferences, and skills are acquired from others by thinking about food preferences—about the culinary cultures of different ethnic groups. Why do Yemenite Jews prefer very spicy food with a lot of schug (a mixture of crushed hot peppers, coriander, garlic, and various spices), whereas Polish Jews cannot tolerate such hot food, enjoying instead sweet gefilte fish, a dish that makes many Yemenite Jews shudder with disgust? These are big questions, and trying to answer them fully would take us beyond the scope of this book, but we can provide part of the answer. It is that the type of food to which children are exposed during early life helps to form their adult food preferences, and hence to determine the culinary culture and preferences that they, in turn, will transmit to their children.

It is surprising how early some food preferences are learned—it's a lot earlier than most people think. Just how early it is has been shown nicely by some experiments with European rabbits. Rabbits are notoriously prolific breeders, but they are not what we think of as devoted mothers. After giving birth, they leave their pups in sealed burrows, returning to nurse them for only five minutes or so each day. The pups are weaned when just under four weeks old, by which time their again-pregnant mother, who mated within a few hours of giving birth to them, is busy preparing a nursery burrow for her soon-to-be-born next litter. Subsequently, as they explore the world outside their burrow, the young pups have little direct help from their mother. Yet, although their world contains plants of differing nutritional value and possible toxicity, the youngsters know what is good and safe to eat. They know because their mother has given them useful information about food long before they leave the nest.

In experiments that showed how young rabbits acquire information about food, a group of European scientists fed lab-living pregnant females a diet containing juniper berries. This is a food they would eat naturally in the wild, and it did them no harm. When the pups of juniper-fed does were weaned, although they had had no direct contact with juniper food, the youngsters clearly preferred it to normal lab food. This was true even if the newborn pups were taken away from their mother at birth and given to a foster mother who had never eaten juniper, and whose own youngsters showed no preference for it. Clearly, the offspring of juniper-eating mothers had acquired information about juniper food from her while in the womb, presumably because chemical cues had reached them through the amniotic fluid and placenta. Remarkably, not only had they received such information before they were born, they had retained it for the four weeks until they were weaned and had to make their own food choices.

Young rabbits' food preferences are not determined solely by what happens before birth. The experimenters also looked at what happens during the suckling period. They took pups from normally fed mothers immediately after they were born, and gave them to juniper-fed mothers for the nursing period. At weaning, when given a choice, these pups preferred juniper food. This means that although they were nursed for only a few minutes each day, the pups got a taste for juniper from their foster mother. It was not clear whether the nursing mother's influence was through her body odor or through components of her milk, but in rats there is experimental evidence showing that what is in mother's milk can affect the food preferences of her young. Whatever the exact route of transfer, it is clear that both in the womb and while suckling, rabbits get information from their mother about what she has been eating.

What is true for rabbits also seems to be true for humans (figure 5.3). Recently it has been found that the six-month-old babies of women who had had a lot of carrot juice during the last three months of pregnancy preferred cereal made with carrot juice to that made with water. The same was true if the babies' mothers had had the carrot juice only during the first two months of the breastfeeding period. Babies whose mothers had drunk just water showed no such preference. Clearly, for some mammals, including humans, food preferences begin to form very early, when the young are still in the womb, and are then enhanced by the tastes and smells that are transferred to them during suckling. The amniotic fluid, placenta, and milk do more than provide food materials—they also transmit information in the form of traces of the substances that the mother has eaten. This information helps to determine the preferences that become evident in the eating habits and culinary culture of the next generation.

Information about eating habits can be transmitted in other ways. In rodents it is sometimes transmitted through the mother's saliva and breath, as she licks her offspring and they sniff at her mouth. Another channel of transmission is the feces: many young mammals eat their own and their

Figure 5.3
Food traces transmitted in a human mother's milk (top) influence later food preferences (bottom).

mother's feces, a seemingly unhygienic behavior known as coprophagy. Coprophagy allows the animals to make the most of their food, extracting all the remaining useful components from it. In plant-eating animals it also helps to ensure that the microorganisms needed to break down plant cellulose are transferred from mother to offspring. But in addition, the feces can be sources of information about food. They certainly seem to be so for young rabbits. Just before the youngsters have to face the world on their own, their mother deposits a few fecal pellets in the nest, and the youngsters eat them. In the series of experiments described earlier, the scientists found that when the fecal pellets of normally fed mothers were replaced with those from mothers on the juniper diet, the young rabbits had a strong preference for juniper food. So it seems that by leaving feces in the nest, a mother gives her soon-to-be-independent young some additional, up-to-date information about what she has been eating.

There is a lot of evolutionary logic in the existence of these channels through which information affecting food selection is transmitted at a very early age. If youngsters had to find out what is good to eat entirely through their own efforts—through trial-and-error learning—they would probably make some very costly mistakes. The development of early, maternally induced food preferences prevents this. The information received from the mother is likely to be about foods that are nutritious, nonpoisonous, and common. It is much better for a naive youngster to find and eat these foods, and only later, when it has had more experience of life, expose itself to potentially dangerous new types of food.

Information that is transferred through the placenta, milk, and feces may make the young prefer the same food as their mother, so it can contribute to the formation of family food traditions. But food preferences, like other types of behavior, change during an animal's lifetime. A large part of an animal's behavioral repertoire is acquired later, through various learning processes. We will come to these shortly, but before doing so we want to do as we did with the epigenetic and genetic systems, and try to characterize this BIS.

It is immediately obvious that the behavior-affecting information that is transmitted through placenta, milk, and feces is holistic, not modular. The substance transferred is itself one of the building blocks that enables the mother's behavior to be reconstructed by her offspring. When the substance is transferred, the young may inherit (reconstruct) her behavior; when it is not transferred, the mother's behavior cannot be reconstructed (unless the information is communicated by another route). This BIS is therefore far more like the inheritance of self-sustaining loops or structural inheritance than it is like the genetic system. And like them, the number of transmissible variations for any one aspect of the phenotype—for any single type of behavior, such as a food preference—is probably quite small. Nevertheless, the number of combinations of various preferences and tendencies that individuals in a population can display may be very large.

There are two other properties of this and other behavioral inheritance systems that we want to emphasize, because they make BISs very different from the genetic system. The first is that although the transfer of information is usually from parents to offspring, it need not be. For example, substances in the milk of a foster mother can be transmitted to her adopted offspring. The second concerns the origin of variation: with BISs it is difficult to talk about blind or random variation, because the information inherited by the offspring has been acquired and tested by their mother, and variations in this information are the result of her development and learning processes. It is a change in the parent's behavior that generates a new behavioral variant that may be reconstructed in the next generation, and there is rarely anything blind about a behavioral change.

Inheritance through Nonimitative Social Learning: On Opening Milk Bottles and Getting at Pine Seeds

Behavior-affecting information is inherited not only or even mainly through substances transferred from the mother. Young birds and mammals also get information by observing and learning from the activities of their parents and others with whom they interact. Although they could probably learn asocially, through their own trials and errors, most young birds and mammals are not left to fend for themselves—the world is too complicated and dangerous, and the young are too ignorant. Instead youngsters associate with and learn through others, usually (although not always) through their parents and relatives.

Before considering how behaviors are passed on by learning through or from others, we need to say something about the nature of what is transmitted with this type of inheritance, because it worries some people. With the genetic and epigenetic systems, something material is passed from one generation to the next: information is carried in DNA, in chromatin, or in other molecules or molecular structures. The same is true with the BIS we have just been dealing with—molecules of substances that influence behavior are transmitted from parent to offspring. But now we have come to inheritance systems in which nothing material is transmitted. It is what an animal sees or hears that matters. Does this make any difference? For our purposes, we believe it does not. In all cases, information is transmitted and acquired, and in all cases the information has to be interpreted by the recipient if it is to make any difference to it. An animal can receive information through its ears and eyes, as well as through its DNA and chromatin, and the interpretation of this information can affect behavior, just as the interpretation of DNA information can. So from our point of view, information transmitted through observational learning is not essentially different from any other types of inheritance. All provide heritable variations which, through selective retention or elimination, may lead to evolutionary change.

Let's return now to social learning in young animals. Like learning that is mediated by the transfer of behavior-affecting substances, early social learning is usually rapid and has long-term effects. The habits acquired early in life are often hard to change, and some behaviors that are learned very easily while young are much more difficult to acquire when older. There seems to be a special "window of learning" for some types of behavior—a window that is wide open early in life and gradually closes as the individual matures. The learning that takes place during this circumscribed period early in life is known as "behavioral imprinting," because learning is so rapid and the behavior is so stable that it seems as if the stimuli that induce it leave a persistent "imprint" on the youngster's brain.

A well-known example of this is the "filial" imprinting that can be seen in farmyard birds. For several days after hatching, young chicks, ducklings, and goslings very devotedly follow their mother around, learning as they do so her shape, color, calls, and actions. Consequently, they can later recognize and respond to the specific appearance and activities of their own mother. Mistakes can be made, however (figure 5.4). The photograph of the famous Austrian ethologist Konrad Lorenz, marching through a meadow with a file of goslings faithfully following him as if he was their mother, is familiar to most zoologists. In fact this odd behavior was first investigated and described scientifically in the late nineteenth century by the Scottish biologist Douglas Spalding, who discovered that ducklings and other chicks respond to the sight of the first large moving object they see by following it and forming an attachment to it. Spalding also found that the learning process occurs during a very narrow time frame—the first

Figure 5.4
Filial misimprinting: goslings follow the ethologist they saw soon after hatching, rather than their natural mother.

three days after hatching. In natural conditions, of course, the first large moving object that chicks see is almost certainly their own mother, and there is a lot of adaptive logic in becoming imprinted on her features. After all, she knows where there is good food and safety. It is only in the rather unnatural conditions of experiments that the youngsters acquire some rather useless information by following and becoming attached to the experimenter or a shoe dragged along by a piece of string.

Learning by newly hatched chicks is a good example of imprinting: it is rapid, happens during an early and very limited period, occurs without any immediate reward, and leads to a normally adaptive pattern of behavior. But there are many other types of imprinting. One that has been well studied is sexual imprinting, which is important for finding an appropriate mate. Youngsters become imprinted on the image of the parents who care for them, and in adulthood this image is the model for their choice of mate. It makes good biological sense that this should happen, because it normally means that the animal courts someone of its own species, but it has made life very difficult for those trying to breed endangered species in captivity. If appropriate measures are not taken—for example, by hiding the human carer and exposing the young to a dummy looking like an adult of its own species—things can go seriously wrong. The young become imprinted on humans, and when they mature they show complete indifference to adults of their own species, preferring instead to court and sometimes even attempt to mate with their human keepers. Before the problem was recognized, many attempts to maintain populations in captivity failed because the animals misimprinted on their human foster parents.

There is a lot more that could be said about imprinting, because it has been studied for many years, but even the little we have said here should be enough to show how it can function as an inheritance system. In all types of imprinting, youngsters are exposed to species-specific, group-specific, and lineage-specific stimuli, and the information they acquire through learning leads to the production of typical behavioral responses when they are older. This in turn can eventually lead to the reconstruction of the stimulus on which the individual was imprinted. For example, when a chick becomes sexually imprinted on a particular model (a parent with blue feathers), it will tend to choose a blue-feathered mate; therefore its offspring will in turn be exposed to and become imprinted on a blue-feathered parental image, and so the cycle will continue.

Imprinting is one way in which information that has been acquired by learning can be transmitted from generation to generation, but not all social learning is as early or as rapid as with imprinting. Many behaviors can be learned throughout life, some taking a very long time to learn. The complex behaviors seen when meerkats hunt their various types of prey or when bowerbirds build their beautiful bowers are learned and perfected over a long period, several years in the case of bowerbirds. So, how do they acquire their skills? Do the young imitate experienced adults? The answer is that in most of the cases that have been studied, they do not. Nevertheless, they do learn from others. What seems to happen is that by being close to and watching an experienced individual, an inexperienced one (a "naive" one, in the jargon of behavioral biology) has its attention drawn to something in its surroundings that it hadn't taken much notice of before. The naive individual may also observe the outcome of an experienced animal's activity. As a result of what it sees, it develops similar behavior. For example, if an animal sees another eagerly eating a type of food it doesn't know, it may itself try the food, but the way it obtains it and handles it is not an imitation of the experienced individual's activity. It finds out how to deal with the food through its own trials and errors.

A famous example of this type of social learning is the spread of the habit of opening milk bottles that occurred in English tits. In the United Kingdom and some other parts of Europe, milkmen deliver bottles of milk to peoples' homes, leaving them on the doorstep. At some stage tits began to take advantage of this, and learned to remove the bottle cover and eat the rich cream underneath. By the 1940s, the habit was already widespread in most parts of England, "infecting" not only more and more great tits and blue tits, but also several other species. In some places the tits learned not only to open the milk bottles but also to identify the milkman's cart, flying behind it and attempting to open the bottles even before the milk reached its destination (figure 5.5).

The rapid and extensive spread of the bottle top–removing habit received a lot of attention from biologists, since it was clear that the habit was

Figure 5.5
Thieving tits.

"cultural"—it was due not to a new genetic mutation, but to a new invention and its social dissemination. Close observation of what was happening showed that the birds did not learn by imitation: they did not copy the actions of the bottle top removers, since some opened bottles one way, some another way. What naive birds learned through watching experienced ones was that milk bottles are a source of food. They learned how to open the bottles by individual trial and error, each developing its own, idiosyncratic technique.

In the case of these thieving tits, naive individuals, who might be of any age, were probably observing several experienced ones. So you could argue that it is not surprising that they did not imitate the actual actions of the experienced tits—they had too many "tutors," each using a somewhat different technique. Had they had a single tutor, with a single technique, they might have imitated its actions. However, there is evidence that even when youngsters do learn almost exclusively from one individual, usually it still does not involve imitation. A good example of this has been described by Ran Aisner and Yosi Terkel, two zoologists from Tel-Aviv University.

In the 1980s, Ran Aisner found something unusual on the floor of a Jerusalem-pine forest—a lot of pine cones from which the scales had been totally stripped. It looked as if an animal had been after the pine seeds under the scales, but what was it? The obvious candidate for this type of cone-stripping behavior, the squirrel, does not live in this part of Israel, so some other animal must have been the culprit. Some careful detective work soon revealed that the nocturnal, tree-dwelling cone stripper was the black rat. Black rats, we should add, do not normally live in the Jerusalem-pine forests, nor do they normally eat the pine seeds in pine cones; they are omnivores, who live on the fringes of human habitation. Their spread into a new habitat, the dense pine forest where almost the only food available is pine seeds, was therefore very curious.

Stripping closed pine cones is not easy, so the efficient way the cones were stripped fascinated the two Israeli scientists. It showed that the rats must have acquired a complex new skill. Aisner and Terkel therefore studied the distribution and development of the stripping behavior. They found that whereas adult rats from the pine forest population could strip pine cones rapidly and expertly, moving from the lower part of the cone spirally upward, rats from other populations had no idea how to get at the seeds. Cross-fostering experiments showed that the skill was acquired by learning, not by inheriting particular genes: when pups of nonstripping mothers were adopted by stripping mothers, they learned how to strip the cones, whereas pups of an experienced stripping mother who were adopted by a nonstripper never acquired the skill. Pups could learn from a skillful mother, irrespective of their genetic relationship. To learn the skill, pups need to be allowed to go on stripping cones that have already been partially stripped by their mother. They learn to complete the job by following the direction of stripping that the mother used, and in this way they eventually develop an efficient technique for getting to the seeds. It is the initial condition produced by the mother's behavior that ensures that the efficient spiral method of stripping will be adopted. The pups do not imitate their mother's behavior—they do not copy what she does—but her presence and her tolerance toward them when they snatch seeds or partially stripped cones from her are necessary for them to learn the stripping technique for themselves.

Socially learned changes in feeding behavior have enabled Israeli black rats to adapt to a habitat that is different in many ways from their traditional one. But their adaptation involves more than just eating a different food. By living in trees, the rats provide a new learning environment for their young. These young, just like their parents, have to perfect tree-climbing and learn tree-nesting skills. In this way, merely by associating with their mother and experiencing their parent's environment, the youngsters' learned behavior becomes similar to that of their parents, and later will be transmitted to their own young.

Israeli black rats and British tits have shown how socially mediated learning can lead to the formation of new habits that are passed from one generation to the next, thus forming new traditions. Although these two examples have been studied more closely than most others, they are not at all exceptional. The investigation of animal traditions has been an active area of research in recent years, and it has become clear that behavioral traditions, mediated through social learning, affect all aspects of bird and mammal life—their food preferences, courtship behavior, communication, parental care, predator avoidance, and choice of a home. Inheriting behavior through social learning is not uncommon.

What can we say about the nature of the information and its transmission with this type of behavioral inheritance system? First, as with the previous type of BIS, in order for a habit or a skill to be transferred, it has to be displayed; if it is not displayed, it is not transferred. Second, the information is holistic—it cannot be deconstructed into discrete components that can be learned and transferred independently of each other. Third, in no sense is the origin of new variation random or blind. A new behavior can be initiated by a lucky or curious individual who learns by trial and error, or by observing individuals of another population or species, and once acquired it may be transmitted to other members of the group through social learning. But what is learned and transmitted depends on the ability of an individual to select, generalize, and categorize information relevant to the behavior and, no less important, to reconstruct and adjust the behavior about which it has learned. The receiving animal is not just a vessel into which information is poured—whether or not information is transferred depends on the nature of that information and the experiences of the receiving animal. Neither the transmitting nor the accepting animal is passive in this type of learning.

A fourth characteristic of nonimitative learning is that, as with the previous BIS, information can be transferred not only from parents to offspring but from any experienced individual to any naive one. The milk bottle–opening behavior seems to have spread even to different species, although we cannot be sure that they did not invent it for themselves. Finally, with this inheritance system, as with the last one, the number of different variants of transmissible behavior is probably not very large if we are looking at any one habit: black rats either know or don't know how to strip pine cones, tits are either aware or not aware of the food potential of milk bottles. Nevertheless, the number of combinations of different transmissible habits and practices that the individuals in a population can show may be very large.

Imitative Learning: On Singing Whales and Birds

Our third type of BIS involves imitation: a naive individual learns not only what to do but how to do it. It copies the actions of another. There is a lot of argument about how common this is in the animal world, but it is generally agreed that one type of imitation, vocal imitation, occurs in some birds, dolphins, and whales. These animals learn what song to sing by imitating the song of others. As a result, different populations may have different dialects, just like the dialects found in human populations. Birdwatchers have known about these regional variations in bird song for many years. More recently, similar variation has also been found in the songs of killer and sperm whales, where members of a group are united by a dialect that is clearly different from those of other groups.

Studies of bird song have shown that learning through imitation is an important part of song development: the patterns of sounds that a youngster hears are reconstructed in its own song (figure 5.6). Since this usually happens during an early and limited period of life, this imitative learning

Figure 5.6
Imitation by a songbird.

is known as "song imprinting." There are fewer studies of whale songs, but the few that there are suggest that here too the young probably imitate the sounds made by adults. In contrast to nonimitative social learning, the "students" learn to reproduce the vocalizations of their "tutors"—they are not just reacting to the environment in a similar way.

Vocal imitation certainly occurs in birds, and possibly in whales and dolphins, but it is less clear how much motor imitation—imitation of movements—there is in the nonhuman world. It is undoubtedly important in human development (babies are great imitators of movements as well as sounds), and both vocal and motor imitation have probably been extremely important in human evolution. Being able to imitate sounds must have been crucial for the evolution of language, and the ability to imitate movements was probably one of the things that led to the evolution of our unique aptitude for culture, especially with respect to toolmaking and tool use. There is some evidence that motor imitation also occurs in chimpanzees, rats, dolphins, gray parrots, starlings, and a few other species, but in general it seems to be relatively rare in nonhuman animals. We have to be cautious, however. Distinguishing between imitative and nonimitative learning is not always easy, and there have been very few experiments that would detect motor imitation unequivocally, so we cannot really evaluate how frequent and important it is in the animal world.

The way in which information is acquired and transmitted through imitation is not the same as with the other two behavioral systems we have described. Consequently the evolutionary effects are different. We can see this if we characterize this final type of BIS in the same way that we did the earlier ones. Imitation is similar to the other BISs in that information is not transmitted in a latent, encoded form—the behavior has to be displayed in order for it to be inherited. But in contrast to the other two BISs, with imitative learning information is transferred in a modular manner—unit by unit. For instance, it is possible to alter a unit in the song of a bird and transmit the altered song to the next generation. Similarly, when a dance is imitated, some parts can be altered and imitated without affecting other parts of the dance. This modular structure allows many variant patterns of behavior to be formed. Nevertheless, we have to remember that not every variant has the same chance of being transmitted. Transmission depends on the difficulty of copying the units (the notes or actions), the length and rhythm of the sequence, and its functional effectiveness. In addition, the number of "teachers," their skill, and how often the behavior is performed will all affect the speed and efficiency of learning. Unlike a photocopier or DNA polymerase, which copies information irrespective of its content, transmission through imitation is not independent of function and meaning.

The variations that occur in imitated behavior are usually not blind to function either. A brand-new behavior may be the result of individual trial-and-error learning, or stem from a new group activity, or it may even be learned from another species. Whatever its primary origin, however, before it is passed on it is adapted and reconstructed so that it becomes adjusted to the animal's general lifestyle and is easier to perform. If we compare a new behavior to a new mutation, it's as if the "mutation" was thoroughly edited before being transmitted to the next generation.

Imitation allows a lot of heritable variation in a behavioral sequence, because what is imitated can be changed bit by bit. The sequence of actions may be quite long, but because the rewards come from carrying out the whole sequence, not the parts, variations in the individual parts are not constrained. What matters, from an evolutionary point of view, is how effective the imitated sequence is as a whole. In theory, therefore, through the selection of variations in an imitated behavior, it should be possible to build up a quite complex tradition. Yet this seems not to have happened. Many songbirds and parrots have a wonderful ability to imitate, but as far as we know this has led to nothing very intricate or elaborate. It looks as if just having the potential to generate a lot of transmissible variation is not enough to produce sophisticated traditions comparable to those of human culture. So what is missing?

To appreciate the missing component we need to recognize that although the number of potential variations in behavior that are possible with a modular system is huge, many are neither functional nor useful. If all were produced, the useful combinations would be buried among a mass of useless ones. But as we know, unless an animal has a brain disorder, it does not produce random combinations of actions. Animals must therefore possess some kind of internal filter—some set of principles or rules—that allows only behavioral variations that have a reasonable chance of being useful to be formed. All animals must have rules that restrict the variations that are generated, and animals that imitate must also have rules or "guidelines" that restrict what is copied. Humans, for example, usually do not imitate blindly—our decision to imitate another individual is often guided by our belief in the meaning, relevance, or usefulness of the observed activity. Imitation by humans is intentional: it is directed by perceived goals and inferred reasons. Only if there is a fairly sophisticated understanding of other minds can the modular system of imitative learning open up truly revolutionary (evolutionary) possibilities.

Although we have emphasized the uniqueness of imitation as a system for transmitting behavioral information, it shares many features with the previous BISs. In all behavioral transmission, variation is targeted and culturally constructed. It is targeted in two senses: first, there are simple rules that organize perceptions, emotions, and learning processes. For example, all animals, including humans, group things into fairly distinct categories, even though the world is not so sharply organized: we distinguish between color categories, shape categories, and so on. Similarly we anticipate that what has happened many times before will continue to happen in the same way: if smoke has always been associated with fire, the next time we see smoke we (and some other animals) anticipate fire. Second, the type of information an animal may acquire by learning is structured by the past evolutionary history of its lineage: some things are learned easily in one species, but not in another. Most humans can easily distinguish between individuals by sight, but are poor at distinguishing between them on the basis of smell, which is something that dogs have no trouble with. We most easily form, remember, and transmit information that corresponds to our general and species-specific evolved biological biases. All behavioral variation reflects these biases and the organizational rules of the animal's mind, and behaviors are constructed to be compatible with the individual's own preexisting habits and those of the social group of which it is part.

One of the striking things about almost all BISs is the active role that the animals play in acquiring and transmitting information. Sometimes the role is very direct, as with active teaching, but often it is indirect. Darwin was one of the many biologists who recognized long ago that animals actively participate in shaping the environment in which they live and are selected. In recent years, interest in "niche construction," as it is now called, has grown. Both theoretical and empirical studies have shown that it plays a significant role in social learning and the evolution of animal traditions. We can see why if we think about the tarbutniks. In their imaginary world, as the burrow-digging habit that protects them against predators spreads, individuals will concurrently be selected for their ability to live in these burrows. Consequently they may change the time of day at which they are active, or alter their food habits. Similarly, in the real world of the black rats in the Jerusalem-pine forest, their change in diet means that they spend most of their time in the trees, building their nests and caring for their offspring there. Through having learned how to extract pine seeds from pine cones, the rats have constructed for themselves an environment that is very different from that of other black rats. If this tree-dwelling habit persists over several generations, any variations, whether socially learned or genetic, that make the rats better adapted to tree living will be selected. The rats may end up evolving gray squirrel–like habits, which will be mutually reinforcing, since all involve a style of life that is tree-oriented. In this way, a new habit can result in animals constructing for themselves an alternative niche in which they and their offspring are selected. Animals are therefore not just passive subjects of selection, because their own activities affect the adaptive value of their genetic and behavioral variations.

Traditions and Cumulative Evolution: Evolving New Lifestyles

Earlier we defined cultural evolution as a process involving a change through time in the nature and frequency of socially transmitted patterns or products of behavior in a population. The examples we have given so far have shown that socially learned and transmitted changes in behavior patterns, skills, and preferences certainly do occur, although of course most behavioral innovations are ephemeral and do not manage to become established, let alone spread, in a population. Nevertheless, occasionally new patterns of behavior do spread through social learning, and this leads to new traditions and cultural change. The question we now have to ask is whether this is of any evolutionary significance.

Many people believe that cultural evolution in animals is very limited and fragile, and only exceedingly simple traditions are formed. Yes, they say, animals do learn socially, and they even admit that it is quite common. Yes, they agree, this does lead to the formation of habits and to some simple traditions, although these cannot be very stable. But, no, they insist, it cannot lead to any complex cultural adaptations. For them, what is really interesting about animal traditions is how genetic evolution led to the ability to form them. Whenever the question of the evolutionary significance of animal traditions is raised, it is the genetic aspects that dominate discussions.

The evolution of the genetic basis of the ability to construct cultures is a very important topic, and we will discuss it in later chapters. Here we just want to make the point that there is no real justification for the assumption that the culture of animals is limited in scope and complexity, and therefore cannot be a significant independent agent of evolution. To make such a judgment, we need to know how common animal traditions are, and unfortunately we do not know this. Studying animal traditions is not yet a very fashionable area of research. This type of work is difficult and takes many years, whereas research grants are usually given for only short periods, so funding is a problem and often limits what people can do. Yet, in spite of this, many more traditions are now recognized than were even dreamed of a few years ago, and new traditions are regularly reported. There are now enough long-term studies to show unambiguously that the diversity of animal traditions is enormous, involving many different species and many different aspects of life. Recently, scientists described thirty-nine cultural traditions in nine African populations of the common chimpanzee. The researchers believe that this is still an underestimate, because the studies have been made over a relatively short time and the understanding of many aspects of chimpanzee behavior is limited.

One of the reasons why the number and range of animal traditions may have been underestimated is that it is commonly assumed that all hereditary variation is genetic. Usually this assumption is neither questioned nor verified. When it is, the usual finding is that the inherited behavior has genetic, ecological, and traditional aspects. The fact that there is genetic variation affecting behavior does not mean that other factors, such as social learning, are less important. In the same way, when social learning is found to be an important cause of differences between groups, it does not exclude genetic differences.

It is interesting to ask why, in spite of the growing evidence that animals have traditions, people are still reluctant to acknowledge that significant cultural evolution occurs in animals. We think the main reason is probably associated with our awareness of the complexity of cultural evolution in humans, and the relationship it has to human values. If we think about the culinary culture of Polish or Yemenite Jews, this complexity is obvious: the culture includes ways of preparing, cooking, and serving the food, and is closely associated with other aspects of life, such as religious and secular rituals. It is clearly the outcome of cumulative evolution in which habits acquired in the past are preserved and become the foundation on which additional habits are established. In this way, a complex culture has been built up. If we now compare this human culinary culture with the food traditions of nonhuman animals, we are struck by the relative simplicity of the latter. Even the Israeli black rats, which developed such an elaborate method of stripping pine cones, have acquired just one new technique and preference. The cumulative evolution required for cultural complexity does not seem to exist in animals. That is why the critics claim that, in animals, evolution along this axis, the axis of culture, is very limited. Many doubt that the simple traditions of nonhuman animals deserve to be called culture at all.

Fortunately, there are some long-term studies that show that cumulative cultural evolution does occur among nonhuman animals. We will look more closely at one of these, the research on cultural evolution in the Japanese macaque monkeys on the small island of Koshima. The study began in the 1950s, when Japanese primatologists started to provide food for the macaques they wanted to study. They used sweet potatoes to lure them from the forest to the sandy seashore, where they were easier to observe. This trick had unexpected results. Before eating the potatoes from the beach, one female, Imo (which is Japanese for potato), then one and a half years old, started washing them in a nearby stream, thereby removing the soil from them. The new habit spread to other monkeys. Some time later, they began washing the potatoes in the sea rather than the stream. They also started to bite the potatoes before they dipped them into the salty water, thus seasoning them as well as washing them.

Imo's inventiveness did not end with potato washing. A few years later, she solved another problem. The macaques were being fed on the beach with wheat, which was difficult to collect and eat because inevitably it became mixed with sand. Imo's solution to this problem was to throw the mixed sand and wheat into the water, where the heavier sand sank while the wheat floated, making it easy for her to collect it. The new habit spread, first from the young to adults, and then from mothers to children. Adult males, who interact much less with the young, were the last to learn, and some did not learn at all.

The habit of bringing food to the sea and collecting the food that their human observers threw into it had other effects. Infants that were being carried by their mother when she washed the food became used to the sea, and started playing and bathing in it. Swimming, jumping, and diving became popular. Another new sea-related habit developed when hungry older males began eating fish that the fishermen had discarded. The habit spread to others, and now when there is nothing better available, they collect and eat fish, limpets, and octopus from pools.

What has happened since the scientists first started feeding the macaques on Koshima Island is that a new lifestyle has developed. The original potato-washing tradition triggered another—separating wheat and sand in the water—and these two in turn helped to trigger the tradition of using the sea for playing and swimming. Each habit reinforces the others: the pleasure of swimming brings the monkeys to the shore and reinforces the tendency to wash foods, while washing food in the sea increases the likelihood that monkeys will discover the pleasures of swimming. Each transmissible habit varies little, but a whole new lifestyle has evolved through one modification in behavior producing the conditions for the generation and propagation of other modifications.

Clearly, cultural changes in animals can be cumulative, but the result is not linear evolution with a consistent increase in complexity in one direction. Instead, what we see is that cultural variation in one domain influences the chances of generating and preserving cultural variation in another, and this in turn can affect another domain, and so on. One habit can stabilize other habits, so eventually there is a network of habits that together construct a new lifestyle. The elements of the new lifestyle become more stable as mothers begin to transmit the behaviors to their young, because early learning has particularly potent and long-term effects. Of course, the persistence of behaviors also depends on their adaptive value—even if a new habit is popular initially, it will disappear if it reduces the chances that its practitioners will survive and reproduce. The persistence of a new habit also depends on the ease with which it is learned and transmitted, and this in turn depends on the extent to which it is integrated with the habits that are already established in the population. In the case of the Japanese macaques, the newly acquired habits have certainly persisted: sadly, Imo is now dead (figure 5.7 is our tribute to her genius), and for the past quarter of a century the macaques have been given

Figure 5.7
Imo's restaurant.

sweet potatoes only about twice a year, yet the potato-washing culture that Imo initiated remains.

The lesson from the Koshima study is that cultural evolution in animals can be complex, gradual, and cumulative, and involve several different aspects of behavior. We believe that many other complex heritable behaviors will also be found to have a strong "traditional" component. Usually we will not see a linear sophistication of just one aspect of behavior, but rather we will recognize that through social learning a new web of behaviors has been constructed, just as happened in the Japanese macaques.

Dialogue

I.M.: Although I hadn't thought about it before, it is obvious to me that what you have just been describing—the transfer of behavior and even whole packages of behaviors—must occur in intelligent creatures like mammals and birds. But I must confess I found some of the things you wrote about a little disquieting. You described several channels through which information is transferred to the very young, and said that this information affects their behavior later in life. Does this mean we are slaves of our early education, and that a lot of our behavioral destiny is determined when we are six months old? Are you implying, just as some psychologists claim, that the basic structure of personality is determined in the first few years of life? I am not convinced. I have a Yemenite friend who is a great admirer and connoisseur of gefilte fish, and I need not mention in present company that Polish Jews and even Brits sometimes eat the spicy Yemenite food you call schug with great relish!

M.E.: Of course education is important at every stage. But early education and early input often have strong effects, which are difficult to alter. But, we stress, *difficult*, not impossible! Early learning creates preferences and biases; it does not determine them. If the preferences and biases are positively reinforced, the behavior consolidates easily; if they are not, then another behavior develops. For example, suppose that a mother had a lot of carrot juice at the breastfeeding stage, but later her carrot juice–preferring infant is given carrot juice that has gone off and makes it sick. The child will probably then develop an aversion to carrot juice, and prefer something else. Yet if fresh carrot juice had been readily available, the child would probably prefer carrot juice to any other drink. Carrot juice is not something it can't live without, and it is not an addiction, but the preference is there and may be reinforced if it drinks more fresh juice. Clearly, the individual has a choice. If there were suddenly a social or medical taboo on carrot juice, a lot of people who were reared on it and love it would stop drinking it. It is quite obvious that the information that the individual acquires during its early life is updated as it develops. So it is not surprising that there are some gefilte fish admirers among the Yemenites, although certainly they did not acquire their liking for it with their mother's milk! It is even less surprising that Polish Jews in the Middle East end up preferring the local, generally popular, spicy food. But often they had to "get used to it," and it took time for them to acquire the taste.

With some types of imprinting, such as sexual imprinting, altering the imprinted preference may be more difficult. For example, we mentioned that it's not easy to alter sexual imprinting in birds once it is established, and there is a lot of evidence for this. It is also difficult to alter imprinting on a habitat. Mauritius kestrels almost died out because monkeys that were introduced onto the island stole and ate the eggs from their nests, which were in tree cavities. It was only when a pair of kestrels nested on a cliff ledge, which was not a traditional nesting site, that things got better. They reared their brood successfully, because the new nesting site was safe from monkeys, and the cliff-reared offspring became imprinted on the new site. As adults, they too nested on the predator-free cliffs. Had kestrels not somehow found this solution, the species could have become extinct, just because of the conservativeness inherent in social learning.

I.M.: So there is inertia in behavioral systems and that can lead to extinction?

M.E.: Yes. Information transmitted through social learning, like information transmitted in other ways, can sometimes be a curse when the environment changes. But socially learned information is usually not as fixed as nonacquired information, such as genetic information, so there is more chance of adaptive alteration.

I.M.: Is there really? You have now reminded me of something that bothered me about those early-learned preferences that you described. What is the difference between a preference that is acquired early in life and has long-term transmissible effects, and a genetic predisposition to the same behavior?

M.E.: If you take two animals, one with a genetic predisposition to prefer a certain food (e.g., it has a genotype that leads to the development of a nervous system which produces an automatic association between a particular taste and pleasure), and a second individual who has the same preference but in this case it was acquired while in the uterus or during suckling, then you will probably see no difference in their behavior. In both cases, the preference will disappear if the animal has an aversive experience with the food. If you looked at the brain using imaging techniques that show what parts are active when a certain behavior occurs, you would also probably see no difference. In both cases similar areas would light up when the individuals are offered the preferred food. The difference between them has more to do with the future—with what happens in the next generation. When you have a genetically determined preference that biases the development of the nervous system, the information is heritable, irrespective of the parent's environment or actual experiences. It doesn't matter whether she ate the food or not, the preference will be passed on. In the case of early-learned information, transmission of the preference depends on the mother eating the particular food during pregnancy or when her young are suckling. If she does not, there will be no transmission, and consequently no preference and no corresponding eating habit in her young. The preference will disappear. For an acquired preference to be maintained over generations, it has to be satisfied, and that will depend on the conditions of life. When conditions are stable, it is very difficult to see any difference between a genetic predisposition and early-learned behavior. If conditions keep changing, the difference becomes apparent.

I.M.: But this means that an acquired preference can disappear within a generation. For cumulative evolution, surely you need some degree of stability. What has been acquired in previous generations must be a reliable foundation for acquiring new habits. If an acquired behavior can disappear within one generation, how can there be cumulative evolution? There will always be something that prevents the flow of acquired information from generation to generation. Animals cannot write their history and learn it from books!

M.E.: Lack of stability is not the insurmountable problem you think it to be. It is true that, just as with human habits, most animal habits are ephemeral, lasting for only part of the lifetime of an individual, for at most one or two generations. But this is not always so. Some cultural traits can be stable for many generations. In the case of the Koshima macaques, it is already more than fifty years (six generations) since they washed sweet potatoes for the first time, and with some chimpanzees in West Africa, the tradition of cracking open nuts with a stone is at least 400 years old. We know this because an early seventeenth-century Portuguese missionary wrote about it. And there are probably a lot of other cases of long-term cultural stability which we have simply failed to identify as cultural. You are probably wondering where this stability comes from. Well, it could be the result of many different things, but we think that it is due mainly to the interaction and dependence of one behavior on others.

Think again about the macaque monkeys of Koshima. Their cultural system became stabilized because links were formed between different types of behavior. These links mean that even if one particular aspect of behavior disappears, there's a good chance that it can be easily reconstructed. Let's say you stop giving the monkeys sweet potatoes and wheat, but they still go to the beach because they like swimming and are used to water. Now if after a few generations you start giving sweet potatoes again, the chances that they will wash them are quite high, simply because they are often in the water and are likely to take their food with them. So they are not back to square one, where inventing potato washing is much less likely. The presence of one type of behavior increases the chances that a related behavior will develop. Sometimes the feedback between behavior and environment is ecological. For example, when birds hide the seeds they like in the ground so that they have food in times of scarcity, they increase the chances that this behavior will be perpetuated in subsequent generations. This is so because they do not find all the seeds they hide, and some germinate and develop into the plants that produce the very seeds the birds cached. In a sense the birds plant the seed supply of the next generations. So the chances that future generations will repeat the behavior are enhanced, because the birds' activity has stabilized the relevant aspect of the environment.

I.M.: But the genetic system is more stable. We know that culture changes rapidly. Look at what happened with humans in the last century!

M.E.: This leads me to a second answer to your question about the instability of cultural traits. Instability does not have the same consequences for traditions as it does for genes. The problem with genetic instability is that most changes are blind mutations, and are usually deleterious. You need relatively high fidelity in the genetic system; otherwise the lineage will degenerate as useful information is lost. With cultural, learned traits, on the other hand, most change is not blind: it is a functional variation on a theme. Not all (or even most) cultural changes are for the worse, because there are so many internal learning filters and social filters to pass through before anything is transmitted to the next generation. You don't need stability in the genetic sense, provided changes are *functional*, not degenerative. This is really a more general point, which also applies to some cellular epigenetic variations. When variations in information are targeted and constructed, when there are processes that direct the changes and then filter the information prior to its transmission or acquisition, fidelity in the strict sense is not necessary. You still need fidelity, but it is fidelity of a different kind. You need *functional* fidelity—the new information has to be at least as functional as the previous information. Since information is often updated, sometimes getting increasingly more sophisticated, adaptive evolution through BISs or EISs can be very rapid.

A last point to notice is that natural selection can lead to the genetic stabilization of traits that were initially purely cultural. If an environment is very stable, and a rapid reliable behavioral response is very important, the speed and reliability of learning may be improved by genetic changes that lead to biases in the development of the behavior. For example, if you have an early-learned food preference, and the food is such that it is absolutely vital, there may be an advantage in having a genetically based bias toward it. There will be selection of genes that stabilize the learned preference.

I.M.: OK. I can see that traditions can be stabilized or adaptively modified in the ways that you have suggested. But is this really enough to produce a new species? In your tarbutnik story you said that this might happen.

M.E.: In the 1970s the German ethologist Klaus Immelmann suggested that cultural differences resulting from sexual imprinting and from imprinting on the natal habitat could be important in the formation of new species, especially in birds. You can appreciate his idea if you think what might happen if a few birds start using a different part of their environment. Their young will become behaviorally imprinted on the new habitat, and they will try to find a similar habitat as adults. If they do, then just for reasons of proximity they are likely to find a mate who also prefers that habitat. Sexual imprinting may also be affected. For example, if the new habitat is acoustically different, males may adjust their song so that it is different from that of males in the parent population. If females prefer the dialect of the local males because they were imprinted on it when young, they are even more likely to mate with a neighbor. In this way the population in the new habitat will begin to be reproductively isolated from that in the old. This is the beginning of speciation.

I.M.: That's theory! Is there any evidence?

M.E.: There are no observations or experiments that unambiguously show purely cultural speciation. But the same is true for most of the suggested mechanisms of speciation in animals. However, although direct evidence is lacking, there are experiments that provide good circumstantial evidence for the importance of behavioral imprinting in speciation. The animals that have supplied it are African parasitic finches—birds that, like some cuckoos and cowbirds, lay their eggs in the nests of others, who are therefore tricked into rearing their offspring for them. Robert Payne and his colleagues, who have been studying parasitic birds for over thirty years, have shown experimentally how cultural transmission could promote rapid speciation. They took eggs of a parasitic species, and put them in the nests of a species unaccustomed to parasites. The new hosts incubated the eggs and looked after the parasite nestlings with the same loving care they showed to their genetic offspring. Not surprisingly, the young who grew up in the nest of the new host became imprinted on the song of their foster father. When adult, the males sang his song, and females were attracted to the males who were singing it, preferring the song of these males to that of those reared by the normal host.

So within one generation, the parasites of the new host species had become at least partially reproductively isolated from the species from which they originated. Their reproductive isolation was the result of imprinting, which made females prefer males who sing their foster parents' song. Imprinting also made the females lay their eggs in the nests of their foster parents' species, so their offspring were exposed to the same type of imprinting stimuli as they themselves experienced, and they should continue to parasitize the new host. Eventually, if the experiment were to go on for long enough, reproductive isolation might be stabilized through natural selection of morphological variations in the parasite that lead to even more devoted parental care by the hosts (although the parasitized host would almost certainly change too). But that is speculation.

I.M.:   I have another problem with imprinting. You said that many birds and mammals become sexually imprinted on the image of their parents. What prevents incest in these species? If they like a mate who is similar to their parents, why do they not mate with the parents or with their very close kin, who are likely to look just like the parents? I thought that too much incest creates dangerous genetic problems, but sexual imprinting seems to me to be a sure recipe for incest!

M.E.:   You are right: mating with close kin, known as inbreeding, usually does lead to genetic problems. It leads to increased homozygosity—to individuals having more genes for which both alleles are the same. This means that detrimental recessive alleles are likely to be expressed in the phenotype. As to your question about "incest," there are several answers. First, in most species, the young do not stay in the group in which they were reared. Juveniles of either or both sexes leave and look for mates elsewhere. In most mammals it is the young males who disperse, while the females remain in their natal group, but in birds it is usually the other way around—females disperse and more of the males remain. So incest with siblings does not occur. There are other reasons why incest doesn't happen. The English biologist Patrick Bateson found that, at least in the lab, Japanese quails prefer as mates individuals who are similar, but not identical, to their parents; there is a taste for mild novelty! This seems ideal: on the one hand, the young ensure that they will have mates similar to their parents in behavior and appearance, and that increases the likelihood that their mates are compatible with them; on the other hand, the taste for novelty ensures that genetic and behavioral diversity is introduced.

I.M.:   Clever quails! But I want to go back to the actual transfer of information, rather than its consequences. Tell me, in social animals, what happens if the young are exposed to individuals who are not their parents? Are they affected by them?

M.E.:   It depends on what you are looking at and how much contact there is between the young and these other individuals. For example, a lot of song imprinting happens when the young are out of the nest, so the singing of neighboring males as well as that of the father affects the song the young learn. In other cases, imprinting is earlier, and it is the parents, the usual carers, who transmit the learned information.

I.M.:   I don't want to be tiresome, but I remember hearing that in many species of birds and mammals the parents have "helpers"—older offspring or "friends" who help them to rear the young. Are these helpers transmitting behavioral information to the offspring? Do the offspring become imprinted on the helpers' behavior?

M.E.: It may sound quite unbelievable, but we don't know! The question has not been investigated. To the best of our knowledge, no one has tried to see if, in families with helpers, birds develop preferences and behaviors similar to those of their past helpers. We would certainly expect to find such an influence.

I.M.: How much influence? As much as that of parents?

M.E.: Less, probably, because there is less contact and fewer opportunities to transfer information, but this would really depend on the specifics of the social system. In many species of birds and mammals, you have not only helping behavior but also actual adoption. In these cases the adopting parents may transfer almost their whole package of preferences and behaviors to the young, just as natural parents do. This may mean not just information relevant to food preferences, sexual tastes, and so on, but also to a style of parenting, which in the case of an adopter or helper may include those behaviors that enhance the tendency to adopt or help. So the young who are exposed to this parenting style may themselves, when they are adults, become adopters and helpers.

I.M.: Are you implying that this will lead to the spread of adopting or helping behavior? I would have thought that such behavior must undermine the efforts of individuals to rear their own offspring, so it would disappear! Natural selection should get rid of such altruistic behavior.

M.E.: Not necessarily. Adopting or helping need not harm an animal's ability to rear its own young, and sometimes it may enhance its reproductive success. It may be that by caring for the young of others a helper is practicing caring skills, and by doing so it will later become a better parent. Often adoption is not a substitute for having genetic offspring, because the adoptees are added to an existing litter or brood, and their presence does not affect the survival of the adopting parents or their foster sibs very much. Even if there is a small decrease in the reproductive success of helpers and adopters, if the transmissibility of their parenting style is high, then, yes, adopting and helping behavior will spread. This is so because the helper or adopter transmits its parenting style not only to its genetic offspring but also to its cultural ones (the adopted or helped young). Since helpers and adopters may help and adopt young from lineages that practice more "selfish" parenting styles, they will infect these lineages with their seemingly altruistic behavior, and it will spread.

I.M.: Do you think that is why adopting and helping behaviors are so common?

M.E.: It may be one of the reasons.

I.M.: Information can be transmitted by peers and other influential individuals in a group, even if there is no helping or adopting. Does that make any difference to cultural evolution?

M.E.: If there is a lot of transmission through nonparents—let's follow convention and call it "horizontal" transmission—the rate and pattern of evolution will be different from that expected when information is transmitted almost exclusively through parents. For example, potato washing among the macaques would have spread a lot more slowly if there had been no horizontal transmission. Bad habits, for example eating a fruit with strongly addictive but health-endangering effects (something equivalent to smoking in humans), also spread more rapidly when transmission is horizontal. There will be selection against the bad habit of course, but whether it is eliminated or spreads, and to what extent it spreads, depends on the strength of selection against it on the one hand, and the rate of its transmission on the other. Since horizontal transmission of information increases the number of individuals who can receive it, you can have really bad habits persisting. Smoking is definitely one! However, such bad addictive habits are rare in animals, who have to survive in rather demanding environments and cannot afford ill health. Most really bad habits will be eliminated by selection rather rapidly. If they are not, the group with the bad habits will soon perish. But some mild bad habits may persist.

I.M.: But wouldn't horizontal transmission, even of habits with neutral rather than bad effects, undermine the results of cumulative selection? Learning new behavior from animals in a different lineage, ones with very different habits, could lead to the adoption of their way of life, and that would mean the animal's own cultural adaptations were gone for good! Sharing behaviors would undermine the cultural adaptations of different lineages.

M.E.: Not necessarily. Animals are not passive recipients of behaviors. As we have been emphasizing all along, they *develop* their behavior—newly acquired behavior is reconstructed and adjusted to preexisting biases and patterns of behavior, many of which are acquired through early social learning from their parents. A new behavior has to fit into this behavioral package. If it does not, the individual will usually not accept it, or will modify it until it does fit with its existing set of habits and practices. Yes, horizontal transmission will lead to some degree of cultural homogeneity, sometimes maybe even strong uniformity, but in general there is no reason to believe that it will always undermine useful and well-established cultural adaptations. Often it may assist their spread.

I.M.: Would you say that the greatest difference between genetic and cultural evolution is that the latter is more rapid?

M.E.: No. Cultural evolution usually is faster, although in some circumstances it can be very slow, and there may well be periods of cultural stasis. However the most interesting feature of cultural evolution is that the variation that underlies it is more channeled and therefore more adaptive. There is a constant updating of behaviors and preferences as individuals develop, so many of the variations that are transmitted through behavioral inheritance systems are generated by what we earlier called "instructive" processes. They are the outcome of an individual's learned response to external conditions: a young mammal receives information about its mother's new food source through her milk, and develops a preference for it; an individual macaque discovers how to wash newly available sweet potatoes, and others learn from her; an animal in a new habitat discovers a call to which its young respond more readily, and the young learn it; a bird discovers a safe new nesting site, and through habitat imprinting its young learn to use it too. As an individual's behavior develops and changes adaptively, it is transmitted to others through social learning. A Dawkins-type division of labor between replicator and vehicle is inappropriate, because the development of a variant behavior and its transmission go together. You can't separate heredity and development. And since you can't separate heredity and development, you can't separate evolution and development. The changes induced during development, mainly through learning, play a big role in behavioral and cultural evolution. There is an awful lot of Lamarckism!

I.M.: There is one last question, which is in fact the one I wanted to start with. I have already said that I don't find it too surprising that behaviors are transmitted from generation to generation in intelligent creatures like birds and mammals. But it would really surprise me if you told me that insects or something similar have traditions. Do they?

M.E.: For insects there is a lot of circumstantial evidence showing that learned information is transmitted between generations. Most of this transmission is through the transfer of behavior-affecting substances that lead to food preferences, and to preferences for mating and laying eggs on particular hosts.

I.M.: Are there actual examples of socially mediated behavioral inheritance in insects?

M.E.: This has not been studied sufficiently, but there are a few examples. One of the best known is the cultural transmission of alternative social organizations in the fire ant, *Solenopsis invicta*. In some lineages of these ants the nests have several small queens, whereas in others there is just a single large queen. Cross-fostering experiments have shown that a queen's adult phenotype and her colony type are determined largely by the social organization of the colony in which she matures. The differences between the two colony types and queens probably depend on the level of a queen-produced pheromone—a chemical that is released into the colony. This pheromone affects the maturation of larval queens directly, and may also act indirectly through its effect on the quantity or quality of the food given to them by worker ants. The outcome is that the existing queens' phenotypes and colony organization are perpetuated.

A rather different kind of tradition has been found in several species of butterfly and other insects, which can become imprinted on a new food plant during their larval stages and will then use the same plant for egg-laying when adult. The changed behavior is maintained over several generations. There are also species of cockroach in which the young follow their mother as she forages for food at night, just like ducklings following a mother duck. It seems likely that the young learn what and where to eat, and that they in turn transmit this information to their offspring, but this needs further study.

I.M.: Why is it not studied? It may have practical implications!

M.E.: It is probably for the same reason that made you say that you would be surprised if traditions are found in insects. Insects are seen as "soft automata," so people don't expect them to have social learning. We agree that socially mediated learning may have lots of practical implications. It might be possible to exploit social learning to condition insects to prefer particularly damaging weeds, for example. In the conservation of mammals and birds, it is already clear why social learning is important. If you try to reintroduce a species into an area where it has become extinct, you have to teach the animals a lot before releasing them into the wild, especially if they are social animals. Their genetic inheritance is not enough; they need the information that is transmitted through the behavioral system as well.

I.M.: One very last question: what about humans? How does this behavioral inheritance concern humans? I mean beyond carrot juice and such things.

M.E.: It sounds like a simple question, but the answer is horribly complicated. We believe that with humans there is a qualitative leap. Not only has the behavioral inheritance system become extremely important but another mode of transmission has evolved and taken over: there is a massive transfer of information through symbols (like in our linguistic system), and this is an entirely new dimension. We look into this in the next chapter.

6　The Symbolic Inheritance System

When an evolutionary biologist looks at her own species, *Homo sapiens sapiens*, she sees a contradiction. On the one hand she recognizes that in their anatomy, physiology, and behavior humans are very similar to other primates, especially chimpanzees. She can see how alike humans and chimps are in the way they express basic emotions, in their highly developed sociality, in their ability to improvise, and in some of their ways of learning. It is easy for her to see why Jared Diamond called our species "the third chimpanzee" because as an evolutionary biologist she discerns the homologies that suggest a common ancestry. Yet, on the other hand, she also sees that humans are very different from other primates: this species of chimpanzee writes music and does mathematics, sends missiles into space, builds cathedrals, writes books of poetry and of law, alters at will the genetic nature of its own and other species, and exhibits an unprecedented level of creativity and destruction, rewriting the past and molding the future. In these respects, *Homo sapiens sapiens* is totally unlike any other species.

What is it that makes the human species so different and so special? What is it that makes it human? These questions have been answered in many ways, but in our opinion the key to human uniqueness (or at least an important aspect of it) lies in the way we can organize, transfer, and acquire information. It is our ability to think and communicate through words and other types of symbols that makes us so different. This view is not new or original, of course. The idea was explored more than half a century ago by the German philosopher Ernst Cassirer, and recently it has been discussed in depth by the neurobiologist Terrence Deacon. Like Cassirer, we choose the use of symbols as a diagnostic trait of human beings, because rationality, linguistic ability, artistic ability, and religiosity are all facets of symbolic thought and communication. This is what Cassirer wrote:

... this world [the human world] forms no exception to those biological rules which govern the life of all the other organisms. Yet in the human world we find a new characteristic which appears to be the distinctive mark of human life. The functional circle of man is not only quantitatively enlarged; it has also undergone a qualitative change. Man has, as it were, discovered a new method of adapting himself to his environment. Between the receptor system and the effector system, which are to be found in all animal species, we find in man a third link which we may describe as the *symbolic system*. This new acquisition transforms the whole of human life. As compared with the other animals man lives not merely in a broader reality; he lives, so to speak, in a new *dimension* of reality. (Cassirer, 1944, p. 24; Cassirer's italics)

Cassirer goes on to suggest that rather than defining man as the "rational animal," we should define him as the "symbolic animal," because it was the symbolic system that opened the way to mankind's unique civilization. The symbolic system—the peculiar, human-specific way of thinking and communicating—may have exactly the same basic neural underpinnings as information transmission in other animals, but the nature of the communication (with self and with others) is not the same. There are special features that make symbolic communication different from information transmission through the alarm calls of monkeys, or through the songs of birds and whales.

What symbols are, how they form and develop, and how they are used are among the most complex issues in the study of man, but for us there are some minor consolations. The most obvious is that there is no need to resort to thought experiments to convince anyone that symbolically represented information is passed on from one generation to the next. It is something we all take for granted. Those of us living in the Western world know that most of the people we meet will have at least a nodding acquaintance with the Bible, and will share the long cultural heritage of which it is part. And everyone will readily agree that our symbol-based culture is changing through time: we only have to think of what has happened to technology during the last hundred years to be convinced. However, before we look at cultural change in our species, we will try to explain in general terms what symbols and symbolic systems are, and how they provide a fourth dimension to heredity and evolution.

Mr. Crusoe's Great Experiment

It would be nice to start with an accepted, intelligible, general definition of symbols, but unfortunately any definition we could offer at this point would either be misleading and limited, or incomprehensibly long and awkward. We will therefore take a less formal approach and try to explain the peculiarity of symbols and symbolic systems through examples.

One way to start is to take a sign—a piece of information that is transferred from a sender to a receiver—which looks like a symbol but clearly is not one, and compare it to a very similar sign that certainly is a symbol. This may help to pinpoint the defining features of symbols. Since there is no real-life example that is not in some way partial, we will make use of another thought experiment. This one is not entirely original; it is a rather free adaptation of one suggested for other purposes over a hundred years ago. Its author was Douglas Spalding, the brilliant Scottish biologist mentioned in the previous chapter, who is rightly regarded as one of the fathers of modern ethology. We will be using the original Spalding story in chapter 8, but meanwhile here is our version of it. Before beginning, we need to make the usual writer's disclaimer: all characters in this story are entirely fictitious, and any resemblance to real people or animals is purely coincidental.

Imagine a Robinson Crusoe who, soon after settling on his island, caught a few of the local parrots and started teaching them various English phrases (figure 6.1). He used the well-known method of teaching through positive and negative reinforcement—the carrot-and-stick method—which rewards good behavior and punishes bad. He quickly discovered that the parrots learned much more quickly when taught in pairs, competing for the tidbits

Figure 6.1
Mr. Crusoe's great experiment.

he used as rewards. Since Mr. Crusoe was feeling rather lonely, he first taught his parrots to say "How do you do?" when they saw him or one of the other parrots first thing in the morning. The parrots on the island had a great talent for vocal imitation, so they soon learned the greeting. Mr. Crusoe then taught them to say the words "fruit," "veg," "grain," "water," and "coconut milk" when they saw these types of food and drink. After they had learned these words, he trained them to say "found fruit," "found veg," "found grain," and so on when they found the corresponding foods, and "give fruit," "give veg," "give grain," and so on when they wanted Mr. Crusoe, a mate, or parent to give them those foods. He also taught the parrots to name several of their natural enemies—"eagle," "snake," and "rat." These animals all prey on parrots' eggs and nestlings, but each uses a different method of attack, so each requires a different type of defense. Mr. Crusoe taught the parrots to call out the correct name for each predator when they saw it, and rewarded them if they acted appropriately when they heard the calls of other parrots, even if they themselves did not see the predator. They were thus trained to use English words as alarm calls, warning relatives and neighbors about the dangers.

Not only did Mr. Crusoe train the parrots to use various words, he also amused himself further by doing some selective breeding. He allowed only those parrots who could say all the words with the best English accent and in the right circumstances to breed. And perhaps most important, each year he bred only from those families in which the youngsters learned the words from their parents and other parrots rather than from himself. In this way he gradually established the calls as part of the parrots' human-independent habits.

Now imagine that after over forty lonely years of intense training and selective breeding, Mr. Crusoe dies, but the parrot population thrives, and the various alarm calls and food calls are passed on from parents to offspring, from mate to mate, and among neighbors. They become, as Mr. Crusoe always intended, part and parcel of the local parrots' behavioral repertoire. Although the parrots still use several non-English calls and gestures, some of the English calls have now supplanted the traditional parrot ones.

Picture now an unsuspecting English ethologist coming to this island fifty years later, knowing nothing of Mr. Crusoe and his great experiment. When she hears the clearly adaptive and appropriate English calls of the parrots, she is of course amazed. For a fleeting moment she believes that her secret prejudice is now justified, and English (the only language she can speak) really is God's universal primary language, the ancient language of the Garden of Eden. But after discovering Mr. Crusoe's diary, with its full documentation of his training and selection experiments, she reverts to more conventional ways of thinking and begins to analyze the phenomenon she is observing. Being interested in the symbolic nature of language, the ethologist asks herself whether the repertoire of English words and phrases that the parrots use so adequately and clearly is the much sought-for example of a simple and true protosymbolic system—a missing linguistic link. Are the parrots on the road to symbolic language?

There are several features of the parrots' word calls that the ethologist recognizes are quite similar to the way words are used by humans. First, the calls are arbitrary (in the sense that the birds would have learned Hebrew words and phrases if only Reuben Krutnitz had reached the island instead of Mr. Crusoe). Second, the calls are clearly referential; each call refers very specifically and accurately to a particular thing or situation in the world, and evokes a typical, appropriate response. Third, they are conventional, in that all the parrots "agree" about what each call refers to.

All of these things, as the ethologist realizes, are true of human language, but she knows that a symbolic language is that and a lot more. The parrots' repertoire is poor, but it is not the poverty of the repertoire that worries her. After all, she knows that when people from groups speaking different languages first try to communicate, they too have a very poor repertoire of words, and so do very young children. What strikes the ethologist is the *rigidity* of the parrots' call system. She traps a few eloquent parrots and teaches them the word "cracker" for a new food item. They learn it successfully and rapidly. However they never join the words "found" or "give" to the newly learned word "cracker" to say "found cracker" or "give cracker" without being taught. They do not generalize the property of the word "give" or "found" and apply it to a new item. Each phrase, like each word, has to be learned from scratch, as a single unit. The birds do not grasp the *relation* between the words; they do not map the relation between objects and acts to the relationship between words in a phrase. For the parrots, each call, whether it is a word or a phrase, is a unit.

Why is this so important? Think about a child with a limited vocabulary. There is a stage when the child's use of words and phrases seems rather similar to the parrots': each utterance is tied to a certain situation and to a certain response, and is learned as a unit. But very soon the child goes beyond this. She begins to combine words flexibly: she will use the already learned word "give" with many of the other words she has learned and with the new words that she is learning at present. The word "give" does not lose its meaning, because it still refers to a certain action directed at the child and it is still a "tool" that leads to the fulfillment of a desire, but it has acquired a new kind of mobility. It can be transferred between situations and contexts. The child can say "give cookie," and she can easily apply the verb to the newly learned word "teddy," saying, without being taught, "give teddy." The particular wish expressed by the word "give" has become generalized to many different demand situations, so the word "give" maintains a certain general meaning, full of possibilities. Of course, the expression and potential satisfaction of a particular desire depend on the combination of words, but the word "give," once learned, need not be learned anew for every possible combination. Its use can be extended even further: it can later be used metaphorically for saying things like "give hope."

Take another example: think of what goes on when our learned parrots shriek "snake." This call, "snake," has a truth value—it is related to a situation (snake around!) and the parrot transfers information about this situation, information that can be true or false. (It is usually true, except for rare cases of mistake or deceit.) But when a child who has reached the stage of combining words says "snake," this does not necessarily mean that there is a snake around. In fact it does not refer to any one single situation. She can say "want snake" (referring to a toy); she may say "find snake" (in a game); she can say "feed snake"; or she may say that she likes or is angry with a snake (whether or not it is actually there). It is as if the child at this stage treats the word "snake" (and any other word) as an analytical unit, as part of the analysis of an actual or imagined situation which is reflected in the stringing together of words. This unit, the word "snake," always retains its reference, but not its situation-specific truth-value, its emotional value, or its effects on the speakers' and hearers' actions. The truth-value, the emotional value, or action value has been transferred to the sentence level. This gives the unit, the word, great freedom to move from context to context, because it is no longer constrained by referring to one particular situation or desire, or to a particular behavioral response.

This stage, even if it is at first very modest, requiring only a very limited stringing together of words and no special rules for ordering them, depends on the ability of a child to understand that the relation between words reflects the relation between parts of a perceived and experienced situation (including the child's own desires), and vice versa. It is not just that the words reflect the relations between objects, actions, and goals; the organization of words *points* to such relations. The child can go not only from an experienced situation to words but also from word combinations to situations. This has enormous implications. The realization that words refer to parts of a situation helps the child to recognize that aspects of her experience can be differentiated and named. Moreover, existing words help her to focus her attention on the elements of the situation she experiences, to break it into parts. And the correspondence between different aspects of her early experience (for example, between sweet taste and pleasure, and between mother and pleasure) may lead her to use words metaphorically, as, for example, in "sweet mummy." The ability to assemble words into phrases and sentences, and to use them metaphorically, allows the child to create fictions—imaginary objects and situations.

Combining words has to be subject to rules, simply because the number of possible combinations becomes enormous, even when the vocabulary is small, and the meaning of different combinations becomes very ambiguous. What these rules are and how they develop is not our concern here, but we should note that the rules of language (grammar, especially syntax) allow us to generate and understand an infinite number of varied meaningful sentences. We see this even when we consider relatively simple phrases like "man bites cruel dog," "dog bites cruel man," "cruel man bites dog," "cruel dog bites man," and so on, which may be true or imaginary, and which we understand unambiguously if we understand the individual words and the rules of grammar (in this example, the rules are reflected in the word order). A related and important property of language is that even if a word has never been heard or seen before, the words that are already known and the grammatical structure in which the unknown word is embedded often give a good hint about what the word may refer to. This is especially true if the word is playing a role in several different sentences. Words can therefore be said to refer to each other. The way that words refer to each other is most clear when we think about what a dictionary is—it is where we define words by means of other words. To sum up, we can say that *words act as symbols because they are part of a rule-governed system of signs that are self-referential.*

Language is, of course, a lot richer than we have suggested, because it involves sounds, gestures, intonations, and so on. But we will leave it for now because, although it is so central to human nature, language is really a rather special symbolic system, and we need to think about symbols in general. So think of a picture—a picture of Jesus on the Cross, or of the Virgin Mary holding the infant Jesus in her arms. No one who is familiar with Christianity will doubt that these pictures are pregnant with symbolic meaning. But why? In contrast to strings of words, there is nothing arbitrary about these pictures (assuming that they have been painted in a more or less realistic style); they depict a terribly suffering man, a woman

200
Chapter 6

Chapter 6 200

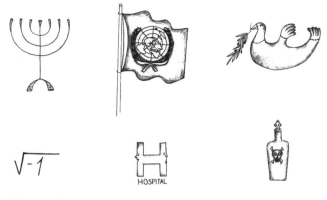

Figure 6.2
Symbols.

holding a baby. Also, unlike a sentence or paragraph, it is not very easy to break a picture into parts: the organization of a picture is more holistic than that of a sentence, and the parts of the picture are more interdependent. Yet, in spite of these important differences between linguistic utterances and pictures, we still think of pictures as symbols, because we interpret them within a shared framework of religious or artistic practices in which they have a role or function. In the examples we used, the pictures are part of a complex of Christian religious practices, where they play an actual or potential role by symbolizing suffering, redemption, and so on. The choice of these particular images is, in a sense, conventional: non-Christians have other ways of representing similar ideas and emotions. The pictures are part of an organized communication system involving religious symbols and practices, and where there are symbols there is, by definition, a *symbolic system*.

So, in summary, we can say that signs—the pieces of information transferred from sender to receiver—become symbols by virtue of being a part of a system in which their meaning is dependent on *both* the relations they have to the way objects and actions in the world are experienced by humans, and the relations they have to other signs in the cultural system. A symbol cannot exist in isolation, because it is part of a network of references. However, the extent to which the interpretation of a symbol depends on other symbols is not the same in all systems. For example, a picture within the system of religion or art may convey something meaningful to a spectator even if the cultural system of which it is part is unfamiliar, whereas a mathematical notation, like the mathematical symbol

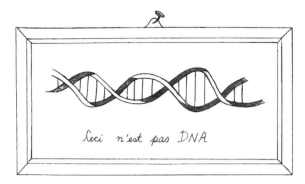

Figure 6.3
Ceci n'est pas DNA (after René Magritte).

$\sqrt{-1}$, derives its meaning entirely through relations with other symbols in the mathematical system. In all cases, however, the systems allow people to share a fiction, to share an imagined reality, which may have very little to do with their immediate experiences. This is true of stories, of pictures, of rituals, of dances and pantomime, of music, indeed of any type of symbolic system we may think about. All symbolic systems enable the construction of a shared imagined reality.

Symbolic Communication as an Inheritance System

Earlier we suggested that our ability to communicate through symbols is at the root of many of the things that make us so different from other animals. Humans have a unique method of transmitting and acquiring information. What we now want to do is look at symbolic communication from a narrow angle, that of the system that provides a fourth dimension to heredity and evolution. We want to try to characterize this special inheritance system in the same way that we did the genetic, epigenetic, and behavioral systems, and see how similar it is to each of them. There is at least a superficial resemblance between the way we transmit information through speech and the way animals use various calls and songs, so does the symbolic system work in the same way as the behavioral inheritance system? Or is it more like the genetic system? DNA is called the "*language* of life," and our characteristics are said to be "*written* in our genes," so there must be obvious similarities between the two systems. What are they? What features does the symbolic system share with other systems of information transmission, and what is it that makes it so different and special?

There is one important property that the symbolic and genetic systems share but is missing from behavioral inheritance. Symbols and genes can transmit *latent* information, whereas information has to be used before it is transmitted or acquired by behavioral means. It's easy to see this if you think about how a song or a dance is transmitted to others. Let's consider three cases: transmission through the genetic system, transmission through the behavioral system, and transmission through a symbolic system. For the genetic example, we can use fruit flies of the genus *Drosophila*, which have very beautiful songs and dances. The songs are sung by males, which produce them by vibrating their wings. The dances, in which males are the more active partners, include wing movements, wafting scents around, circling around each other, touching, and licking. If it all sounds very sexy, it's because it is—it's a courtship dance. Each species has a characteristic song and dance, which enables the flies to identify their own species. These songs and dances are innate, and quite a lot is known about their genetics, but the important point is that they will be inherited even if the parents never perform them (perhaps because an evil experimenter has cut off their wings). The same is true of the songs of some species of mammals and birds, which are also innate. In other bird and mammal species, however, if individuals are to inherit a song (we know less about dances), it has to be sung in their presence. It is only through hearing the song that individuals can obtain the information that will enable them to reproduce it. In other words, for a pattern of behavior to be transmitted through the behavioral system, it has to be displayed; there is no such thing as latent information which can skip generations.

This is certainly not so with transmission through the symbolic system. Humans can transmit a song or a dance to others even if they are tone-deaf and have two left feet. There is no need for us to sing a note or dance a step, because we can transmit the information required to reproduce a song or dance using disks or films, or even through written or oral instructions. Symbolic information does not have to be acted upon in order for it to be transmitted. Provided that the culture that can interpret it remains intact, it can remain unactualized for generations. Information for building the Third Temple has been transmitted among the Jews for almost two thousand years, but the Temple has not been built. And grandma's soup recipe may be passed down through the family for several generations before someone gets around to making the soup again.

The symbolic and genetic systems are similar in that both can transmit latent information, but the symbolic system can do far more than this. Since symbols are shared conventions—socially agreed-upon signs—they can be changed and translated into other corresponding conventions. Theoretically, their translatability is unlimited. An instruction in English that is given in Roman letters can also be given in Morse code, or in semaphore, or in computer code. Symbols can even be "translated" between systems: the idea of Jesus on the Cross can be expressed in language, pictures, dance, and mime. "Danger" can be expressed by a word, a picture, a tune. A story can be transmitted orally, after learning by rote; it can be transmitted through song and pantomime; it can be transmitted in writing; and these days it can also be transmitted through films, TV, and computer games. So although symbolic information is like genetic information in that it is encoded and is translatable, the translatability of symbolic information is much broader than that of information in the genetic system. Since we can "translate" symbols from one form to another, and separate and combine different symbolic forms and levels by following general principles of coherence, vast amounts of meaningful symbolic information are readily generated.

Some types of symbolic information are more likely to be generated, transmitted, and acquired than others. Like information communicated through the behavioral system, a lot of new symbolic information is targeted; it is not the outcome of uncorrected mistakes like most new genetic variants are. New symbolic information is organized and categorized mentally, and tested and adjusted to fit existing ideas, habits, and culture before it is passed on. And with symbolic information, an additional element comes into the editing and testing processes, because symbolic construction is not always concerned with present realities. It is often fictional and future-oriented. Symbolic systems can readily generate fictions such as the Third Temple, the unicorn, the communist utopia, the square root of minus one, and the nth dimension. Moreover, future goals, future scenarios, and future plans are part of the background against which newly generated information is edited. Obvious examples of such future-oriented constructions are utopias, myths, and the new plans invented in a manufacturer's research and development unit. The construction and selection of future-oriented ideas can work at different levels—in the head of a creative individual, in the actions she performs to test them (which then feed back into her ideas), in the group within which the ideas and their implementations are presented, and in the larger society.

Different aspects of the symbolic system have different structures. Language, at the sentence level, is modularly organized: its units (words) can be changed one by one, just like the units (nucleotides) in a DNA sequence, so an enormous number of variations can be generated and transmitted.

A picture or dance has a more holistic organization, with the parts being more interdependent, although the elements in it can still be combined in many different ways. Moreover, symbolic systems are hierarchically organized: words are components of phrases, phrases are components of sentences, sentences are components of stories, and stories (along with pictures, dances, songs, books, and other artifacts) are themselves units within a larger cultural system such as a religion. With symbolic systems we therefore have to think about the way that symbols can join to form ever-larger hierarchical structures, and about the ways in which the symbolic elements can be moved and reorganized to produce new meanings. This makes thinking about the generation and transmission of variants in the symbolic system quite difficult, because there are so many interacting factors and levels to consider.

There is another complicating factor. Symbolic information is often transmitted from adults to nonrelated young (as in a school), from young to adults, and among individuals belonging to the same age group. In this it resembles the behavioral system of other animals. But there is a significant difference: active instruction is important with symbolic transmission systems. In other animals, social learning usually does not involve intentional teaching, but for humans it is essential, because the symbolic system itself, and not just the local culture that it produces, has to be culturally acquired. For example, although people argue about the role of learning and the type of learning that is involved, no one doubts that a lot of learning is necessary before a child can understand and use language. The need for overt learning and instruction is seen even more clearly with other types of symbolic system: we are *taught* the symbolic system of literacy, we are *taught* mathematical symbols and rules, we are *taught* how to understand and participate in the rituals of our culture. The framework needed to interpret symbolic information has to be learned.

Cultural Evolution and Symbolic Communication

We can summarize the last section by saying that although the symbolic system of transmitting and acquiring information has properties that it shares with other inheritance systems, it is also clearly different from any of them. Inevitably, therefore, human cultural evolution, which is based largely on information transmitted through symbolic communication, has characteristics that make it very different from other types of biological evolution.

In the previous chapter we defined culture as the system of socially transmitted patterns of behavior, preferences, and products of animal activities that characterize a group of social animals. Cultural evolution was described as a change, through time, in the nature and frequency of these socially transmitted preferences, patterns, or products of behavior in a population. Nonhuman animals transmit behavioral information in a variety of ways: often it is through vocal signs, as in the communication systems of birds and whales; in other cases it may involve a complex combination of vocal, visual, tactile, and olfactory signs. When communicated across generations, these animal signs may form a culture. As with human culture, the process of acquiring information is an active one that involves the reconstruction and transformation of the information. However, animals do not have a symbolic culture, because their communication signs do not form a self-referential system. Human culture is unique in that symbols permeate every aspect of it, and even acquired behaviors such as food preferences or songs, which in other animals are transmitted by non-symbolic means, are usually associated with symbolic communication in humans. And things such as ideas, artifacts, and institutions are almost entirely based on symbols. This does not mean that all human communication depends on symbolic systems: the ways of acquiring and transmitting information that we described in the previous chapter are also important in human societies. Nevertheless, the defining feature of human culture is its dependence on the symbolic system and the great weight that symbols have in it.

The consistent, long-term differences in the cultural habits and beliefs of different human societies show that the symbolic system provides very effective ways of transmitting information. Not all cultural variants fare equally well, of course. Some customs and beliefs persist with little change, others disappear, and still others are modified. The question we now have to ask is how we should view these processes. Should we think in terms of Darwinian evolution, or is what we see Lamarckian evolution, or is it something totally different? What is the best way of looking at cultural change?

The nongenetic aspects of human variation and heredity have been incorporated into evolutionary thinking in various ways. Scientists like Luca Cavalli-Sforza and Marcus Feldman, Robert Boyd and Peter Richerson, have built mathematical models that describe how the frequency of cultural practices that are transmitted by nongenetic means (not necessarily symbol-based) change over time. These models show that when you have the essential ingredients of Darwinian evolution—cultural innovations (variation), cultural transmission (heredity), and differential multiplication and survival (selection)—the result is cultural change. However, as the French anthropologist Dan Sperber has noted, some of the models assume that the transmission of cultural ideas is a copying process, whereas in fact in most cases it is a reconstruction process in which the receiver actively acquires and transforms the information she receives according to her own cognitive and cultural biases. Consequently, because they do not focus on the central process of reconstruction, most of the mathematical models can provide only limited information about the spread of cultural variants.

The forbidding mathematical form of some of the models, and their ability to describe only general patterns of cultural change rather than throwing light on questions about its amazing diversity and sophistication, may explain why these models have not received much public attention. In contrast, two very different and more ambitious approaches have become very fashionable and are widely debated in both the scientific and public arenas. According to one, that of the evolutionary psychologists, in order to understand human societies and cultures we have to recognize the evolved genetic basis of human behavior. This approach is explicitly genetic and usually takes a "selfish gene" viewpoint to explain many of the most fundamental facts about human behavior. The second approach seems almost the complete opposite of this: it views cultural evolution as the outcome of competition between "memes"—cultural units that replicate and are selected in a way analogous to but separate from selfish genes. We start by looking at this latter approach, that of the memeticists.

The "Selfish Meme" View of Cultural Evolution

The meme was christened by Dawkins in 1976 in his first book, *The Selfish Gene*, where he described memes as the "new replicators" (the old ones being the genes, of course). Although Dawkins was not the first to discuss units of cultural transmission, thanks to his easily remembered term *meme* and his discussion of it in the context of selfish genes, the idea became more widely known. It seemed to suggest a simple and useful way of understanding complex cultural processes, and some people felt that just as genes had broken through the mess of hereditary phenomena and made them intelligible, so memes would do the same for culture.

After an initial lag period, Dawkins's meme concept began to flourish, and there are now books, an Internet journal, websites, and academic conferences devoted to memes. However, Dawkins's discussion of the meme concept is still one of the clearest, so we will start with what he said. He defines the *meme* as "a unit of cultural inheritance, hypothesized as analogous to the particulate gene, and as naturally selected by virtue of its 'phenotypic' consequences on its own survival and replication in the cultural environment" (*The Extended Phenotype*, p. 290). According to Dawkins, the meme is a unit of information residing in the brain, embodied as neural circuits. This neural "genotype" of the meme has phenotypic effects:

The phenotypic effects of a meme may be in the form of words, music, visual images, styles of clothes, facial or hand gestures, skills such as opening milk bottles in tits, or panning wheat in Japanese macaques. They are the outward and visible (audible, etc.) manifestations of the memes within the brain. They may be perceived by the sense organs of other individuals, and they may so imprint themselves on the brains of the receiving individuals that a copy (not necessarily exact) of the original meme is graven in the receiving brain. The new copy of the meme is then in a position to broadcast its phenotypic effects, with the result that further copies of itself may be made in yet other brains. (Dawkins, 1982, p. 109)

In this paragraph Dawkins makes a clear genotype/phenotype type of distinction, a replicator/vehicle distinction. The organism and the cultural products it creates (books, pictures, music, etc.) are the vehicles of the replicators, the memes. These memes, the information-containing entities, reside in the brain, and by virtue of their phenotypic effects they can spread by being copied into other brains (figure 6.4).

Memes compete to get into our brains and be passed on again. Usually it is the catchiest tune, the best idea, the most effective tool, or the most useful skill that succeeds. But memes, like genes, are "selfish" replicators. A "selfish" gene may undermine the survival and reproductive success of its carrier (for example, by making it behave altruistically), because this enhances the success of other individuals carrying copies of itself. The gene benefits; its vehicle does not. Similarly, memes can "selfishly" undermine the survival and replication of their vehicles, yet still increase in frequency. For example, the meme for cigarette smoking continues to proliferate, even though it harms its vehicles, because it has strong socially contagious and addictive effects. The meme thrives; the individuals harboring it do not. Memes are often described as "viruses of the mind."

A clear albeit extreme version of the selfish meme concept has been developed by Susan Blackmore, whose memes are ideas, instructions, behaviors and information, which are passed from person to person by imitation. She writes:

From the meme's-eye view, every human is a machine for making more memes—a vehicle for propagation, an opportunity for replication and a resource to compete

Figure 6.4
Cultural transmission according to the meme theory. In the upper part of the figure, the meme for the fancy shoe fashion spreads from the central figure to the minds of the observers on either side. As a result (below), all end up having the meme and displaying the fancy shoe phenotype.

for. We are neither the slaves of our genes nor rational free agents creating culture, art, science and technology for our own happiness. Instead we are part of a vast evolutionary process in which memes are the evolving replicators and we are the meme machines. (Blackmore, 2000, p. 54)

The meme idea that Blackmore and others have espoused is a seductively simple way of explaining the evolution of human behavior and culture in Darwinian terms. However, we believe the arguments on which it is based are flawed. The flaw stems from the distinction that is made between replicators (memes) and their vehicles (human brains, human artifacts, and humans themselves are all given this role). According to Dawkins's definition, a vehicle is an entity that *cannot* transmit its acquired variations from generation to generation: a change in an amoeba (a vehicle) cannot be transmitted unless it originates from a change in its DNA sequence (the replicator); or, using Dawkins's cake analogy, a change in a cake cannot be transmitted, whereas a change in the recipe can be. In general, no change in the vehicle will be passed on unless it stems from a change in the replicator. The problem for the meme concept is that if the developmental processes that vehicles undergo result in the generation of variations that are heritable, then the distinction between genelike replicators and phenotype-vehicles breaks down. Since heritable variations in behavior and ideas (memes) are *reconstructed* by individuals and groups (vehicles) through learning, it is impossible to think about the transmission of memes in isolation from their development and function.

The point will become clearer if we look at some of the examples of memes given by Dawkins and others. Consider the transmission of the memes for the skill of opening milk bottles in tits, for a new style of clothes, for a way of caring for babies, and for the set of ideas induced by the picture of Jesus on the Cross. If a meme is an information-containing entity analogous to a gene, then for each example it should be possible to identify something that is copied and transmitted. Yet in no instance is anything copied, except in a very broad and loose sense. In each case the organism (or group) actively reconstructs the pattern of behavior, or the pattern of emotions and ideas, through learning. And learning is not blind copying—it is a function- or meaning-sensitive developmental process.

Look at what happens when the milk bottle–opening behavior of tits is "copied." The activities of tits who know how to get at the milk draw the attention of naive tits to something (milk bottles) that they had not previously regarded as a source of food. As a consequence, the naive tits eventually reconstruct the bottle-opening behavior through trial-and-error learning. However, although this particular behavior, milk bottle-opening, can be named and delimited for some analytical purposes (such as studying its rate of spread, or geographic distribution), there is no "meme for milk bottle-opening" that is transmitted from tit to tit. The milk bottle–opening behavior and its transmission are tied together. The reproduction of the presumed "meme"—the milk bottle–opening circuits in the brain—is a developmental consequence of the reoccurring social and ecological interactions that cause tits to reconstruct the bottle-opening behavior. If we are to explain the reconstruction of bottle-opening behavior and the factors that impinge on its spread, we have to understand the properties and logic of this social-ecological system.

The same kind of reconstruction processes can be seen with human examples of behavioral transmission. There is a form of severe mental illness in which, among other things, the sick mothers do not touch their babies. This early deprivation has devastating long-term effects on the children, who show the same psychopathology when they grow up. Again, the daughters do not touch their babies, so the cycle is repeated. In this way the behavior is transmitted from generation to generation in the female line, but it is quite clear that no "meme for not-touching-babies" is being passed from mothers to daughters. What happens is that the daughters' interactions with their sick mothers lead to the reconstruction of the same pathological maternal behavior. "Not-touching-babies" is both a cause and symptom of the syndrome. It cannot be isolated as an autonomously transmitted "meme." It is a part of a psychophysiological system of interactions.

With the meme for a new fashion in clothes, we can again see that it spreads through a process of reconstruction, not through development-independent and learning-independent copying. It is true that acquiring the preference for a particular style of dress may seem a rather trivial and dispensable part of development, and we are often most impressed by the apparently contagious nature of the spread of fashions. But even if the reasons for following a fashion remain largely unconscious, adopting a style of clothes is still a consequence of development and learning in a social setting. What is reproduced when we adopt a fashion is not just a particular consumer behavior, but also complex social factors that are related to class, economic status, cultural icons, and so on.

As a final example we can take the transmission of the ideas embedded in a picture of the Crucifixion. Here, the element of cultural construction is far more significant and dominant than in the other cases. What has to be reconstructed in each individual is a hugely complex cultural-religious package, which is formed through a lengthy developmental process. The learning that is necessary involves several levels of social organization, from the child and its family, through the local community, to the organized church and society. To talk about a "Crucifixion meme" leaping from brain to brain tells us very little about this cultural phenomenon. In fact, it misses the very thing that it tries to explain—culture!

A big problem for the meme concept is that when patterns of behavior such as milk bottle–opening or not-touching-babies are transmitted to others, the copying mechanism is not independent of what is copied. It is not the same as with DNA replication or copying with a photocopier, where what is copied is irrelevant to the copying process. This is obvious if you think about transmitting a nursery rhyme (a meme) to a child. How well we transmit it and how well the child acquires it (for how long it will be remembered and how true it is to the standard version) will depend on the content of the rhyme, its melody, how many times we and others repeat it, when we learned it, our own and the child's musical talent, the child's motivation, and many other factors. In other words, transmission and acquisition involve learning processes that are sensitive to the behavioral-developmental history of both the "student" and the "tutor." Of course it can be argued that we can teach a child many different nursery rhymes, as indeed we do, and that provided these rhymes conform to some very general structure, they can be learned. So in this sense we can say that such learning by imitation involves processes that are not sensitive to content. It is certainly not as content-sensitive as transmitting a behavior such as milk bottle-opening by tits. This is probably why memeticists like Susan Blackmore focus on imitation as the major mechanism of meme transmission. However, when we imitate, what is copied is the "phenotype" of the meme: if we introduce a change into the nursery rhyme that we teach a child, the mistake will be perpetuated. So even mechanical imitation is not equivalent to the replication of genes, which would be unaffected by a phenotypic modification. And if imitation is not mechanical, if what is to be imitated is evaluated and controlled by the imitator, then imitation is a context- and content-sensitive process, not mere copying (Figure 6.5).

The learned, developmental dimension to the generation and reproduction of most cultural information makes it very difficult to think of cultural evolution in terms of distinct replicators and vehicles. There are no

Figure 6.5
The transmission of a nursery rhyme and the origin of a new variant. Notice that the variations in the song are not entirely random: they make sense, and the rhyme and rhythm remain the same.

discrete unchanging units with unchanging boundaries that can be followed from one generation to the next. Although the meme concept may appear to provide an intellectually manageable theory of cultural evolution, it does so by focusing on the selection of copied ideas and behaviors, and ignoring the much more difficult and messy issues of their origins, social construction, and interactions. It tells us nothing about the generation, implementation, and processes of transmission and acquisition of new cultural information. Surely we are likely to gain far more insight into the process of cultural evolution if, instead of thinking about the selection of supposedly distinct cultural variants or memes, which spread by virtue of their ability to "replicate" themselves more than others, we try to understand the complex processes that generate and mold cultural changes. The distinctive feature of human culture is its potent constructive power, which includes the ability to design and plan the future, and its coherence and internal logic. Symbolic communication allows humans to communicate ideas and artifacts that are constructed to deliberately shape their future within a very complex social and political system. Thinking of the spread of human habits and ideas in terms of the replication of selfish memes obscures these unique aspects of human evolution.

Evolutionary Psychology and Mental Modules

If interpreting human cultural evolution in terms of memes is unhelpful, is there a better way of thinking about it? Many evolution-oriented sociologists, psychologists, and anthropologists would say that there is—that we should think about human behavior and culture in terms of our genes. They see culture as a colorful and thin veneer spread upon genetically selected, innate, human-specific, psychological mechanisms.

This view is most obvious in the currently fashionable approach of the more radical evolutionary psychologists. They assume that the human mind is made up of a set of largely autonomous "mental modules" or "mental organs," likening the brain to a collection of minicomputers, each dedicated to a particular task. We have specialized modules for choosing mates, for language, for recognizing people, for number, for detecting cheats, for parental love, for a sense of humor, and so on. We are told that these modules were fashioned by natural selection during crucial periods of human evolution, particularly during the Pleistocene, when our ancestors were hunter-gatherers in the African savannah. Each module is dedicated to processing a specific type of information, and generates behavior that enables a person to do something that is likely to be adaptive. Modules show some organizational autonomy, have a high speed of operation, and are inaccessible to consciousness. When they go wrong, the processes they control are selectively impaired.

The main point of the evolutionary psychologists' arguments is that our uniquely human behavior is *not* the product of our greater general intelligence; rather, it is the result of highly specific neural networks that were constructed by Darwinian selection of genetic variations. In the past, this potent selection acting on distinct facets of behavior led to the evolution of correspondingly distinct mental modules. When the psychological mechanisms that these supposed modules determine produce behavior that is maladaptive, the evolutionary psychologists assume that it is because they evolved in the Pleistocene or more remote past. Then, they claim, the behavior *was* adaptive; it is only in modern society that it is not. So our sweet tooth *was* adaptive in our evolutionary past when high-energy food was in short supply; it is only in today's affluent societies that satisfying our cravings for sweet things has become self-damaging.

Although people quibble about the details of when and where the evolution occurred, the gene-based evolved-module view of human behavior does make biological sense and is gaining adherents. That does not mean it is right, of course. An alternative is to see human behavior and culture as consequences of hominids' extraordinary behavioral plasticity coupled with and enhanced by their powerful system of symbolic communication. According to this view, an important aspect of cultural evolution is the extremely variable ecological and social environments that humans construct for themselves. Evolutionary psychologists tend not to take such alternatives seriously, dismissing them as relics of the old-fashioned social sciences approach. We believe that in so doing they fail to recognize the power and subtlety of cultural evolution. An amazing amount can be accomplished through cultural transmission on its own, without any genetic change. To show just how much cultural practices can adjust and adapt themselves to existing genotypes, we will use yet another thought experiment, this time one that was first used by Eva Jablonka and Geva Rechav in 1996.

The Literacy Module

Imagine that, despite all our efforts, we manage not to destroy the world and ourselves, and that 500 years from now our planet not only sustains most of the present forms of life but humans have also managed to construct a better world for themselves. Most people have enough to eat, a home, health care, and freedom, and almost all healthy people are literate. They can read and write because they all grow up in an environment in which they are exposed from birth to a flow of words, and to visual and tactile linguistic symbols that stand for things, ideas, and relations. These symbols are produced and displayed by the complicated computer-like machines and other communication devices that have become a necessary part of everyday life (Figure 6.6). As a result, children acquire the ability to read without any formal instruction, in much the same way as many of today's children learn to read through being constantly exposed to modern communications technology. Children in the year 2500 also readily learn how to write, since by then writing requires only simple button pushing. Without doubt, people in the middle of the third millennium take their literate lifestyle very much for granted.

Now imagine that a scientist from another planet arrives on Earth, having been given the task of trying to work out how literacy evolved. She soon finds out that all healthy children with a normal upbringing acquire the ability to read and write early in life. They do so almost without formal instruction, although at varying speeds. The systems of symbols used by different populations are not identical, but most are acquired with more or less the same ease, so there is no great variation between populations.

Figure 6.6
Acquiring literacy in the year 2500.

The alien realizes that literate behavior is extraordinarily complex. When humans read and write, information from several sources has to be integrated within the framework of a system of rules of which they are not consciously aware. The alien starts looking at human genetics and neurology, and finds that there are specific defects, known as dyslexia, that affect literacy. Dyslexia tends to run in families, and the best evidence suggests that there is a strong genetic component involved. There are different types of dyslexia, but only some of them are associated with defects in other mental capacities such as spoken linguistic proficiency or general intelligence. So it seems that genetic variations can affect literate behavior directly, not just through their effect on general intelligence. However, the alien discovers that as well as the genetic component there is also a large learned element in literate behavior. She finds that socially deprived children who have not had the usual exposure to the behavior and technology that lead to literacy can still learn to read and write when older, but they do so less easily than children who have had a normal upbringing. Older children and adults need formal instruction. When the alien uses brain-imaging techniques to investigate the neural basis of literacy, she finds that reading and writing show a fuzzy, somewhat variable, but non-random localization.

From the facts about literate behavior that she has collected—its complexity, the ease with which it is acquired at a very early age, and the genetic, neurological, and developmental data—the alien concludes that it is a sophisticated adaptation, underlain by a distinct, genetically evolved "literacy module." It seems obvious that it is the product of lengthy past selection of genetic variations influencing literate behavior.

But then the alien researcher consults the historical and archaeological literature. There she finds that reading and writing are very recent cultural practices, and there was no direct genetic selection for literacy during human evolution. So she abandons her first hypothesis, and concludes that the "literacy module" is not a distinct, separately evolved structure after all, but is constructed during the early development of each individual. Because the behavior is so complex, she argues that a combination of various preexisting cognitive adaptations came together to form literate behavior. No selection of genes was necessary—the process was one of cultural evolution.

We are using this thought experiment because the reasoning that led the alien to suggest at first that literacy is the result of genetic selection having produced a "literacy module" is just the same as the reasoning that leads evolutionary psychologists and linguists to conclude that there is a "language module" in the human brain which was selected as such during hominid evolution. Language, it is said, is universal and human specific; its structure is very complex; it is acquired early and without conscious effort; there are brain defects that are to varying extents language specific; and some genetic variations are related to linguistic defects. This is all taken as evidence for a species-specific, evolved mental module for language. But as the literacy thought experiment shows, we must be very careful about inferring genetic selection for such a faculty, for although it may be the outcome of direct genetic selection, it need not be. We must also consider the alternative or complementary possibility—that what we see is the outcome of cultural-historical evolution and developmental construction.

Those who believe that there is a genetic language module in the human brain provide supporting evidence from several sources, but for many of the other mental modules postulated by evolutionary psychologists (e.g., the module for detecting social cheats, the sex-specific module for choosing a mate, the module for sex-specific creativity, etc.) there are no neurological or genetic data to support the claim. Their proponents rely on inferences drawn from more general sociological and psychological findings. For example, the American psychologist Leda Cosmides and anthropologist John Tooby suggested there is a module for identifying social cheats because, in psychological tests, most people make fewer mistakes when they reason about breaking social rules than they do when they have to apply similar reasoning to nonsocial rules. Cosmides and Tooby's argument is that, as our ancestors began cooperating with each other for mutual benefits, there was strong selection for being able to detect anyone who took all the benefits of cooperation without contributing anything. This led to the genetic evolution of a cheat-detecting module—an adaptive way of thinking that is used instead of logical reasoning (at which we are rather poor) in many social situations.

In the case of mate choice, the American psychologist David Buss based his argument for a sex-specific module on answers people gave in a questionnaire about their choice of sexual partner. He found that people from different cultures answered in broadly similar ways, showing similar sex-specific preferences. Men usually preferred young and beautiful women to older, wealthy ones, whereas women preferred older, wealthy men to young, poor ones. The evolutionary explanation here is that both sexes have been selected in their evolutionary past to prefer a sexual partner with qualities that enable them to produce and raise children. For men, this is a woman who is fertile, well-fed, and disease-free (qualities indicated by youth and beauty); for women it is a man with the resources (money and power, which usually come with age) to provide for them and their child.

The English psychologist Geoffrey Miller gives sexual selection an even greater role in human cultural evolution. He sees much of culture as a set of adaptations that evolved for use in courtship. Cultural products, he says, are indicators of their producers' intelligence and creativity, so they are valuable clues when it comes to selecting partners who will be good parents. This idea enables him to explain why men publish more books, paint more pictures, and compose more music than women. It is because men have had to compete more intensely than women do for sexual partners. Women have to be highly discriminating in their choice of partner because they invest so much in producing children, and they have preferred highly intelligent, creative men. These men got more mates, so the genes that made men creative spread. Today, men display their sexually selected creativity through their books, pictures, and music.

Can these arguments be correct? Theoretically (although probably in a rather different world) they could be, but this does not mean that they are likely to be right in our deeply cultural world. We believe that it is necessary to be very cautious before accepting evolved mental modules and other such ideas wholesale and uncritically. We have already illustrated the dangers through the literacy thought experiment, but we want to highlight the major difficulties by looking more closely at two of the most common features that make the evolutionary psychologists postulate the existence of an evolved mental module for a behavior. The first is universality and invariance; the second is the ease with which the behavior is acquired or applied.

Universality and invariance mean that everyone develops the behavior, whatever their social environment and psychological idiosyncrasies. This, the proponents of modules argue, probably means that an invariant genetic program, which everyone shares, drives the behavior. But there other ways of explaining the fact that enormous differences in social organization, learning opportunities, and individual psychology often have no effect on the acquisition of a behavior, even though they may affect the particular variant acquired (e.g., the particular language learned).

An alternative possibility is that no one has identified the initial conditions that cause the behavior's apparent invariance because we *all* experience these conditions. In the literacy thought experiment the shared initial conditions that produced universality and invariance are clear: *all* children were exposed to a complex literate environment. Literacy would not have been universal without this. But there are less obvious examples: for some time people believed that the way newly hatched ducklings respond to the calls of their mother was not learned, but was the outcome of an evolved genetic program. It was thought to be an instinct—an experience-independent, inbuilt, adaptive response. Ducklings recognize the calls of their own species even if they have not heard the calls of the incubating hen prior to hatching, so it certainly looks as if the response is genetically inbuilt. But it is not necessarily so. Experiments made by the American developmental psychologist Gilbert Gottlieb showed that, in at least one species, ducklings have to learn the call. It seems that during the development of their vocal system, while they are still in the egg, the birds exercise their sound-making apparatus, and consequently hear their own self-produced vocalizations. This is what tunes their perceptual system and makes them respond to the call of their mother when they hatch. In this case, therefore, because the initial condition—the embryonic auditory experience—that is necessary for the development of the behavior was not at first identified, a wrong conclusion (an explanation in terms of an evolutionarily selected response) was reached. Could not the same be true for human universal behaviors—the ones for which we are supposed to have genetically selected modules?

If the argument that universality and invariance indicate a distinct genetically evolved module is weak, what about the argument about the ease with which a behavior is acquired or applied? When people learn a pattern of behavior very quickly and early in life, often without much explicit instruction even when it involves complex rules (of which they are frequently unaware), does this mean that there must be an evolved genetic module for the behavior?

It was the great discrepancy between the small amount of learning and the complexity of the response that led Eric Lenneberg and Noam Chomsky to argue for the existence of an innate linguistic capacity in humans. It is certainly difficult to see how very young children can learn to apply the rules of grammar correctly, as they obviously do, just from the limited and inconsistent exposure to language that they get. It does look as if there must be some preexisting neural preparedness for producing such a complex behavior. But this is not the only possible explanation. Although the speed and ease of learning may indicate that there are some preexisting specifically selected neural mechanisms, the same properties could also be due to a culturally evolved system that is well adapted to the brain, and therefore makes learning easy. For example, think how difficult it was 1200 years ago for someone in Europe to divide one number by another. Say they wanted to divide 3712 by 116, or as it would have been then, MMMDCCXII by CXVI. Using the Roman notation system, they would have needed an abacus or a set of tables to accomplish this task, or more probably would have hired a specialist to give them the answer (XXXII). Today, with our Arabic notation system (and the useful zero), it takes the average ten-year-old only minutes to get the answer 32. If we knew nothing of the cultural change in the number notation systems and judged simply by the ability to learn to do sums quickly and correctly with just pen and paper, we might well deduce that during the last 1200 years a great mathematical mutation had occurred and been incorporated into our maths module through natural selection.

Such a genetic change could not happen so quickly, of course, but this is not the point. What the example shows is that the speed with which behavioral practices are acquired or applied does not depend solely on genetic evolution. Ways of learning and doing things have been structured by cultural evolution, and this cultural tailoring also determines the speed with which behavioral practices, possibly including linguistic practices, are learned and accomplished.

These arguments do not mean that we advocate a purely cultural-evolution explanation of the language capacity. In fact we think that it is reasonable to assume that language has evolved through the coevolution of genes and cultural linguistic practices, and we say more about this in chapter 8. However, we do believe that one needs to be alert to the great power of cultural evolution and the way in which it can rapidly and efficiently adjust behavioral practices to the idiosyncrasies of the developing brain. We should always consider and test this possibility as an alternative or an addition to any gene-selection hypothesis that is offered. Some behavior may be a purely cultural creation, even though it appears to be universal. For example, it is not very difficult to think how human cultural and social evolution have created conditions that lead men to write more books, paint more pictures, and compose more music than women. Nor is it difficult to think of aspects of cultural evolution that lead women to sometimes prefer wealth to youth in a potential mate. The symbolic system is very powerful, and certainly quite capable of constructing and reconstructing a suite of variations that can lead to behaviors that seems to be an almost universal, invariant part of human nature.

From Evolution to History

As we see it, neither memes nor modules provide a fully satisfying description or explanation of human behavior and cultural evolution. What is missing from both memetics and evolutionary psychology is development. Memeticists and evolutionary psychologists confine the role of social, economic, and political forces to the *selection* of cultural variants; their importance in the process of innovation and in the consolidation of innovations is commonly overlooked. Memetics and evolutionary psychology have little to say about how cultural constructions actually begin: they tell us almost nothing about the ways in which social, political, and economic forces transform societies and culture through the plans and actions of people. As Mary Midgley has recently remarked somewhat acidly, current theories about what changes the world tend to assume that it is "certainly something far grander than a few people worrying in an attic." Yet, as we stressed earlier, the one aspect of human cultural change that makes it totally distinct from any of the other types of evolution we have discussed is that humans are aware of and can communicate about their past history (whether real or mythical) and their future needs.

Forward planning by individuals, communities, companies, and states is involved in the selection, dissemination, and often also in the introduction of cultural novelties. Human beings share information about their imagined futures, choose between existing and latent variants, and construct the present in anticipation of what is to come. The ability to plan the future and communicate about it accentuates the importance of the social and cultural environment at every stage of the development and spread of cultural innovation. The rational and imaginative faculties of humans have to be understood and factored in if we are to understand the origin and generation of new cultural variations. Memetics and evolutionary psychology have little to say about these aspects of culture. They stress the selection of variants, rather than the conditions that are important for their development.

The developmental aspect of cultural evolution is especially significant in humans, but it is not unique to them. It is important in the construction of animal traditions as well. In the previous chapter we emphasized how through their behavior animals can construct the ecological or social conditions in which information is transmitted and new variants are generated. We argued that sometimes, as in the case of the Koshima monkeys, this can have cumulative effects, with one new habit resulting in a cascade of behavioral and ecological changes that stabilize and reinforce the original habit and each other, eventually leading to a new lifestyle. Such processes of cumulative, self-reinforcing cultural construction are even more obvious in human societies. Each aspect of a human symbolic culture is part of an interacting network of behaviors, ideas, and the products of behavior and ideas that is shaped by many different forces. This makes it very difficult to think about any artifact or behavior as an evolutionary unit in isolation from the culture in which it is historically embedded. Not only is the survival of an innovation dependent on the existing culture but so too are its generation and reconstruction, and all three are interdependent. Because symbolic systems are by definition self-referential, if innovations are to survive they must conform to the system. For example, when a new law (new Halacha) is introduced into the Jewish religious system, it must not contradict previous laws and must be shown to be intimately related to them, preferably directly derived from them. This gives the cultural system as a whole great stability, and constrains change.

Although many aspects of symbolic cultures are quite stable, whether or not an individual element within a social system is faithfully transmitted is often unimportant. If you think about it, it is easy to see why high fidelity would frequently be detrimental, particularly in complex and ever-changing societies. It is not a good idea to stick faithfully to your parent's style of dress, way of speaking, and type of automobile. What matters is not the fidelity of transmission, but the *functional adequacy* of any change in a cultural element. Usually a changed element must continue to play a role similar to that of the original one, and remain integrated with other aspects of the culture. Whether a new variant in dress style, speech habits, car, or anything else is preserved and regenerated accurately will depend on the constraints that are imposed by the wider aspects of cognition and culture. The important general point we want to reiterate is that the selection, generation, and transmission or acquisition of cultural variants cannot be thought about in isolation from one another; neither can they be thought about in isolation from the economic, legal, and political systems in which they are embedded and constructed, and the practices of the people who construct them.

Our view of human behavior and cultural evolution obviously differs from that of the memeticists and the evolutionary psychologists, who take an essentially neo-Darwinian point of view and ask how a cultural entity or behavior has been selected—who benefits from it. Evolutionary psychologists are usually looking for the benefit to the individual (or the gene), whereas the memeticists' bottom line is that the beneficiaries are the cultural activities and entities themselves. Our approach is more Lamarckian, because we see things in a developmental-historical way. We believe that in order to understand why a particular cultural entity exists or changes, one has to think about its origin, its reconstruction, and its functional preservation, all of which are intimately linked with each other and with other aspects of cultural development. It is necessary to ask not only who benefits and what is selected but also how and why a new behavior or idea is generated, how it develops, and how it is passed on. As we see it, trying to identify the beneficiary whose reproductive success is increased by some facet of culture is usually impossible, because commonly there is no single beneficiary, and often cultural evolution is not primarily the result of natural selection.

To illustrate the difference between our approach to cultural evolution and that of those who seek neo-Darwinian explanations, we will take a relatively trivial cultural change and see how it might be interpreted. Our chosen "cultural entity" is the punishment for stealing sheep. At different times sheep stealers in England have been hanged, imprisoned, sent to Australia, or fined. The form of the punishment has definitely changed (evolved) through time, so in what terms should we think about this change? If we think in terms of simple neo-Darwinian selection of individuals and societies, it is not easy to find reasons why one form of punishment should replace another. The evolutionary psychologists would tell us that we have a genetically evolved module in the brain that was formed through past selection for controlling the activities of people who break social rules. They would probably not attempt to explain the historical changes in the form of the punishment, other than to suggest that additional modules (such as a module for defending one's property or a module for searching for new resources), whose parameters were triggered by the social environment, would have contributed to the changes.

Memeticists, on the other hand, would probably focus on the spread of ideas about how sheep stealing should be punished, and argue that some ideas replaced others because they fitted in better with the current "memeplex" (the name some of them use for the set of memes in the brain). This approach, which assumes that the only beneficiary is the meme for the type of punishment, seems quite relevant in this case, because the memeticists' way of thinking acknowledges the importance of the wider culture in determining a preference for one type of punishment rather than another. But memeticists would have little to say about *why* a new form of punishment was invented in the first place, or *how* social interests (for example, policies about colonization, or plans to have a more centralized and controlled judicial system) have influenced the invention and dissemination of new forms of punishment.

Neo-Darwinian explanations do not incorporate the fact that the invention, regeneration, and preservation of ideas about punishment are all linked to the network of interactions that forms the wider cultural system.

Answering the Darwinian question, Who benefits?, gives us only a very limited understanding of historical and cultural change. In our opinion, changes in the form of punishment can only be understood by looking at the Lamarckian questions as well. We have to ask, What are the mechanisms that generate variations in the form of punishment? How, when, and in what circumstances are such variations generated? How do they develop? The generation, acquisition, development, and selection of variants must *all* be considered if we are to understand changes in cultural practices.

In summary, what we are saying is that cultural evolution cannot be explained in purely neo-Darwinian terms. If we are to begin to understand how and why cultures change, we need a far richer concept of the environment than is used in Darwinian theory, and a different concept of variation. We have to recognize that the environment has a role in the generation and development of cultural traits and entities, as well as in their selection, and that new cultural variants are usually both constructed and targeted.

Dialogue

I.M.: Early in this chapter you talked about linguistic, mathematical-rational, and artistic-religious systems as if they are separate symbolic systems. Are they? I think that one can make a very good case for the hypothesis that all existing symbolic systems hang heavily on language. In other words, linguistic ability is what transformed nonsymbolic, visual, and gestural systems of parrot-like communication into symbolic ones.

M.E.: This is, in fact, quite a popular hypothesis. Whether you accept it or not depends on how you envisage the evolution of our linguistic ability. If you think that it emerged almost fully formed, quite quickly, and that once in place language reorganized all aspects of cognition, then of course you can suggest that the other modes of symbolic communication were shaped almost entirely by language. Our view is that linguistic ability evolved gradually over a long period, coevolving with other modes of symbolic communication (visual, musical, and gestural). During this evolution, some division of labor between the different symbolic systems occurred, and they became more specialized, particularly the system that has eventually become mature language. We say more about all this in chapter 8. Language is, of course, an enormously powerful faculty, and we agree that once it appeared, it profoundly affected cultural evolution.

I.M.: I still find it difficult to think about a symbolic culture that is non-linguistic. Can you give an example of such a culture, or at least a plausible scenario for one?

M.E.: The Canadian neurophysiologist Merlin Donald has suggested that before the symbolic-linguistic culture of *Homo sapiens* evolved (a stage he calls the stage of Mythic culture), there was a stage of culture which was, in his terms, mimetic (not to be confused with "memetic"). It was a relatively long stage, which characterized *Homo erectus*, who had the ability to mime and reenact events. Intentional communication and representation occurred through gestures and sound. But it was not simple imitation, because it involved the representation or reenacting of a situation or a relationship. A ritual dance depicting a hunt was, according to Donald, a mimetic presymbolic representation. Teaching was central to this stage, with adults demonstrating and the young imitating. We find Donald's suggestion that there was a mimetic, prelinguistic stage in hominid evolution very plausible, but would argue that his mimetic stage is already a symbolic culture, albeit a very limited one.

I.M.: Donald's mimetic stage doesn't seem to be very different from what you see with a pack of wolves, in which the animals wind themselves up behaviorally before they go hunting. Yet if I understand you correctly, you are claiming that the symbolic system is separate and distinct from this sort of behavior. It seems to me you are being a bit ambivalent about this fourth dimension of yours. On the one hand you stress the relative autonomy of the new dimension, and the difference between it and the behavioral inheritance systems of other animals. Yet on the other hand you seem to shy away from the autonomy that you have been arguing for. There is a sense in which the world of ideas does have autonomy from that which generated it, just as biology has autonomy from physics, and psychology from genetics. What threatens you about this autonomy? Why not think about the evolution of ideas, or the dreaded memes?

M.E.: You are confusing the autonomy issue by creating the wrong type of dependencies. Of course culture is to some extent independent of the psychology of any particular individual. But culture is not external to individuals. Humans are not just biological and psychological creatures; they are also cultural agents. The problem with the kind of autonomy posited by the memetalk is that the active biological-psychological-cultural *agent* disappears. It cannot. Ideas are generated, edited, and reproduced as part of the development of groups and individuals, and these sociocultural developmental processes impinge on the transmissibility of the ideas and the precise content and form of what is transmitted.

I.M.: But there can also be a nonconstructed aspect to transmission, a purely copying aspect. You admitted that there are some interesting similarities between the genetic and symbolic systems. What about these similarities? Surely they are significant.

M.E.: One obvious similarity is the modular organization of the genetic and some of the symbolic systems, notably language, which allows incredibly rich variation. But the more important similarity is that both the genetic and the symbolic systems can transmit latent information (non-expressed genes, unimplemented ideas) rather than information that is actually used. It may be the modular organization of information and the ability to transmit unused information that underlie the very rich evolution that occurs through both the genetic and symbolic systems. When the transmission of variation is decoupled from its actual display and expression, it is possible to have a reservoir of variation that can be used in the future. It is really paradoxical, because it is the transmission of currently nonfunctional, nonexpressed variations—their future potential rather than their immediate use—that allows the genetic and symbolic systems to have such large and diverse evolutionary effects! But, we stress again, the differences between the two systems are far greater than the similarities. Even meme enthusiasts would accept that.

I.M.: I must admit that I, too, find the meme idea attractive. It is wonderfully simple, yet not intuitive, so it demands a conscious intellectual effort to understand it. I really wish it would work. Let's take something that seems like a *non*meme—the transmission of gender behavior in those gerbil lineages you described at the end of chapter 4—and see if it can be described in terms of memes. If I can make the meme idea work in this awkward case, it will surely work in any case. With the gerbils, what you have in each generation are interactions between the hormonal state of the mother, the sex ratio of her litter, and the developmental processes of the embryos. Aggressive mothers have litters with more males, which means that the females in the litter are exposed to more testosterone, which means that they grow up as aggressive mothers whose litters have more males, and so on. Obviously, as part of this developmental package, there are changes in female brains which are perpetuated. So why not focus on this component of the system—on the changed brain circuits—the meme?

M.E.: Yes, changes in the nervous system are reconstructed in each generation. But so are the hormone concentrations during development, and a host of other physiological variations. From our point of view this is a good example, because it shows how unwise it is to decide to focus on just one aspect of the physiology of the animal. The nervous system does have a privileged position when we think about learning, but it is rather strongly associated with other systems, and there are various feedbacks and bidirectional regulation going on. The gerbil case shows the tight association between the nervous system and other systems. It is bad biology to think about the nervous system in isolation.

I.M.: OK, maybe I was stupid to try and use that case, because the various aspects of the animal's physiology really do form a very tight package. But let's look at a looser package, at the more autonomous aspects of cultural systems. I can see that the idea of memes replicating through the imitation of actions runs into problems if one takes the replicator-vehicle distinction seriously, because the vehicle can transform the meme. So let me limit myself to a special type of imitation or copying—to cases where you really do automatically copy the instructions rather than the product, the recipe not the cake, the notes not the music. Why not call this type of meaning-free, automatic copying "replication," and the entities "replicators" or "memes?"

M.E.: You can do that, but you still run into a lot of trouble. For example, a story transmitted through writing would undoubtedly be a meme. However, if the same story is passed on through oral tradition, it would not be a meme, because its transmission involves various types of active learning processes, and directed variations are introduced as it is adjusted to the local way of life. So this would mean that a meme has to be defined solely by its very specialized mode of transmission. But an even bigger problem is that the transmission of ideas, patterns of behaviors, skills, and so on involves several types of concurrent and interacting learning processes. Focusing on one aspect will not lead us very far. It is the non-automatic and nonrote aspects of symbolic transmission—those aspects that involve directed, actively constructed processes—which are the most dominant and interesting in the generation and construction of cultural variations. And these aspects are the ones that are so often ignored or dismissed.

I.M.: So you think that cultural evolution cannot be modeled using the tools of epidemiology or population genetics? You mentioned that simple models of cultural evolution do exist.

M.E.: Some things can be modeled. For example, you can describe and follow the changes in the frequency and nature of a habit, such as eating gefilte fish, through generations. But if you want not only to describe the pattern of change of this artificially isolated "unit" but also to understand it, you will need to study psychology, anthropology, and sociology. The same, by the way, is true for animal culture. Little will be achieved without an understanding of the psychology and sociology of animals.

I.M.: You know, I am beginning to wonder if cultural evolution is a useful term at all! You defined cultural evolution as a change in the frequency and nature of socially transmitted behavior patterns from one generation to the next. Yet, if all these socially transmitted behaviors are reconstructed, readjusted, and modified to fit the ideas and practices of individuals and groups, all within immensely complex social systems, what does this definition mean? What is "inherited?" The frequency of exactly what changes? And what exactly is selection? The greater the role you give to these instructive and constructive processes, the less clear it is to me that we are really speaking about anything at all similar to our notions of Darwinian evolution.

M.E.: You probably didn't realize it, but we modeled the definition of cultural evolution in animals that we used in the previous chapter on Dobzhansky's definition of biological evolution. Dobzhansky was one of the founding fathers of the Modern Synthesis, and he defined evolution as a change in the genetic composition of populations over time. Of course his definition, like our derived one, is very general and schematic, and of course evolution involves a lot more than a change in the frequencies of some variant units or processes. Nevertheless, during cultural change the frequency of socially learned behaviors does increase and decrease, although these changes are shaped by the interactions within an entire ecological and social system. We certainly recognize that this definition is very superficial: it tells us nothing about the actual processes involved— nothing about the role and nature of the changes, nothing about the generation of variation, nothing about the communication and reconstruction of behavior as part of a social and cultural whole. It tells us very little about what constructive cultural evolution is about, in the same way that Dobzhansky's definition of gene-based evolution tells us very little about the actual processes of genetic evolution. We agree that if one's notion of evolution is based on *neo*-Darwinian thinking—on selection acting on discrete units that are not altered during the process of transmission and that are random with respect to the factors that affect their generation and subsequent chances of spread—then calling cultural historical changes "evolution" seems to be a misuse of the term.

I.M.: You seem to be saying that cultural evolution is neither a simple variation/selection process, nor a process that is guided by inner developmental laws, but some kind of construction process in which whole sets of sociocultural systems are changing. When you are describing cultural evolution, you imply that every case is so deeply embedded in the local social and cultural systems that each is peculiar to its own location in history. Everything is unique so there can be no generalizations. So what can you say? It seems to me that you can provide a very rich and "thick" description of each cultural change, but no generalities. What justifies calling these historical changes "evolution?"

M.E.: Your argument about the uniqueness of each case is not peculiar to cultural evolution. A close-up look at the evolution of any plant or animal can lead to a very similar conclusion. Each instance of genetic evolution occurs in a unique ecological and social setting, with organisms affecting the properties of their niche, the mode of their selection, and all the other factors that can impinge on their evolution. The close-up view is excellent for highlighting the unique features of the process of change, but it does not allow you to see the general features.

I.M.: And what are the general features of your so-called cultural evolution?

M.E.: What we are interested in are complex cultural practices that did not emerge in one single step, but are the result of cumulative historical processes. With such complex cultural practices, we can see no alternative to the assumption that the historical changes involved some form of selective retention of cultural variants that allowed the elaboration of the cultural practice. Of course we need to understand how the cultural changes happen and how they become established. These processes have to be based on some valid theories of cognitive and social psychology, and an understanding of the dynamics and logic of social systems. There are some good theories in each of these domains, but they are not integrated into a general theory of historical cultural change. We believe that it may be possible to construct such a general theory or set of theories. Of course in each particular case you will still have to relate the theories to the unique and contingent social and cultural circumstances.

I.M.: I wonder, and I am really rather skeptical! The level of contingency and uniqueness in the case of cultural change seems to me much greater than in the other types of evolutionary systems that you have described. But since we are into sociology, let me ask you a sociological question. These instructive and constructive aspects that you keep mentioning mean, as you admit, that you are talking about some form of Lamarckian cultural evolution. Is Lamarckism accepted in the cultural domain?

M.E.: Today many people (although not everyone) agree that cultural evolution has at least some Lamarckian components. This line of thought has a long tradition, stretching from Herbert Spencer through Peter

Medawar to Steven J. Gould. People like Medawar, Gould, and many others have recognized that its Lamarckian aspects make cultural evolution a very different kind of evolution, because adding the inheritance of acquired information to neo-Darwinism transforms the evolutionary process. Acquiring and transmitting information through symbolic systems involves both internal and external construction processes, which alter the dynamics of evolution. Basic concepts such as "transmission," "heredity," and "units of variation" have to be rethought, because the Lamarckian approach requires that you treat inheritance as an aspect of the development not just of individuals, but of the social and cultural system.

I.M.: It looks as if the role of what you call instructive and constructive processes is getting greater and greater as you move through your four dimensions of heredity. It seems to me that with cultural evolution you are not just Lamarckians, but threefold Lamarckians! Symbolic variation is directed in three ways—it is targeted, it is constructed, and it is future-oriented! Nevertheless, you mentioned that not everyone accepts that cultural change is Lamarckian. What are the arguments against Lamarckism?

M.E.: Labels and "isms" are never precise, so it is not difficult to find definitions of Lamarckism that do not fit the processes of cultural change. The American philosopher of science David Hull, for example, thinks that even in memetics Lamarckian notions are a conceptual mistake. This is not because he thinks that memes are not "acquired." On the contrary, Hull thinks that memetics is *about* the inheritance of acquired memes. But for him Lamarckian evolution requires that acquired phenotypic characters are transferred to replicators, so that in the next generation the acquired character will be manifest through the effects the altered replicator has on the development of the phenotype. Since, according to Hull's understanding of the term "meme," memes are like genes and not like phenotypes, acquiring memes does not count as the Lamarckian inheritance of acquired characters, simply because a meme is not a character. This is a very Weismannian view of Lamarckism. For most biologists, however, when inheritance is what Ernst Mayr called "soft inheritance"—that is, when the hereditary material (or process) is not constant from generation to generation but can be modified by the effects of the environment or the organism's activities—it qualifies as Lamarckian inheritance. Hull's position is a minority one, even among memeticists. And as we have already said, claiming that cultural evolution has a Lamarckian aspect is not a grave sin among biologists. Many of them do accept a constructive, Lamarckian, element in cultural evolution. ·

I.M.: I suppose that the evolutionary psychologists, who attempt to link human behavior and culture firmly with genes, are not very keen on these Lamarckian processes. I must confess that this mental module view of human psychology surprised me. It seems rather like postulating a simple causal relation between genes and phenotypes, just one step removed. I find it difficult to believe that people still think like that. I thought that the terrible consequences of eugenics had taught scientists a lesson. I cannot believe that scientists today really attribute differences in the cultural output of men and women to genetically evolved psychological mechanisms.

M.E.: Some do. But you must understand that these evolutionary psychologists do not claim that the differences between the achievements of different human populations are due to genetic differences. They are not racists. They are talking about *a universal human nature*, which all people share. They believe that the behavioral and psychological attributes that are common to all cultures are the result of genetically selected, more or less distinct, mental modules. And this does mean that they explain observed sociopsychological patterns, such as differences between the creative output of men and women, in this way. If you point out to them that the creative output of women has grown at least a hundredfold in the last century, they will stress that even in today's "equal" society the inequalities in output still persist. They will then go on to tell you that in many species of animals, the males have extravagant and "creative" courtship displays involving large colorful tails, elaborate dances, and so on, while the females do not. So, using the principle of parsimony, they will argue that the same evolutionary process—selection for impressive male displays by choosy females—explains the greater cultural output of human males. The strategy of this argument is to draw a very tenuous biological comparison between ourselves and other animals, a comparison that ignores the dynamics and sophistication of cultural-social evolution. Of course, as decent people they believe that the unjust and objectionable aspects of the present human situation should be altered by changing social conditions, but they claim that it may be difficult to do so, given our genetic predispositions.

I.M.: But some of these modules—these psychological adaptations—make sense. It seems to me that since social awareness is so central to our lives, the evolved-module explanation of why people reason better about social rules than about equivalent nonsocial rules may be valid.

M.E.: Maybe, but we have reservations. There are other ways of looking at it. If a lineage has been social for a long time, as our primate lineage obviously has, it is likely that natural selection will have constructed an intelligence that is biased to pay attention to, learn rapidly about, and manipulate social situations and relations. You can then argue that as humans acquired greater reasoning ability, they applied it to their preexisting social awareness. In other words, the primates' general bias to be particularly well attuned to social relations is inevitably reflected in a bias in reasoning when the species evolved greater general intelligence. But this is not what Cosmides and Tooby claim. Their claim is that the cheat-detecting module is human specific, and evolved in the Pleistocene as a distinct cognitive module, not as the combination of primates' *general* social awareness plus better *general* reasoning ability. We do not claim that what we have just suggested is the correct explanation, just that this possibility has to be seriously considered.

I.M.: So you prefer your tangle of constructions, with as few modules as possible, and with little independence for cultural units. Don't you think that you can borrow some useful ideas from these other approaches?

M.E.: We think that the memeticists' emphasis on the role of perceptual, emotional, and cognitive biases in cultural evolution is very important. We agree that a lot of cultural evolution involves the invention of cultural practices that are adjusted to the biases of our brains. The change in number systems that we discussed—the change from Roman to Arabic numerals—may well reflect such adjustment. We accept that in some cases it is useful to focus on psychological biases that lead to the spread of certain ideas and behaviors in an almost autonomous manner. We also accept that, as the evolutionary psychologists suggest, there may be some genetically selected biases that make it easier to learn certain things, and these have to be included in an evolutionary explanation of culture. But we think that in all cases, even the extreme ones, it is ultimately the agent—the individual and social group—that generates and constructs ideas and practices. The focus should therefore be on the social and cultural construction processes.

I.M.: You have been describing a terrible tangle of interactions, but it's still not quite interactive enough, I am afraid. Where do you go from here? You've got your four dimensions. Something like evolution can occur in *each* dimension, I grant you that. But how are they related? We are not made up of four neat and separate dimensions; we are a messy complex! And it is the complex that evolves!

M.E.: That's the last part—putting Humpty Dumpty together again. This is what we have to try and sort out in the next three chapters. First though, we will summarize some of what we have said so far about our four dimensions of heredity.

Between the Acts: An Interim Summary

In part III we shall look at the interactions of the four systems of information transmission that we have described in the previous six chapters. Since we have covered a lot of ground, a summary and comparison of the salient properties of the four systems may be helpful. To avoid excessive repetition, we have done this in the form of two tables. The first table describes the way in which information is reproduced and varies for each of our four dimensions of heredity and evolution. It shows (1) whether the organization of information is modular (the units can be changed one by one) or holistic (the components can't be changed without destroying the whole); (2) whether or not there is a system dedicated to copying that particular type of information; (3) whether or not information can remain latent—unused but nevertheless transmitted; (4) whether information is passed solely to offspring (vertically) or to neighbors as well (horizontally); (5) whether variation is unlimited and capable of indefinite variation, or limited, in that only a few distinct differences can be transmitted.

Tables such as this are always something of an approximation, because in biology things rarely fall into discrete categories. We have said before that the different epigenetic systems overlap, and also that putting transmitted substances that affect the development of an animal's form with epigenetic inheritance systems, while putting those that affect its behavior in a different category, is somewhat arbitrary. Tables also have to make do with words such as "sometimes" and "mostly," rather than spelling out the details. For instance, we say that the direction of transmission for the genetic system is "mostly vertical," which is shorthand for "genetic transmission is vertical in eukaryotes, except on the probably rare occasions when DNA is transferred from individual to individual by various vectors, or directly through ingestion; in bacteria and other prokaryotes horizontal transfer may be quite common and evolutionarily important, although we do not have enough data to know how frequent it is."

The reproduction of information

Inheritance system	Organization of information	Dedicated copying system?	Transmits latent (nonexpressed) information?	Direction of transmission	Range of variation
Genetic	Modular	Yes	Yes	Mostly vertical	Unlimited
Epigenetic					
Self-sustaining loops	Holistic	No	No	Mostly vertical	Limited at the loop level, unlimited at the cell level
Structural templating	Holistic	No	No	Mostly vertical	Limited at the structure level, unlimited at the cell level
RNA silencing	Holistic	Yes	Sometimes	Vertical and sometimes horizontal	Limited at the single transcript level, unlimited at the cell level
Chromatin marks	Modular and holistic	Yes (for methylation)	Sometimes	Vertical	Unlimited
Organism-level developmental legacies	Holistic	No	No	Mostly vertical	Limited
Behavioral					
Behavior-affecting substances	Holistic	No	No	Both vertical and horizontal	Limited at the single behavior level, unlimited for lifestyles
Nonimitative social learning	Holistic	No	No	Both vertical and horizontal	Limited at the single behavior level, unlimited for lifestyles
Imitation	Modular	Probably	No	Both vertical and horizontal	Unlimited
Symbolic	Modular and holistic	Yes, several	Yes	Both vertical and horizontal	Unlimited

Whatever the shortcomings of tables, they do help to highlight patterns of similarities and differences. When viewed in this way, it is obvious how alike the genetic and the symbolic inheritance systems are. In both, variation is modular; both can and often do transmit latent information; and in both, variation is practically unlimited. These properties give the two transmission systems enormous evolutionary potential by providing vast amounts of heritable information that can be sifted and organized by natural selection and other processes. The first table also shows that whereas with the genetic and epigenetic systems the direction of information transmission is mainly vertical, from parents to progeny, with other systems there is a significant amount of horizontal transmission to peers or neighbors. In fact there is a kind of jump in the direction in which information flows, with horizontal transmission becoming much more common as we move to the behavioral systems and socially mediated learning. This introduces a bias that can alter the effects of selection in significant ways. It also means that a new set of considerations (about why, how, and when horizontal transmission is occurring) have to be included in evolutionary thinking about changes based on information transmitted by behavioral or symbolic means.

Whereas the first table focuses on the nature and reproduction of information, the second one summarizes the more Lamarckian aspects of information generation and transmission. It shows whether newly generated variation (1) is blind ("random") or is targeted to specific activities and functions; (2) passes through developmental filters and is modified before transmission; (3) is constructed by direct planning; (4) can construct a different environmental niche. From the table it is clear that as we move from the realm of genetics to epigenetics and on to behavior and culture, the instructive aspects of variation generation and transmission become more dominant and more diverse. Although with the genetic system there is a small amount of targeting through the various types of interpretive mutations, and there is some filtering through selection among cells during gametogenesis, the role of instructive processes is relatively limited. With epigenetic systems, targeting is much more pronounced. Many epigenetic variations are induced, and the regulatory features of the genetic and cellular networks determine whether and how chromatin marks, self-sustaining loops, and RNA silencing are affected by external signals. Which variants are reproduced in the next generation of cells or organisms depends on the properties of the developmental system and on selection between cellular variants.

With behavioral transmission, both targeting and construction are even more apparent. Variation is targeted in the sense that the evolved biases

Targeting, constructing, and planning transmitted variation

Inheritance system	Variation is targeted (biased generation)?	Variation subject to developmental filtering and modification?	Variation constructed through direct planning?	Variations can change the selective environment?
Genetic	Generally not, except for the directed changes that are part of development and the various types of interpretive mutation	Usually not, although expressed genetic changes may have to survive selection between cells prior to sexual or asexual reproduction	No	Only insofar as genes can affect all aspects of epigenetics, behavior, and culture
Epigenetic	Yes, a lot of epigenetic variations are produced as specific responses to inducing signals	Yes, selection can occur between cells prior to reproduction; epigenetic states can be modified or reversed during meiosis and early embryogenesis	No	Yes, because the products of cellular activities can affect the environment in which a cell, its neighbors, and its descendants live
Behavioral	Yes, because of emotional, cognitive, and perceptual biases	Yes, behavior is selected and modified during the animal's lifetime	No	Yes, new social behavior and traditions alter the social and sometimes also the physical conditions in which an animal lives
Symbolic	Yes, because of emotional, cognitive, and perceptual biases	Yes, at many levels, in many ways	Yes, at many levels, in many ways	Yes, very extensively, by affecting many aspects of the social and physical conditions of life

of the mind restrict what can be learned. It is constructed through individual trial and error and various types of social learning, which are limited and channeled by the nature of social interactions. With variation transmitted by the symbolic system, there is a quantum leap in social complexity, with families, professional groups, communities, states, and other groupings all influencing what is produced in art, commerce, religion, and so on. Construction plays an enormous role in the production of variants, yet because symbolic systems are self-referential, the rules of the systems are powerful filters. The ability to use symbols also gives humans the important and unique ability to construct and transmit variants with the future in mind.

The final column in the second table indicates the extent to which the various inheritance systems are instrumental in constructing the niche in which selection takes place, a subject about which we shall have more to say in part III. Organisms can engineer the environment in ways that affect the development and selection of their descendants, as well as their own lives. Even bacteria and blue-green algae (Cyanobacteria), the oldest inhabitants of our planet, can be thought of as ecological engineers, since the products of their metabolism diffuse into the environment and transform it, changing the selective regimen of their neighbors and descendants. Moving to higher levels of organization, we can recall how the Israeli black rats constructed a new niche for themselves and their offspring by changing their diet and beginning to live in Jerusalem-pine trees. A much more famous example is the beavers' dam, which can have long-term effects on a lineage. The "inherited" dams provide the environment for new generations of beavers, and modifications to dams can accumulate over many generations, although they affect only individuals in the immediate vicinity. With human symbolic culture, the ability to construct the selective environment is far greater, often extending over many generations and affecting distant individuals and communities. The effect of such ecological and social construction can be enormous. Jared Diamond has argued that some of the most important patterns of human migration, colonization, and domination during the last fifteen thousand years are the outcomes of domestication, which made certain plant and animal species a necessary part of the human niche. When you think about it, it becomes clear that there is almost no aspect of the human world that is not engineered, including our own cognition. The use of written symbols, for example, inevitably alters our thoughts and perceptions, because being able to read and write in effect extends our memory span and our reasoning ability.

All four heredity systems allow the construction and transmission of information that reflects the interactions of the organism with its environment. Developmentally acquired or learned information—information that is likely to be useful to future generations—can be passed on. In today's unicells, fungi, plants, and lower animals, evolution is based on information transmitted through the genetic and epigenetic systems, and some of this information is acquired and targeted. With animals that have behavioral transmission, the capacity to generate and pass on acquired adaptive information is far greater. Some learned behaviors can create traditions, which, as we shall argue in chapter 8, interact with and guide evolutionary changes through the genetic system. In these animals, epigenetic inheritance continues to be important in development, although the transmission of epigenetic variants between generations is probably of little significance when so much information is transmitted behaviorally. With the emergence and elaboration of the symbolic systems, even the genetic system has taken an evolutionary back seat. Throughout human history, adaptive evolution has been guided by the cultural system, which has created the conditions in which genes and behavior have been expressed and selected. Soon, if some of the promises that accompanied the genome project are fulfilled, the dominance of the symbolic system will be even greater. We shall have the capacity to change our genes directly—to make genetic "educated guesses" that will affect future generations. Clearly, the different dimensions of heredity and evolution have different significances in different groups, and equally clearly, they all interact. In the next chapters we look at the nature and outcomes of these interactions.

Imagine an entangled bank, clothed with many plants of many kinds, with birds singing in the bushes, with various insects flitting about, with worms crawling through the damp earth, and a square-jawed nineteenth-century naturalist contemplating the scene. What would a modern-day evolutionary biologist have to say about this image—about the plants, the insects, the worms, the singing birds, and the nineteenth-century naturalist deep in thought? What would she say about the evolutionary processes that shaped the scene?

Undoubtedly the first thing she would say is that the tangled bank image is very familiar, because we borrowed it from the closing paragraph of *On the Origin of Species*. The nineteenth-century naturalist who is contemplating the scene is obviously Charles Darwin. The famous last paragraph is constantly being quoted, the biologist would tell us, because in it Darwin summarized his theory of evolution. He suggested that over vast spans of time natural selection of heritable variations had produced all the elaborate and interdependent forms in the entangled bank.

Our modern-day evolutionary biologist would almost certainly go on to say that she thinks Darwin's theory is basically correct. However, she would also point out that Darwin's seemingly simple suggestion hides enormous complications because there are several types of heritable variation, they are transmitted in different ways, and selection operates simultaneously on different traits and at different levels of biological organization. Moreover, the conditions that bring about selection—those aspects of the world that make a difference to the reproductive success of a plant or animal—are neither constant nor passive. In the entangled bank, the plants, the singing birds, the bushes, the flitting insects, the worms, the damp earth, and the naturalist observing and experimenting with them form a complex web of ever-changing interactions. The plants and the insects are part of each other's environment, and both are parts of the birds' environment and vice versa. The worms help to determine the conditions of life for the plants and birds, and the plants and birds influence the worms' conditions. Everything interacts. The difficulty for our evolutionary biologist is unraveling how changes occur in the patterns of interactions within the community and within each species.

Take something seemingly simple, like where a plant-eating insect chooses to lay its eggs. Often it will show a strong preference for one particular type of plant. Is this preference determined by its genes, or by its own experiences, or by the experiences of its mother? The answer is that sometimes the insect's genetic endowment is sufficient to explain the preference, but often behavioral imprinting is involved. Darwin discussed this in the case of cabbage butterflies. If a female butterfly lays her eggs on cabbage, and cabbage is the food of the hatching caterpillars, then when they metamorphose into butterflies her offspring will choose to lay their eggs on cabbage rather than on a related plant. In this way the preference for cabbage is transmitted to descendants by nongenetic means. There are therefore at least two ways of inheriting a preference—genetic and behavioral. An evolutionary biologist would naturally ask whether and how these two are related. Can the experience-dependent preference evolve to become an inbuilt response that no longer depends on experience? Conversely, can an inbuilt preference evolve to become more flexible, so that food preferences are determined by local conditions?

Similar questions can be asked about the plants on the entangled bank. The most obvious effects of the insects' behavior are on the survival and reproduction of the plants. Being the preferred food of an insect species may be an advantage to some of them, because it means that their flowers are more readily and efficiently pollinated. If so, those plants that the insects find tasty may become more abundant. Any variation, be it genetic or epigenetic, that makes a plant even more attractive to the insects, or that makes its imprinting effects more effective or reliable, will be selected. Conversely, if the insects' food preference damages the plants, variations that make it less attractive or more resistant to insect attack will be favored. For example, plants often produce toxic compounds that are protective because insects cannot tolerate them. The ability to produce such toxins will be selected. In some species toxin production is an induced response, brought about by insect attack, but in others it is a permanent part of the plant's makeup. Once again, an evolutionary biologist would want to know whether there is any significance in this. When there is an induced response, presumably involving changes in gene activities, does this affect the likelihood or nature of changes in the plant's DNA sequence? Do epigenetic variations bias the rate or the direction of genetic changes? Are the genetic and epigenetic responses related in any way?

How would an evolutionary biologist think about the worms that feature in Darwin's entangled bank? Earthworms must have been one of Darwin's favorite animals, because he devoted the whole of his last book to them. Visitors to Down House, his home for many years, can still see vestiges of his worm experiments in the garden there. Earthworms are a good example of something that is true for many animals and plants: they help to construct their own environment. Darwin realized that as earthworms burrow through the soil, mixing it, passing it through their guts, and leaving casts on the surface, they change the soil's properties. The environment constructed by the earthworms' activities is the one in which they and their descendants will grow, develop, and be selected. An evolutionary biologist therefore wants to know how the species' ability to change its environment and pass on the newly constructed environment to its descendants influences its evolution. How important is such niche construction?

Very wisely, Darwin avoided mentioning human beings when he summarized his "laws" of evolution in the final paragraph of *The Origin*. He realized that suggesting that humans had evolved from apelike ancestors would land him in very deep trouble, and he was going to be in enough trouble as it was. Although he knew full well that his own species is also a product of natural selection, he left discussing it to a later book. He did devote a lot of space in *The Origin* to humans, however. In particular, he described how, through selection, they had changed plants and animals during domestication. Darwin would have been well aware that the naturalist observing the entangled bank was potentially the most powerful evolutionary influence acting on it. Humans could divert a stream, so that the bank dries out and many of the organisms inhabiting it die; or they might introduce new plants or animals, thereby altering the whole web of interactions in the bank. Without doubt, humans are the major selective agents on our planet, and have carried out the most dramatic reconstruction (usually destruction) of environments. Today, in addition to changing plants and animals by artificial selection, humans can alter the genetic, epigenetic, and behavioral state of organisms by direct genetic, physiological, and behavioral manipulations. We are only at the beginning of this man-made evolutionary revolution, which will affect our own species as well as others. Our ability to manipulate evolution in this way is derived from the human capacity to think and communicate in symbols. Through the symbolic system, we have the power of planning and foresight. As the evolutionary biologist knows, this has had and will continue to have effects on all biological evolution.

As she looks at the entangled bank, a modern-day evolutionary biologist would know that explaining how natural selection has produced the complex, interacting living forms she sees is a formidable task. She could recruit the help of specialists, who might enable her to explain bits of the scene: the geneticists could look at the genetic variants in the plant and animal populations, and see how they influence survival and reproductive success; the physiologists, biochemists, and developmental biologists could look at the adaptive capacity of individuals; the ethologists and psychologists could tell her about the animals' behavior, and how it is shaped by and shapes conditions; the sociologists and historians would tell her what role humans have had in developing the bank; the ecologists would investigate the interactions between the plants, animals, and their physical environment. Each of the specialists would probably be convinced that their own findings and interpretations are the most significant for understanding the whole picture, and that the other parts of the study are of marginal significance. This is what usually happens when people look at the isolated parts of a system. A lot of knowledge can be gained from this approach, but eventually it is necessary to reassemble the bits—to put Humpty Dumpty together again. How do the genetic, epigenetic, behavioral, and cultural dimensions of heredity and evolution fit together? What influence have they had on each other?

In these last chapters, we look at these questions. In chapters 7 and 8 we attempt to put the Humpty Dumpty of transmissible information together by seeing how the different inheritance systems interact and influence each other. Since during evolutionary history new ways of transmitting information have evolved, chapter 9 is about the origins of the various inheritance systems. Finally, in chapter 10, we discuss how a view of evolution that includes all types of heritable variation—genetic, epigenetic, behavioral, and cultural—affects practical, philosophical, and ethical issues.

In this chapter we deal with the interplay of the genetic and epigenetic systems, leaving the other interactions for chapter 8. This means that we shall be returning to some slightly tricky genetics and cell biology, which nonspecialists may find rather tough going. We will do our best to make it palatable, but in parts it may be necessary for nonspecialists to do what the distinguished English zoologist Sir Solly Zuckerman said he did when he encountered a complicated mathematical equation—hum through it. The details are not too important, although the general message is. In particular, we hope that readers will get the gist of genetic assimilation, because we believe this concept to be extremely important. It will crop up again in the next chapter, when we look at interactions between genes, behavior, and symbolic communication.

Earlier in this book we used a music analogy to highlight the differences between genetic and nongenetic inheritance, and it may be helpful to use it again to illustrate what interactions between the genetic and epigenetic systems may mean. We suggested that the transmission of information through the genetic system is analogous to the transmission of music through a written score, whereas transmitting information through non-genetic systems, which transmit phenotypes, is analogous to recording and broadcasting, through which particular interpretations of the score are reproduced. A piece of music can evolve through changes being introduced into the score, but also independently through the various interpretations that are transmitted through the recording and broadcasting systems. What we are interested in now is how the two ways of transmitting music interact. Biologists take it for granted that changes made in genes will affect future generations, just as changes introduced into a score will affect future performances of the music. Rather less attention is given to the alternative possibility, which is that epigenetic variants may affect the generation and selection of genetic variation.

A recorded and broadcast interpretation of a piece of music could affect the copying and future fate of the score in two different ways. First, a recorded interpretation could directly bias the copying errors that are made. For example, a copyist might be so influenced by hearing a particular record over and over again that she makes a mistake that reflects this version of the music. The popular interpretation has an extra trill, so she unthinkingly adds it to the score. A second, more indirect effect would occur if a new and popular interpretation affects which versions of a score are copied and used as the basis for a new generation of interpretations. Think of something like traditional folk music, where there is no "master score." Similar, yet nonidentical versions of the music are played and recorded by various bands, each using its own score, its own instruments, and its own interpretation. If a new recorded interpretation becomes very popular, and is played over and over again, it is likely that versions of the score that resemble it will be used, recorded, and copied, and thus become more common. After a long period of such cultural evolution, it will eventually seem as if the beautiful fit between the score and what is heard could never have been otherwise—that the music flows seamlessly from the now dominant version of the score. In this case the recorded interpretation of the music has affected the *selection* of the version of the score, while in the first case that we described the recorded interpretation biased the *generation of variations* in the score. Epigenetic systems could have either or both types of effect on the genetic system: they could directly bias the generation of variations in DNA, or they could affect the selection of variants, or they could do both. We will start by looking at the first possibility—that the epigenetic systems directly bias the production of genetic variation. We will then use the rest of the chapter to explore the ways in which epigenetic variations construct the cellular and physiological niche in which genes are selected.

The Effects of Epigenetic Systems on the Generation of Genetic Variation

Before looking at the interplay of the genetic and epigenetic systems, we need to briefly recapitulate some of the points about genes and their activity that we made in earlier chapters. The most important is that DNA molecules do not sit naked in the cell. They are associated with many different proteins and RNA molecules, which together form the complex known as chromatin. In addition, DNA itself may have small chemical groups (e.g., methyls) attached to some of its bases. These DNA modifications and the components of chromatin influence gene expression: inactive genes usually have more compact chromatin than active or potentially active genes. Following DNA replication, the epigenetic marks—the methyl groups and non-DNA parts of chromatin that affect gene activity—are usually reconstructed, unless the cell responds to external or internal signals that alter its functional state.

We must now add something important to this picture: epigenetic marks affect not only gene activity, they also affect the probability that the region will undergo genetic change. Mutation, recombination, and the movement of jumping genes are all influenced by the state of chromatin, so the likelihood of a genetic change in two identical pieces of DNA is not the same if they have different chromatin marks. In general, DNA is more likely to change in regions where the chromatin is less condensed and genes are active than it is in more compact regions. That's because in active regions DNA is more accessible to chemical mutagens and to the enzymes involved in repair and recombination. It's not unlike what happens with your car, which is more exposed to accidental damage and change when you drive it around than it is when kept parked in the garage. There are exceptions, of course. Just as dead batteries may be more common in cars that are permanently garaged, so some types of DNA change are more common in inactive genes. For example, the base cytosine (C) mutates to thymine (T) more frequently when it is methylated than when it is not, and methylated DNA is usually associated with compact chromatin and inactive genes. Nevertheless, the overall picture (shown in figure 7.1) is that DNA in the regions where genes are active is more likely to change than that in inactive domains.

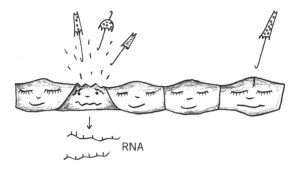

Figure 7.1
The rate of mutation (number of umbrellas) is greater in regions where genes are active than where they are inactive.

We now have to ask whether the influence that chromatin structure has on the likelihood of genetic changes is of any significance in development and evolution. It is not an easy question to answer, because the study of epigenetics and epigenetic inheritance systems (EISs) is young and hard evidence is sparse, but there are some very telling indications that it may be very important. For example, there is an increasing amount of data suggesting that there is an interplay between genetics and epigenetics in the development of cancer. The first sign of cellular abnormality in some tumors is an epimutation—a change in heritable chromatin marks, such as an increase or decrease in the density of DNA methylation. Commonly, genetic changes seem to follow the epigenetic ones, and they may be dependent upon them. Cancer biologists have suggested that what may happen is that epimutations (such as methylation in regulatory regions that are normally unmethylated) turn off one or more of the genes whose products normally help to keep DNA repaired and the cell well behaved. As a result, DNA damage and errors accumulate, and the cell begins to disobey the rules about when to divide. Since both the epimutations and the consequent genetic changes are inherited by daughter cells, the lineage's behavior gets progressively more subversive as new mutations and epimutations are selected and allow cells to evade normal checks and controls.

According to this view of tumor development, genetic and epigenetic events interact, with epigenetic changes (such as increased methylation) leading to genetic changes, and genetic changes (such as mutations in genes coding for chromatin proteins) leading to further epigenetic changes. Whether it is a mutation or an epimutation that initiates the cascade of events is usually unknown, except for some types of cancer that run in families where the inheritance of a faulty DNA sequence is implicated. For most cancers, sorting out the interplay of genetic and epigenetic factors is going to take time. However, although the epigenetic dimension has introduced another complication into the task of understanding cancer, there is a plus side: it offers hope for better diagnosis and treatment. By looking for epigenetic changes such as increased methylation, it may be possible to detect some cancers at an earlier stage and to monitor their progression more easily. Furthermore, since chromatin marks are usually reversible, finding drug treatments that reverse epigenetic changes and halt tumor growth is a real possibility.

The importance of the interplay of the epigenetic and genetic systems in the development of some cancers is now widely accepted, but the role and significance of epigenetic changes in the generation of mutations in germ-line cells is still being debated. In chapter 3 we mentioned Barbara McClintock's suggestion, made more than twenty years ago, that plant cells respond to physiological stresses by reshaping their genomes, thereby producing genetic variations that might enable them to adapt to the new conditions. Her arguments stemmed from experiments with maize in which she showed that mobile elements (also called "jumping genes," "transposons," or "transposable elements") are activated in stressful conditions, and by jumping to new sites they alter genes and gene expression. Today, thanks to the work of molecular biologists, we have a lot more information about how and why the jumping genes of maize and other plants jump. We know that whether or not a transposable DNA sequence excises and inserts itself into a new location, often duplicating itself as it does so, depends among other things on its epigenetic state, which is inherited.

Transposability is correlated with DNA methylation: elements that are capable of jumping are less methylated than those that are inactive, which are normally highly methylated. The methylation marks of potentially active elements vary in a way that depends on factors such as the cell's position in the plant, the sex of the parent cell from which it was derived, and various internal and external conditions. Stresses such as wounding, pathogen infection, or genomic imbalance (having too much or too little of some chromosomes or regions of chromosomes) can lead to substantial changes in methylation marks, followed by vigorous jumping. As mobile elements excise themselves and insert into new locations, they introduce mutations in both coding and regulatory sequences. Active chromosome regions are particularly inviting sites for the transposing elements.

From the work of McClintock and those who followed her, it is clear that the epigenetic changes in chromatin structure brought about by changed conditions can lead to genetic changes, but does this, as McClintock believed, have any adaptive significance? Is it not simply that these transposable elements are parasitic genes, selfishly replicating and moving around the genome, altering the plant's DNA sequences as they do so? This is certainly one way of thinking about them, but it can also be argued that there has been positive selection for this type of global mutation system. It may be that the potentially deleterious effects of transposable elements have been minimized by the selection of silencing mechanisms such as methylation, which keep them quiet most of the time but nevertheless allow them to jump and produce mutations when conditions become life-threatening.

It is not difficult to see the value of such a system, particularly in plants. Plants cannot avoid adverse conditions by moving away, so if they are to have a chance of surviving, they have to respond in other ways. The evolutionary recruitment of epigenetic and genetic mechanisms that enable them to change may have provided them with one way of responding to adversity. Plants have certain features that make exploiting the bursts of mutations induced by the movement of mobile elements less hazardous than it might seem. Many have a modular organization, in which the parts (e.g., the branches of a tree) are semiautonomous, each developing its own reproductive organs. Others form clones—groups of asexually produced and therefore genetically similar offspring, often loosely connected to each other. In addition, the soma and germ line are not rigidly segregated in plants, so it's easier for somatic cells to become germ cells. What all this means is that plants can try out both epigenetic and epigenetically induced genetic variations in their modules, clones, or somatic cells without jeopardizing the survival and reproduction of the whole organism. Some variant clones or modules may be failures, but others may do better than the original, and if so they will contribute most offspring to the next generation. In this way, lineages with an epigenetic response that leads to an increased mutation rate may survive better than others. Transmitting epigenetic variants, and through them the capacity to generate genetic variation when conditions are tough, may therefore be an important survival strategy for plants. Perhaps this is why plants are providing so much of the evidence that epigenetic modifications can be transmitted between generations.

At present, we do not know the full evolutionary significance of the effects of EISs on the generation of mutations, although there are reasons to think that they may have been very important. One intriguing suggestion is that a massive movement of transposable elements following stress-induced epigenetic changes was responsible for the rapid emergence of many evolutionary novelties. At least 45 percent of the human genome is derived from transposable elements, and as much as 50 percent of that of some plants, so jumping genes have certainly played some role in evolution. Mobilizing them when times are bad could undoubtedly produce a lot of new genetic variation. Moreover, when mobile elements move into or out of a gene's regulatory regions, they produce exactly the type of mutation that is likely to be most significant, because they cause changes that affect whether, when, and where the gene responds to the signals that turn it on and off. Such changes can have profound effects on development, and this type of regulatory mutation is believed to have been responsible for many of the major evolutionary modifications to plant and animal organization.

How EISs Have Molded the Evolution of Development

Whatever the magnitude of the direct effects of EISs on evolution turns out to be, no one doubts that the indirect effects have been enormous. This is obvious if we think about complex organisms with cells specialized to do different jobs. It is quite clear that without cell memory, plants and animals with many types of differentiated cells simply could not have evolved. EISs, which are what provides cell memory and enable cell lineages to maintain their characteristics, were one of the preconditions for the evolution of complex development.

Since the different types of EISs are all found in present-day unicellular organisms, it is reasonable to assume that they were also present in the single-cell ancestors of multicellular groups, and that they were necessary for the evolution of multicellularity. This idea is widely accepted. What is not always appreciated, however, is that not only were EISs necessary for the evolution of large and complicated organisms, they also helped to shape the evolution of some of the idiosyncratic and seemingly odd characteristics of their development.

To see how EISs may have shaped the evolution of complex organisms, imagine a primitive multicellular organism with three types of cell: cells that feed, cells that are involved in movement, and reproductive or germ cells. We show such an organism in figure 7.2. Assume that its cells have some epigenetic memory, retaining and transmitting their phenotypes to daughter cells, so that after division each cell usually does the same job. Now imagine what would happen if memory fails, and some of the feeding cells or the movement cells switch jobs, or simply become selfish, dividing vigorously and using up the resources that other cells provide, disregarding their duties to the whole organism. Obviously this kind of thing could happen, because even with present-day EISs, which usually make memory very reliable, cells still occasionally switch types. In our primitive organisms, where EISs have not evolved to be as reliable as those we have today, and where the pathway to specialization is relatively simple, switching to an alternative cell type is quite likely. This may not be too bad if switching is very limited and the organism has lots of cells, but in a small multicellular organism it could cause a lot of trouble.

The first problem resulting from such switches is a general one: the organism is likely to function less efficiently. Multicellular organisms exist only because their cells cooperate, rather than compete. If too many cells neglect or change their duties, the division of labor between them will be disrupted. Cells that disregard their role and use the resources of the rest

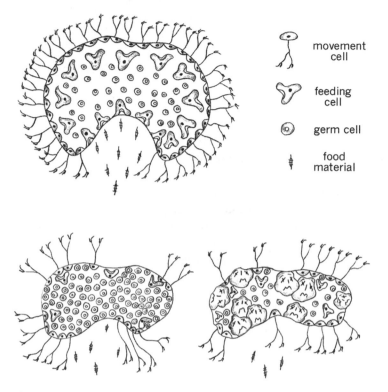

Figure 7.2
A primitive multicellular organism with three types of cell. At the top, the normal organism; below, organisms with inadequate epigenetic memories which have (on the left) insufficient movement and feeding cells, or (on the right) new, function-less cells.

of the body to divide selfishly may destroy the whole, as cancer cells do all too often. So organisms in which too much switching occurs will be less efficient and therefore less likely to survive and reproduce, whereas those that have ways of avoiding improper switching will contribute more to the next generation. Any mechanism that prevents cells from changing their jobs or neglecting their duties will be favored by natural selection. It will be to the organism's advantage for their EISs to be as reliable as possible, and for any cell that deviates from the norm to be destroyed. Through the selection of genetic variants, EISs should evolve so that they are flexible enough to enable the switches necessary for normal development, but not so sloppy as to allow undesirable switching.

Now assume that our primitive multicellular organism has evolved a bit, and with the help of its now fairly reliable EISs, it maintains more cell types. It hasn't as many as the hundred-plus that are found in vertebrate animals, or even the thirty or so types found in plants, but it has more than a dozen. They are formed through a series of switches, in which the original fertilized egg divides to produce cell types A and B, and B divides to produce types C and D, and D produces more D and also type E, and so on. Each cell type is the result of a progressive sequence of epigenetic changes. Which of these cells is going to become a germ cell and give rise to the next generation? In theory, any of them could, provided all the epigenetic marks and other traces of their developmental history are first erased. This is easy to envisage for cell types A and B, which switched only once, but is not quite so easy for cells such as types D and E, which were formed later in the developmental sequence, and needed a series of switches. Reversing the whole sequence might be very error-prone, and could produce mistakes that would jeopardize the development of the next generation and lead to the end of the line. Consequently, anything that prevents errors and improves the capacity of potential germ-line cells to adopt or retain an uncommitted epigenetic state—a state nearer to that that existed at the beginning of development—would be a selective advantage.

We can think of at least three features of development that may be the outcome of selection to prevent cells with inappropriate epigenetic marks from becoming germ cells. First, it may be one of the evolutionary reasons why many epigenetic states are so difficult to reverse. Some cell types have no chance of changing tack and becoming germ cells because, through past selection for stability, the epigenetic changes that produced them have become effectively irreversible. Therefore the problem of their epigenetic legacy is irrelevant, because they cannot become germ cells. Second, the need for germ cells to be free of an epigenetic legacy may be the evolutionary reason why many animals separate off their primordial germ cells very early in embryogenesis. These future germ cells remain physically separate and quiescent, dividing infrequently throughout the rest of development. That way there is little epigenetic memory to erase before the next generation, and little opportunity for epimutations to arise. Moreover, rogue cells that have abandoned their designated job and are selfishly multiplying and invading other regions of the body are less likely to try to take on a germ-cell role if the germ line is physically segregated. A third insurance policy against handing on inappropriate epigenetic variants, even those that arise as epimutations in the germ line, is provided by the extensive reprogramming that occurs during meiosis and gamete production when chromatin is restructured, and in males the gametes lose most of their cytoplasm.

What we are suggesting is that many aspects of development can be seen as evolved mechanisms that prevent the carryover of irrelevant epigenetic information that would destabilize the organization of the next generation. The efficiency of cell memory, the stability of the differentiated state, selection and cell death among somatic cells, the segregation between somatic and germ-line cells in some animal groups, and the massive restructuring of the chromatin of germ cells are all partly shaped by the selective effects of EISs. We stress "partly," because these very complex features of development have several advantages, and their evolution has probably involved other functions.

Genomic Imprints and Gene Selection

In spite of the special treatment that the germ line receives, gametes may nevertheless transmit a legacy of their epigenetic past. This can be seen in cases of genomic imprinting. We mentioned this rather odd phenomenon in chapter 4, where we said that sometimes the appearance or behavior of chromosomes, or the expression of genes, depends on the sex of the parent from which the chromosomes or genes originated. It happens because the chromosomes from the mother and father carry different marks (imprints), and these parental legacies affect how genes in their offspring respond to cellular signals. For example, some genes become active only if they have a "paternal" imprint, others only if they have a "maternal" imprint. Consequently, when a gene is imprinted, two genetically identical individuals, both having one normal and one defective allele, may appear totally different if one inherited the faulty allele from the father and the other inherited it from the mother. We show a simple case of imprinting in figure 7.3.

We believe that the differences in chromatin marks that underlie imprints probably originated as incidental by-products of the different ways that DNA is packaged in male and female gametes. Sperm are very small and mobile, with tightly packed DNA and inactive genes, whereas eggs are much larger, with diffuse chromatin and very active genes. These gametic differences in chromatin structure get carried over into the fertilized egg. Most disappear during early development, so the two parental sets of chromosomes end up similar, but sometimes the differences persist as parental

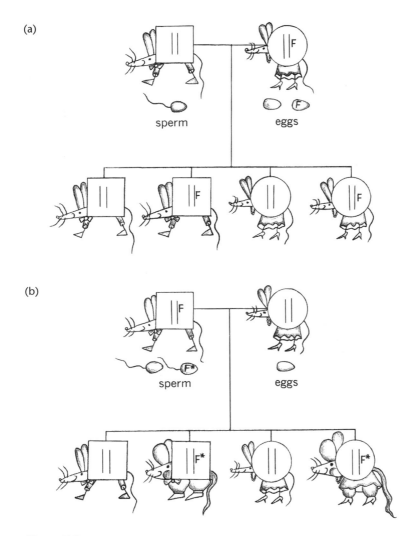

Figure 7.3
Genomic imprinting. The mother in family (a) and the father in family (b) both
have a chromosome carrying an *F* allele. They and their spouses, who carry normal
alleles, are all slim. The youngsters in the two families are genotypically similar, with
two of the four having the *F* allele. However, because the *F* allele of family (b) went
through spermatogenesis, the marks on it were changed and it was expressed in the
two offspring that inherited it, so they became fat.

marks or imprints. Often these make no difference to development: they might, for example, make a gene inherited from the mother slightly more accessible to a regulatory molecule than the gene from the father, so it gets turned on earlier, but this is unlikely to have a significant effect. Just occasionally, however, a difference in parental chromatin structure may be selectively important. If it is detrimental, selection will favor genetic changes that alter the formation of the marks in one or both parents, or ensure that the differences are eliminated early in development. Should differential marking be beneficial, the difference in the marks will be retained and further enhanced through natural selection of genetic variations that affect their establishment and maintenance.

If we look at a few of the examples of genomic imprinting, we can see some of the ways in which imprints have been recruited and modified for different developmental functions. One role that they have acquired in some species is related to sex determination. Several unrelated groups of insects have evolved a very odd sex-determining mechanism: individuals with the usual two sets of chromosomes are females; those with a single set are males. The latter usually develop from unfertilized eggs, so they lack chromosomes from a male parent. In scale insects, however, males develop from fertilized eggs, yet they still end up with only one functional chromosome set. They do so because during early development the chromosomes that they inherited from their father are either eliminated or inactivated. Somehow, parental marks on the chromosomes are recognized and action is taken, although how this strange system works and why it should have evolved are mysteries.

Parental imprints are not involved in sex determination in mammals, but they do have a role in what geneticists call "dosage compensation." Dosage compensation is needed because female mammals have two X chromosomes, whereas males have an X and a Y chromosome. The Y chromosome is what the geneticist Susumu Ohno called a "genetic dummy"—it is small, condensed, and has much less protein-coding DNA than the X chromosome. Consequently, males have only one copy of most X chromosome genes, whereas females have two. In 1961 the British biologist Mary Lyon deduced that the way gene dosage in the two sexes is evened up is through females inactivating one of their two X chromosomes. X inactivation occurs in early development, and usually the decision over which X chromosome is inactivated is a random process—the X chromosome inherited from the father becomes inactive in some cell lineages; in others it is the X chromosome from the mother. Once one of a cell's X chromosomes has been inactivated, the same X chromosome remains inactive in all its daughter cells, so it is a beautiful example of stable epigenetic inheritance. Although X inactivation is usually random, in a few situations, notably in the extraembryonic tissues (the embryo-derived tissues that surround the fetus and provide the route for nourishment from its mother), it is always the father's X chromosome that is inactive. For this to happen, the two X chromosomes must carry marks that reflect the sex of the parent from which they came, and these marks must be recognized by the embryo's epigenetic silencing systems. Why the paternal X chromosome should be preferentially inactivated in cells of extraembryonic

)mosomes have an equal
:ulation. What it shows,
d in some tissues, while

in Australia, David Haig
ut why some organisms
1embryonic tissues. His
1ich the epigenetic and
)oint was the realization
tissues, as they are in
1flict of interest between
1ost of us to think about
carrying offspring from
1 which litters can have
depend on this; all that
)ring from two or more

1ould treat their young.
o obtain as much nour-
· do so at their mother's

It's not getting any easier is it?

... from siblings. A father has no interest in the future welfare of the mother, because it is unlikely that she will ever carry any more of his young, and his offspring's half-sibs have nothing whatsoever to do with him. He wants the best only for his own young. The mother, on the other hand, should give all her present and future offspring the same amount of nourishment—she is equally related to all of them, and it is in her interest that all should do well. Haig suggested that in such an asymmetrical situation, selection will favor any imprints on the chromosomes inherited from the father that make the embryos extract more than their fair share of nourishment from the mother. Through natural selection, the frequency of genes that strengthen 'greedy' imprints on the chromosomes the father gives to his offspring will increase. However, this will lead to selection for imprints on maternally derived chromosomes that counter the paternal greedy-embryos strategy, because greedy embryos endanger the overall reproductive success of the mother by harming her other offspring. What is therefore expected is strong differential marks on genes that are associated with embryonic growth: alleles inherited from the father should have growth-enhancing marks, whereas those inherited from the mother should have growth-suppressing marks. And this is exactly what has been observed with some (although not all) growth-related genes in mice and humans.

It is not difficult to think of other ways in which imprints might be exploited. For example, some years ago we speculated that in organisms with a sex-determination system like our own, fathers might use genomic imprinting to influence their daughters. Females, who are XX, inherit one X chromosome from each parent, and give an X chromosome to all their offspring. Males, who are XY, inherit their X chromosome from their mother and the Y from their father; they give their X only to daughters and their Y only to sons. Now, since a father gives his X chromosome only to his daughters, selection should enhance any imprints on a male's X chromosome that specifically benefit his daughters. Since a mother gives her X chromosomes to both sons and daughters, she doesn't have this opportunity to introduce a sex bias. Whether fathers use this particular way of imposing sex-related differences on their offspring remains to be seen.

From what is already known about imprinting, it seems that sex differences in the chromatin marks transmitted by parents have been exploited in various ways. Although imprints were probably originally by-products of the differentiation of egg and sperm, what we see today is the result of natural selection for many different functions having reshaped the initial chromatin differences, enhancing some, erasing others, and leaving yet others intact. What is true for parental imprints is probably also true for other types of chromatin marks: all are likely to be subject to selective modification and adjustment as conditions change and new adaptations evolve.

Induced Epigenetic Variations and the Selection of Genes

We want to move on now to look at the way in which *induced* epigenetic changes—changes that arise in response to altered environmental conditions—can influence evolution through the genetic system. A good place to start is with some fascinating work that was initiated by Dmitry Belyaev of the Siberian Department of the Soviet (now Russian) Academy of Sciences in Novosibirsk. Belyaev was a convinced Mendelian geneticist, who managed to survive and work in spite of Lysenko's anti-Mendelian ideology and the damage it did to people and science in the USSR. In the late 1950s, he was made director of the Institute of Cytology and Genetics in Novosibirsk, where he started a long-term experiment which involved selecting for tameness in the commercially valuable silver fox. What Belyaev and his colleagues set out to do was therefore more or less the same as our ancestors must have done, albeit less systematically and probably unconsciously, when they domesticated the dogs, pigs, cattle, sheep, and many other animal species that now live alongside us. The experiment, which was continued after Belyaev's death in 1985, was successful. Quite rapidly (on an evolutionary time scale), the scientists established a population of docile foxes, some of which are now quite doglike in the way they are anxious to please their human handlers and compete for their attention.

What is so interesting about the silver fox experiment is that a lot more than behavior was affected by selection for tameness. Within fewer than twenty generations, the reproductive season of the females had become longer, the time of molting had changed, and the levels of stress hormones and sex hormones had altered. There were physical changes too: the ears of some foxes drooped and the way some carried their tail was different; some had white spots on their fur; a few had shorter legs or tails, or a different skull shape. These heritable phenotypic changes appeared quite early in the selection process, and although they affected only a small number of the animals (about 1 percent), they occurred repeatedly. In addition, there were changes in the foxes' chromosomes. Many foxes had tiny additional microchromosomes, with very condensed chromatin and DNA that consisted of many repeated noncoding sequences.

The nervous and hormonal systems are closely related, so it is not surprising that selecting for tameness altered hormone levels and the reproductive cycle. But what about the changes in the foxes' appearance—their droopy ears and curly tail, for example? They seem to be developmentally unrelated to behavior. How could they be explained? There were several possibilities, but the Russian scientists managed to rule out some of them, and concluded that others were unlikely. For example, the new phenotypes appeared too frequently to be the result of new mutations, and the mating scheme allowed very little inbreeding, so these two alternative explanations were rejected. In fact Belyaev's interpretation of his experiment was a little unusual: he attributed the appearance of new phenotypes to the arousal of what he called "dormant" genes. According to Belyaev, animals have a large reservoir of dormant genes—genes which in today's jargon we would describe as permanently inactivated. Belyaev suggested that in stressful situations, such as during domestication, the effects of selection on the hormonal systems cause these inactive genes to become heritably active. The result is a dramatic increase in the amount of variability seen in the population. So, according to Belyaev's interpretation, the new phenotypes that accompanied increasing tameness were the consequence of epigenetic changes rather than genetic changes. Belyaev believed that selection for domesticated behavior had altered the foxes' hormonal state, which had in turn affected chromatin structure and thus activated many normally silent genes in both the soma and germ line. He thought that the microchromosomes might have something to do with this, although as far as we know (unfortunately we cannot read Russian) he never elaborated on this very much.

There are obvious similarities between Belyaev's views and those of Barbara McClintock. Belyaev emphasized the heritable epigenetic effects of stress, whereas McClintock's focus was on genomic effects, but both agreed that stressful environments do more than simply provide a different selective regimen. Belyaev described his ideas about stress in the lecture he was invited to give when the International Congress of Genetics was held in Moscow in 1978. His topic was animal domestication, which he described as one of the greatest biological experiments. He pointed out that since domestication began, not more than fifteen thousand years ago, it had produced a rate of change in behavior and form that was far greater than had ever occurred before in evolutionary history. But, he emphasized, the speed of this change was not solely the result of the intense selection applied. It was also the consequence of stress, which induced changes in the hormonal system that revealed previously hidden genetic variations and made them available for selection.

Genetic Assimilation: How the Interpretation Selects the Score

Epigenetic changes that are induced by stress can do more than just reveal previously hidden genetic variation. They can also guide the selection of genetic variants. In terms of our music analogy, a new recorded interpretation can affect which version of the score is selected and played in the future. To explain what we mean, we want to return to the idea that acquired characters can be inherited.

For as long as people believed that characters acquired through use and disuse in every generation would eventually become inherited traits, explaining why hereditary adaptations are often so similar to induced adaptations was not a problem. One of the favorite examples was the thick skin on the soles of our feet. This is obviously an adaptation for walking over rough ground, and we are born with it. But we can also develop thick skin on our hands or other parts of the body if they are subject to pressure and rubbing. The Lamarckian view of what happened in our evolutionary past was that because the feet were always subject to rough treatment, thick skin that originally was acquired during each person's lifetime eventually became an inherited character that appeared without an abrasive stimulus. Similarly, Lamarckists would explain the observation that some animals have to learn to be wary of snakes, whereas others have an innate fear and avoidance of them, by arguing that after many generations the acquired character, the learned fear response, becomes an inherited one, an instinct.

As enthusiasm for Lamarckian ideas waned, evolutionary biologists had to think of other reasons why inherited adaptations so often mimic physiological and behavioral responses, and at the end of the nineteenth century several people came up with a Darwinian way of turning a learned response into an instinct. We will postpone discussing their ideas until the next chapter, because here we want to look not at behavioral responses, but at how induced developmental or physiological changes can be transformed into inherited characters that appear without an inducing stimulus. A Darwinian explanation for this was provided in the middle of the twentieth century by the British geneticist and embryologist C.H. Waddington.

We mentioned Waddington and his epigenetic landscapes in chapter 2, when describing the intricate networks of genes that underlie every trait. Waddington's epigenetic landscapes had nothing to do with epigenetic inheritance: they were simply visual models that recognized the complexity of the genetic systems involved in developmental pathways. By the time that Waddington began developing them, in the 1940s, it was already clear from the large number of mutant genes that could affect a single character that development requires the correct form and interactions of many genes. Yet, as all geneticists knew, in spite of the genetic complexity and the inevitable hazards of development, the normal phenotype—what geneticists call the "wild type"—is remarkably constant. As Waddington put it, "if wild animals of almost any species are collected, they will usually be found 'as like as peas in a pod.'" In contrast, animals having the same mutant gene often differ markedly from each other. For example, most fruit flies carrying two copies of the mutant allele *cubitus interruptus* have gaps in their cubital wing veins, but the size of the gaps varies; in some the veins are not broken at all, so the wing appears quite normal. Whether and how strongly the wing vein anomaly is expressed depends in part upon the temperature at which the flies are reared. The question is why, if mutants are so variable, is the wild-type phenotype so constant?

In Waddington's terminology (which we have used in earlier chapters because it has become part of the jargon of genetics), the wild-type phenotype is relatively invariant because it is well "canalized" or buffered. Through generations of natural selection for stability, allele combinations have been forged that ensure that any minor perturbations caused by differences in the environment or in genes do not affect the outcome of development. Since mutant strains have never been subject to the natural selection that would stabilize their development, they remain variable. Any small differences in the conditions in which individuals develop or in the other genes they possess may affect the expression of their mutant genotype.

Notice that one of the corollaries of Waddington's canalization concept is that there is a lot of invisible genetic variation in natural populations. Canalization allows genetic changes to accumulate, because they are not "seen" by natural selection. They are revealed only if unusual environmental stresses or exceptional mutations push the processes of development well away from the normal canalized pathway. When this happens, new and selectable phenotypes may be produced. Paradoxically, therefore, while canalization masks genetic variation and prevents phenotypic deviation in normal circumstances, the accumulation of hidden variation increases the potential for evolutionary change when internal or external conditions become dramatically different.

Waddington described some of his ideas in a short article published in *Nature* in 1942. It was entitled "Canalization of Development and the Inheritance of Acquired Characters." In this article he suggested how characters that were originally formed in response to environmental challenges could be converted by natural selection into inherited characters, a process that he later called "genetic assimilation." As an example he used the thickened skin that protects underlying tissues from damage, but rather than using the soles of the feet, Waddington preferred a more picturesque example—the callosities on the underside of the ostrich. These hard patches of skin presumably prevent damage when the bird collapses into its squatting position. Waddington argued that in the ancestors of these animals, the skin thickened only in response to pressure and abrasion, so before this happened, young animals suffered. However, because individuals with genes that enabled their skin to respond quickly and in an appropriate place fared better than others, gradually, over many generations, the adaptation became easier to induce, and the skin thickened quickly following only the slightest pressure or abrasion. In Waddington's terms, selection for the capacity to respond remodeled the epigenetic landscape, making the response more and more canalized, so that only a trivial stimulus was needed. Eventually, as is shown in figure 7.4, a stage was reached

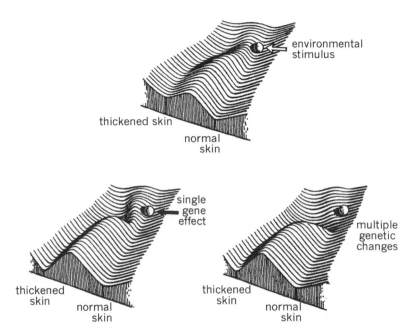

Figure 7.4
Genetic assimilation. At the top, the original epigenetic landscape, in which the main valley leads to normal skin, and a side branch leads to thickened skin. Thick skin is formed only if an environmental stimulus (open arrow) pushes development into the left-hand pathway. Below are two epigenetic landscapes after natural selection has led to genetic assimilation. In both, through selection, the valley leading to thick skin has been tuned and deepened, so that it is an easier path to follow. On the left, a major gene effect (solid arrow) pushes development into this pathway, whereas on the right the selection of variations in many genes has so remodeled the landscape that no stimulus is needed. (Adapted with permission from C. H. Waddington, *The Strategy of the Genes*, Allen and Unwin, London, 1957, p. 167.)

when no external stimulus was necessary at all, either because a genetic switch had been incorporated into the system, or because the genotype constructed by selection had crossed a threshold that enabled the phenotype to be produced without a stimulus. The thick-skin phenotype had been genetically assimilated. The induced (or acquired) character had become an inherited character.

Through the concept of genetic assimilation, Waddington showed how acquired characters can have an important influence on the course of evolution. They can do so because an induced epigenetic change that occurs repeatedly can guide the selection of genes that produce the same phenotype. Support for the idea came from some simple experiments with fruit flies. In one of these, Waddington exposed pupae of a wild-type strain of *Drosophila* to an unnaturally high temperature for a few hours. Following this heat shock, the flies that emerged showed various abnormalities, but Waddington concentrated on those having a "crossveinless-like" phenotype. Crossveinless flies have the whole or part of the tiny crossveins in their wings missing. This can be the result of having a particular mutant gene (*cv*), but Waddington's strain didn't carry this mutant allele, and the heat shock did not induce it. The heat shock didn't change genes; it caused nongenetic changes that upset development. As a result, about 40 percent of the flies had a crossveinless-like phenotype. Waddington selected these flies, bred from them, and gave their offspring a heat shock when they were at the pupal stage. When these offspring became adults, he again selected those with missing crossveins, bred from them, gave their pupae a heat shock, and so on. So in each generation he was heat-treating the pupae and breeding from those adults that had missing crossveins. As he did so, he found that the proportion of flies showing a crossveinless phenotype increased, reaching over 90 percent in fewer than twenty generations.

Selection for the production of a crossveinless phenotype was obviously very successful, but this was not the most interesting part of Waddington's experiment. What was far more exciting was that, beginning around generation 14, some flies in the selected line had missing crossveins *even when the pupae had not been heat-treated*. By breeding from these, Waddington established strains in which the frequency of the crossveinless phenotype was almost 100 percent, without any heat shock at all. In other words, crossveinlessness had been almost completely genetically assimilated—its development no longer depended on the heat treatment. Crossveinlessness, which was originally an acquired character, seen only in an environment that included a heat shock, had, through selection, become an inherited character that was manifest in normal environments.

Waddington analyzed the strain with the assimilated crossveinless phenotype and showed that many genes were involved in producing it. In general terms, his interpretation was that the heat shock had exposed the hidden genetic variation that was present in the original population by affecting the interactions of the many genes underlying wing development. Once epigenetic events had revealed the variation, sexual shuffling and selection had brought together a combination of alleles that produced the new phenotype. Several similar experiments, using other types of inducing stimuli (e.g., briefly exposing newly laid eggs to ether, or altering larval conditions), and selecting for other induced characters (e.g., other types of wing change) were also successful. All could be interpreted in terms of changed conditions unmasking variation in the genes underpinning a developmental pathway, followed by the creation, through selection and sexual shuffling, of a new combination of alleles that led to the alternative phenotype.

At about the same time as Waddington was developing his ideas in England, Ivan Ivanovich Schmalhausen was coming to very similar conclusions in the USSR. His book *Factors of Evolution* appeared in an English translation in 1949. Theodosius Dobzhansky, one of America's leading evolutionists, was instrumental in getting this work translated, and wrote an enthusiastic forward to the English edition. Even so, Schmalhausen's ideas, like Waddington's, had little impact in America. American evolutionists were far more interested in how natural selection led to the integration of genes in populations than in how it molded gene-controlled processes in individuals. In Britain, Waddington's ideas generated rather more interest, but even there the influence of his epigenetic approach was short-lived. As molecular biology became fashionable in the 1960s and 1970s, ideas about development began to be couched in terms of gene action, switches, regulators, feedback loops, and so on. Biologists felt confident that the type of genetic control systems that had been found in microorganisms would soon be found in multicellular organisms too. Consequently, as interest in molecular genetics grew, woolly abstractions such as epigenetic landscapes were increasingly seen as old-fashioned and unnecessary, and quite quickly they fell from favor.

Genetic Assimilation Meets Molecular Biology

Over the past decade, Waddington's ideas have enjoyed something of a renaissance. There are several reasons for this. First, thanks to the more ecologically minded biologists, there has been an upsurge of interest in the role of environmental factors in determining phenotypes. People are now actively investigating the genetic, developmental, and evolutionary basis of phenotypic plasticity—how and why organisms with the same genotype can develop a variety of phenotypes when raised in different conditions. A second reason for the revival of interest in Waddington's work is that as the details of the systems that control gene activity have been worked out, the whole subject of epigenetics has been brought back into the limelight. Once the complexity of the regulatory networks underlying development became clear, it challenged people to think about how such systems evolved, and Waddington's notions began to seem more relevant. A third and very important reason why Waddington's work is being talked about again is that people such as Suzanne Rutherford and Susan Lindquist have been putting some genetic and biochemical flesh on the bare bones of Waddington's explanation of genetic assimilation.

Rutherford and Lindquist and their colleagues have been studying protein folding and misfolding. Proper protein folding is essential for cellular functions in all organisms, and misfolding can cause serious problems. (Recall the horrible effects of conformational changes in prion proteins, which cause diseases such as kuru and Creutzfeldt-Jakob disease.) Protein folding is not always an automatic and simple consequence of the sequence of amino acids in its polypeptide chains. In order to adopt the proper conformation at the right time and in the right place, some polypeptide chains require assistance from one or more members of a family of proteins known as "chaperones." One of these chaperones is Hsp90—heat shock protein 90. As its name suggests, it is one of a group of proteins that were discovered through the way that they and their genes behave when organisms are given a heat shock. Now that more is known about them, "Hsp" seems a bit of a misnomer, because they have a role during oxygen starvation and certain other severe stresses, as well as during a heat shock.

Hsp90 seems to have a dual function. In normal, everyday conditions, it helps to keep a set of proteins that regulate growth and development in a semistable conformation, which enables them to respond to cellular signals appropriately. Without Hsp90, these regulatory proteins are liable to misfold and be incapable of doing their job. The second function of Hsp90 is evident when cells are stressed (e.g., by a heat shock) and the normal folding of many essential proteins is disrupted. When this happens, Hsp90 is recruited to help protect and restore the conformation of the damaged proteins. It is therefore diverted from its usual role of looking after the regulatory proteins.

Rutherford and Lindquist were interested in seeing what would happen to development when Hsp90 was in short supply. Their experimental animal was the fruit fly *Drosophila*, which had been used for a lot of the early work on heat shock proteins. They studied flies in which the amount of Hsp90 was reduced either because one of the two copies of their Hsp90 gene was a mutant allele, or because they had been reared on food containing the drug geldanamycin, which inhibits Hsp90. In both cases they found that some of the flies developed morphological abnormalities, including wing, eye, and leg deformities, faulty wing venation, and duplicated bristles. The spectrum of abnormalities and their severity depended on the strain used. The new phenotypes were heritable, because when individuals with the same type of abnormality were crossed together, some of their progeny inherited the defect. However, it was very unlikely that these heritable defects were the result of new mutations, because there were too many of them and the same strain-specific deformities cropped up repeatedly.

Since the abnormalities were heritable, it was possible to select for the new phenotypes. When the researchers did this, breeding from the individuals with deformed eyes or with wings having a vein defect, their selection was very successful. Within five to ten generations they had obtained lines in which the frequency of the novel phenotype had increased from 1 to 2 percent to 60 to 80 percent. Genetic analysis showed that several different genes contributed to the selected phenotype, so the original strains must have contained a lot of hidden genetic variation that was capable of affecting the selected characters.

To explain their results, Rutherford and Lindquist suggested that normally Hsp90 acts as a kind of developmental buffer, preventing many genetic variations from having any effect on the phenotype. Since it is not too fussy about the precise sequence of amino acids in the proteins it helps to fold into the correct conformation, genetic variations are tolerated so long as Hsp90 is present and doing its normal job. When the supply of Hsp90 is inadequate, however, some Hsp90-dependent proteins don't fold and function properly, and the many developmental pathways for which they are essential get a bit wobbly. Consequently, any variant gene products, which in normal circumstances would have no effect, can push development off course and produce abnormal phenotypes. Hsp90 therefore acts as a general canalization factor, masking variations in many different genes. That is why genetic variation is revealed when Hsp90 is in short supply (figure 7.5).

This is not the end of the story, however. Rutherford and Lindquist also discovered something else: in the lines in which the initial shortage of

Figure 7.5
Unmasking genetic variation. The individuals are identical twins, but the one on the left has normal levels of the chaperone Hsp90, whereas the one on the right has a reduced amount and therefore shows phenotypic abnormalities in some of the features whose development needs the chaperone.

Hsp90 was caused by a mutant copy of the Hsp90 gene, after several generations of selection, the mutant allele was no longer present. The phenotypic abnormality was still there, but the flies they examined all had totally normal Hsp90 genes. One might have expected that if Hsp90 levels were back to normal, as presumably they were in these flies, the deformity would disappear. It didn't. The flies continued to show the abnormality for generation after generation, even in the presence of normal Hsp90 levels.

The interpretation Rutherford and Lindquist gave to their results is essentially the same as that offered by Waddington for his assimilation experiments. They suggested that selection for the deformed eyes or wing vein abnormality brought together several previously hidden genetic variations that affected the developmental pathways of eyes or wings. Once the appropriate alleles had accumulated sufficiently, the new trait was expressed even in the presence of normal levels of Hsp90. The significance of the initially low level of Hsp90 is the same as that of a heat shock—it unmasks cryptic variation, which can then be selected. Eventually, the trait is produced even in the absence of the unmasking agent, because the genes selected have shifted development into a new pathway—the new phenotype has become more canalized. In Waddington's terminology, it is partially assimilated.

Two other scientists in the Lindquist group, Christine Queitsch and Todd Sangster, looked at the effect of reducing Hsp90 levels in another organism, *Arabidopsis thaliana*. This plant is a rather insignificant weed, but it is important because it is the botanical equivalent of *Drosophila*—it is the most genetically researched organism in the plant world. It was therefore the natural choice for studies of the role of Hsp90 in plants. The scientists found that when Hsp90 levels were reduced using drugs, new phenotypes were seen. As with fruit flies, the spectrum of anomalies depended on the strain used, but generally speaking the phenotypes were less monstrous than those seen in flies. Some, such as an altered leaf shape and deeper purple color, looked as if they might even be good candidates for selection if the plants found themselves in new natural environments.

There is an interesting twist in the tale of the variation revealed by heat shock or reduced Hsp90 levels. Apart from its small size and short life cycle, one of the things that make *Arabidopsis* so useful for genetic studies is that it is normally self-fertilized. Genetic strains are therefore very inbred: the two copies of almost all of a plant's genes are identical, and any genetic differences between individuals within a strain are trivial. Consequently, because there is so little hidden variation, it was assumed that the differences between plants from the same strain that were seen after treatment with the Hsp90-inhibiting drugs were caused by random accidents occurring during development. In contrast, in the *Drosophila* experiments, the original flies were probably heterozyous for many genes, so flies within a strain would have differed from each other genetically. Therefore, once a heat shock or lowered Hsp90 level had revealed variation, selection for a new phenotype could bring together combinations of alleles that preserved the selected character even when Hsp90 levels were back to normal. Without hidden genetic differences, selection would have been ineffective. Or at least that is what most geneticists would have thought until recently. Now, however, thanks to some work by Douglas Ruden and his colleagues, things look somewhat different.

Ruden's group looked at selection in an isogenic strain of *Drosophila*. Isogenic strains are constructed using genetic trickery and complicated breeding programs to make flies homozygous for most genes. They are therefore rather like the *Arabidopsis* strains that Queitsch and her colleagues used, with very little genetic variation among flies. The particular isogenic strain that the researchers constructed carried a mutant allele of the *Krüppel* (cripple) gene. Flies with this allele have smaller and rougher eyes, which in certain conditions are prone to form strange outgrowths. For instance, a few flies produce outgrowths when geldanamycin, the Hsp90-inhibiting drug, is added to the food on which they are raised. Ruden and his colleagues therefore did a Lindquist-type experiment in which they added the drug to the flies' food and bred from those flies with the outgrowths. Remarkably, although the food contained the drug for only one generation, and although there was scarcely any genetic variation in the strain, six generations of selective breeding gradually raised the proportion of flies showing the anomaly from just over 1 percent to 65 percent. It remained around that level until the investigators ended the experiment at generation 13. The question is why, if there were no genetic differences between the flies, had selection been so successful?

The clue came from experiments in which the mothers of *Krüppel*-carrying flies had a defective copy of either the Hsp90 gene, or one of several genes that affect the maintenance and inheritance of chromatin structure. Some of the offspring of these mothers developed the eye outgrowth, even when they themselves didn't inherit the defective Hsp90 or chromatin gene. Selecting and breeding from these flies increased the proportion of offspring with the abnormality. These results led Ruden and his colleagues to conclude that variation in their isogenic lines must stem from heritable differences in chromatin structure, not differences in genes. In other words, the flies carried epimutations. The scientists suggested that what happened was that, thanks to her defective chromatin-affecting gene, the chromatin marks in the mother's germ line were altered. When these new marks were transmitted to her offspring, they affected when and where genes were expressed, and because the eye-development pathway had already been made wobbly by *Krüppel*, the inherited epimutations caused eye outgrowths.

There are still many unanswered questions about the mechanisms that produce the variation that has been unveiled by the work described in this section. For example, does Hsp90 affect chromatin structure? Does epigenetic variation contribute to the dissimilar responses in the various *Arabidopsis* strains? No doubt we shall soon be hearing more about the molecular biology behind what has been discovered, but as it stands at present, the rather complicated but beautiful experiments with *Drosophila* and *Arabidopsis* suggest that heritable epigenetic variants, as well as cryptic genetic differences, can be the basis for genetic assimilation. This obviously has very important evolutionary implications. We will come to these shortly, but first we want to look at another aspect of the Lindquist group's work, their studies of yeast prions. This work is interesting because it brings another type of EIS—structural inheritance—into the web of interactions that conceal and reveal genetic differences.

A Revealing Yeast Prion

As we described in chapter 4, the essential thing about prions is that they are heritable architectural variants of normal proteins. There is nothing wrong with the amino acid sequences of their polypeptide chains; they are just folded up in an unusual way. This abnormal conformation is self-propagating: once prions are present, they convert the normal form of the protein into their own prion shape. Because prions often form aggregates and are not available to do the protein's normal job, cellular functions are affected. The results can be disastrous, as they are with the prions that cause kuru and mad cow disease, but the prions found in yeast seem to be fairly benign, and indeed may even be useful.

One of the yeast prions that the Lindquist group has been studying is an altered form of a protein that is involved in translating mRNAs into polypeptide chains. For reasons we needn't go into, this prion is called [PSI$^+$]. Its presence causes phenotypic variation. True and Lindquist showed this by comparing the colony morphology and growth characteristics in many different conditions of seven pairs of yeast strains differing only in whether they did or did not carry the prion. In half of the cases they studied, they found that the prion-containing and prion-free members of each pair reacted differently to their environment. The differences were strain specific, and often the prion-containing strain was better able to tolerate harsh conditions.

Each prion-containing strain was genetically identical to the nonprion member of the pair, so why did they behave differently? The answer lies in the role of the normal form of the prion protein, which is in polypeptide chain termination. This occurs when the ribosomal machinery reaches a stop codon in the mRNA. Recall that the genetic code is a triplet code, in which the nucleotide sequence of mRNA is read in groups of three, with each successive triplet (codon) dictating which amino acid is to be added to the polypeptide chain. There are a few codons that do not code for amino acids, the so-called stop codons. These dictate "end of polypeptide, add no more amino acids." In [PSI$^+$] strains, because the prion form of the protein aggregates, there is not always enough of it available for chain termination. As a result, mRNA translation sometimes goes beyond the stop codon. Such "readthrough," as it is called, adds extra amino acids to the protein that is being synthesized. These may affect the protein's stability or its association with other molecules, so the mistakes in protein synthesis can have phenotypic effects.

Readthrough has another consequence: it allows the synthesis of polypeptides that normally would not be completed at all. Genomes have a lot of duplicated genes, and often the extra copies have mutations. If a mutation leads to an inappropriate stop codon in the middle of the mRNA, it will ordinarily produce a shorter and probably functionless product. This may not matter too much, because unmutated copies of the gene produce the normal protein. However, if [PSI⁺] is present, readthough of the mutant mRNA may allow a functional protein to be formed, albeit one that is slightly different from the normal gene's product. This may affect the phenotype. Cells containing [PSI⁺] can therefore produce a variety of new protein products with potentially new functions (or malfunctions) either because translation goes beyond the normal endpoint, or because stop codons in the middle of mRNA are ignored. Whichever is the case, the new phenotypes are produced without any change in DNA.

Since [PSI⁺] is a prion, it propagates itself and is inherited by daughter cells. Consequently, the reduced fidelity of protein synthesis that leads to phenotypic variability is also inherited, and a lineage with [PSI⁺] retains the capacity to vary. This plasticity might be invaluable in harsh conditions. However, because 1 in every 100,000 to 1 million cells switches spontaneously from the prion-containing to the normal state or vice versa, a population can have some cell lineages with [PSI⁺] and the capacity to vary, and some without. How this can affect evolution through the genetic system is one of the things we discuss in the next section.

Epigenetic Revelations

In trying to explain how epigenetic variations can lead to genetic assimilation, we have gone from the domestication of silver foxes to the molecular biology of yeast, so it may be helpful if we summarize the main points we have made. They are as follows:

• Belyaev's work with silver foxes suggested that there is hidden genetic variation in natural populations. This variation was revealed during selection for tameness, possibly because stress-induced hormonal changes awakened dormant genes.
• Waddington and Schmalhausen attributed the remarkable constancy of many aspects of the wild-type phenotype to past selection for gene combinations that buffer development against genetic and environmental disturbances. In Waddington's terminology, development is "canalized." Because of this, a lot of genetic changes have no effect on the phenotype. However, unusual environmental stresses or mutations can push development away from the normal canalized pathways, and when this happens the genetic differences between individuals are revealed and can be selected.

• Waddington's experiments showed that when variation is revealed by an environmental stress, selection for an induced phenotype leads first to that phenotype being induced more frequently, and then to its production in the absence of the inducing agent.

• Experiments from the Lindquist group showed that selectable variation can be revealed by a shortage or decreased activity of the stress protein Hsp90. Since Hsp90 is a molecular chaperone which helps to maintain the correct shape of a variety of the proteins that are important in development, it may be one of the buffering or canalization agents that enable organisms to tolerate small genetic changes in some protein sequences.

• The work by Ruden's group suggests that, because heritable phenotypic variation can be induced and selected even when there is little or no genetic variation in the strain, some of this variation must be epigenetic.

• In yeast, a prion that upsets protein synthesis can generate phenotypic variation without any change in DNA. Because prions are transmitted to daughter cells, the same spectrum of variations may appear in successive generations.

What is clear from all this is that epigenetic changes, whether caused by environmental or genetic factors, or by random noise in the system, can reveal hidden genetic variation that produces new phenotypes. In normal circumstances, a lot of the genetic differences between individuals are masked, because past selection has made the networks of interactions underlying development indifferent to minor changes. Mutations therefore accumulate in the population, but most of the time do little harm or good. The chaperone Hsp90 seems to be one of the molecules that helps hide genetic variation, through being tolerant of differences in its target proteins' sequences. The buffering capacity of the developmental system is not unlimited, however, and once it is exceeded, genetic differences between individuals become apparent.

The laboratory experiments we described have shown that through selection a phenotype that has been revealed by an inducing substance or stress can be genetically assimilated: development of the induced character becomes independent of the stimulus. But notice that the relationship between the character and the inducing agent is not what you need for it to be regarded as Lamarckian evolution. Strictly speaking, Lamarckian evolution requires that the induced and later assimilated phenotypes are *adaptive* to the conditions that elicited them, but the phenotypes exposed by a heat shock, for example, are not an adaptation to excessive heat. However, there was one of Waddington's experiments in which the response to the inducing agent did seem to be adaptive. Fly larvae were given food that was very salty. This led to a morphological alteration of their anal papillae, which are organs associated with the control of salt levels in the body fluids. The change was therefore probably an adaptation to the new conditions. After rearing the larvae on salty food for many generations, the modified anal papilla phenotype was assimilated, being retained even when the larvae were raised on normal food. Thus, in this case, an acquired character became an inherited one in the traditional sense: the trait was adaptive and genetic assimilation was through natural, rather than artificial, selection.

The genetic assimilation experiments show how Darwinian mechanisms can produce apparently Lamarckian evolution. This is fascinating, but it is not the reason why they are so significant. Far more important is that they show how, when faced with an environmental challenge, induced developmental changes unmask already existing genetic variation, which can then be captured by natural selection. Short-term evolution does not depend on new mutations, but it does depend on epigenetic changes that unveil the genetic variants already present in the population. Once unveiled, these genes can be reshuffled through the sexual process until combinations are generated that readily produce the phenotypes that are most adaptive.

The phenotypes revealed by changed environmental circumstances may or may not be directly related to the conditions that induced them. In the case of the *Drosophila* larvae exposed to salty food, it is likely that the conditions induced epigenetic changes specifically in the suite of genes affecting the anal papillae, because developmental adjustments to these structures are part of the flies' normal adaptive response. In contrast, stresses such as a heat shock may influence a wide range of developmental pathways. For example, if Hsp90 is diverted from its role in development to look after stress-damaged proteins, or if a stress has a global effect on methylation levels, all sorts of genetic differences between individuals might be unmasked. Some of the induced phenotypes might, by chance, turn out to be more appropriate than the existing ones, so through selection they might be assimilated. By disrupting so many developmental processes and revealing such a wide range of variation, even a transient stress could kick-start significant evolutionary modifications. A more persistent stress—one which induces the same epigenetic responses for several generations—is even more likely to lead to evolutionary innovation.

The discovery that some induced epigenetic states are inherited, so phenotypic variants can persist even if the inducing conditions do not, has added a whole new dimension to Waddington's notion of genetic assimilation. Induced heritable epigenetic changes could be important for four reasons. First, they are an additional source of variation, which might be crucial if populations are small and lack genetic variability. Second, most new epigenetic variants arise when conditions change, which is of course exactly the time when they can be most useful. Third, since epigenetic variants are usually reversible, little has been lost through the variants' selection: they can be deselected if the changed conditions are short-lived. The fourth reason, which is perhaps the most interesting in the present context, is that heritable epigenetic variants can do a holding job until genes catch up.

It is easy to see what we mean by a "holding job" if you think about the inheritance of prions or chromatin marks in organisms that are reproducing asexually by fission or fragmentation. In such organisms, because there is no sexual shuffling of genes, new genetic variation can stem only from new mutations. Adaptation could therefore be a very slow process, particularly if it required several genetic changes. However, if EISs augment heritable variation by producing different phenotypes from the same genome, the process may be much more rapid. It would mean that through the selection of epigenetic variants, a lineage might be able to adapt and hold the adaptation until genetic changes took over. For example, in a yeast strain, the [PSI⁺] prions might generate readthrough proteins that allow the lineage to survive in conditions in which prionless strains die out. As numbers in the prion strain increase, the chance of mutations that will enable the proteins to be produced without readthrough also increases. When this happens, the mutants can be selected, and the presence of the prion is no longer necessary. Similarly, selected heritable methylation marks that keep genes turned off could eventually be superseded by mutations that disable the genes. In a sense, the selection of epigenetic variations paves the way for the more stable genetic variants that may follow.

What is true for asexually reproducing organisms is also true for those that reproduce sexually. Epigenetic inheritance augments the probability of genetic assimilation because it maintains a new developmental pathway until it can be established more permanently by the selection of the appropriate combination of alleles. In the next chapter we show how behavioral and cultural inheritance can have similar molding influences on evolutionary change through the genetic system.

We have covered a lot of quite difficult ground in this chapter, and we suspect that even some of our biologist readers have been humming through parts of it. We have therefore summarized the crucial take-home messages in figure 7.6. The first thing to notice is that the figure shows that both genetic and epigenetic variants are subject to natural selection. It also shows that not only do genetic variants affect epigenetic variants (by affecting the marks genes carry, the nucleotide sequences of the RNAs involved in RNA interference, the amino acid sequences of the proteins that form heritable cell structures or have a role in self-sustaining loops, etc.), but epigenetic variants also affect genetic variants. Chromatin marks do so because they bias where and when mutations occur. More importantly though, new epigenetic variants, which are often induced by changed environmental conditions, influence the genetic variants present in a population by unmasking those that were hidden and exposing them to natural selection. When the epigenetic and genetic systems are considered together in the way suggested in figure 7.6, it is clear that thinking about evolution solely in terms of the selection of genes misses out on too much. Yet even the framework shown in the figure leaves out a lot, because it ignores the way the product of the genetic and epigenetic systems—the organism—influences the environment. We come to this aspect of the web of interactions in the next chapter.

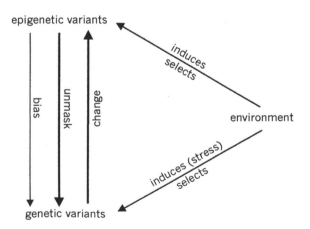

Figure 7.6
The interactions of the genetic and epigenetic systems.

Dialogue

I.M.: I am still puzzled by the role that EISs play in evolution. You suggested that in multicellular organisms EISs can cause various problems, one of which is that rogue cells might decide to switch jobs. They might try to become germ cells, you said, which is a job they wouldn't do very well. You tried to argue that this is one of the evolutionary reasons why somatic cells and germ cells became segregated. Yet you have repeatedly stressed that plants and other creatures do *not* have a segregated germ line, and this is what makes evolution through EISs especially important for them. Once again, you are trying to have your cake and eat it! If EISs are so dangerous, why do plants make so much use of them and still manage without a segregated germ line? And if they are such great agents of evolution, how do creatures with a segregated germ line manage without all the variation EISs provide?

M.E.: You have to recognize that the costs and benefits of EISs have different weights in different organisms. As we pointed out, the reproductive and developmental strategies of plants and animals are very different. Plant cells are surrounded by hard cell walls, so they can't move around and change their position like animal cells can. This makes it more difficult for them to become germ cells, because they cannot migrate into the reproductive organs. Moreover, plants and many other organisms that lack a segregated germ line have a modular organization, so they can afford to experiment with epigenetic variants in the germ cells of some modules, provided others remain unchanged. Nonmodular organisms cannot afford to do this. Finally, plants, fungi, and simple animals do not have a centrally controlled behavioral option for making short-term adaptations to their environment—they can't move to somewhere else; they have to use their epigenetic systems for temporary adjustments. Taken together, these differences suggest that the costs of EISs are not as high in plants as they are in most animals, and the benefits are much greater. That's why the evolutionary uses of EISs and the evolutionary responses to their existence have been different in plants and animals.

I.M.: Would you expect it to be far more difficult for a somatic cell to accidentally change into a germ cell in vertebrates and insects than in plants?

M.E.: Vegetative asexual reproduction is common in plants, and nonexistent in many animals. Gardeners are always taking bits of roots, stems, or leaves and managing to grow whole flowering plants from them, but you can't do that kind of thing with vertebrates and insects. This indicates that cell fate can be changed more easily in plants, and that their somatic tissues can produce germ-line cells rather more readily than animal tissues are able to.

I.M.: What really matters is what happens to the ex–somatic cell's epigenetic baggage when it becomes a germ cell, isn't it? If it isn't removed, then according to your arguments it can bias future genetic changes as well as have its own effects on future generations. But whether epigenetic modifications are removed or inherited will depend on enzymes and other proteins, which means that it will depend on genes. So, isn't it more fruitful to think about the genetic aspects of epigenetics? Surely that would give biologists more fundamental insights into epigenetic inheritance.

M.E.: It depends on what you are interested in. Obviously, understanding the genetics of epigenetic inheritance will give you important insights into the evolution of development, including cell memory and imprinting. We'll say more about this in chapter 9. However, once you accept that there is an independent axis of epigenetic inheritance, seeing evolution in terms of genes alone is not enough. Think about an analogous case, about language evolution. Most of the many questions you can ask about the evolution of languages (like Hebrew and English) make sense only if you assume that they are independent of genetic variation. The same may be true of epigenetic inheritance. If there is an independent axis of epigenetic inheritance, there will be evolutionary phenomena that are coupled with and special to this axis. You can only understand them when you ask questions relating to the epigenetic level. Of course, for some evolutionary questions you have to take both the genetic and the epigenetic dimensions into account.

I.M.: Still, with the evolution of imprinting, selection was for genetic variations—those that gave the best epigenetic outcome. I really liked all those evolutionary stories about imprinting, especially Haig's hypothesis. I would like to play the biologist and suggest one more use for imprints, if I may. Haig explained imprinting in terms of an evolutionary conflict, with each parent marking the transmitted chromosomes in a way that is for its own benefit. And your idea was that in mammals the X chromosome from the father is likely to be marked in a way that will benefit daughters, because the father gives his X chromosome only to female offspring. It seems to me that these two ideas may be related in an interesting way. According to Haig, a father will mark his chromosomes so that they make the offspring who receive them try to extract extra nutrients or care from the mother. But the mother will counterattack—she can mark her chromosomes in a way that will neutralize the marks on the paternal ones. So if the father marks his chromosomes in a way that makes the embryo produce more of a factor that promotes growth, the mother can do the opposite, marking her chromosomes to make it produce less, tipping the balance back in her favor. She has to look after the interests of all her present and future offspring, so she can't let a father gain the upper hand. Now this is where I want to bring in the sex chromosomes, because if a male marks his X chromosome genes so that they make his offspring greedily extract as much as they can from the mother, she has a problem. She cannot countermark her own X chromosome genes (or any others) to overcome those on the father's X chromosome, because she doesn't know whether her X chromosome will end up with the father's X chromosome or with his Y chromosome! If a female's X chromosome is going to be with a Y chromosome, then marking it in a way that makes the embryo nongreedy and restrained will harm her XY sons. So, because she can't mark her own chromosomes to counter the father's X chromosome imprints, there is only one thing she can do—keep his X chromosome inactive in her daughters! This is a drastic strategy, but a very effective one, it seems to me. And it happens, doesn't it? You said that sometimes it is only the father's X chromosome that is inactivated.

M.E.: You may well be right! The idea that the paternal X chromosome is inactivated by maternal factors has been suggested before, although for reasons somewhat different from yours. As we said, in female mammals dosage compensation occurs through X chromosome inactivation: one of the two X chromosomes in the cells of females is permanently switched off, so males and females have the same effective dose of X chromosome genes. What seems to happen is that at fertilization the female gets a partially inactivated X chromosome from the father. In the tissues of the embryo itself, both of the X chromosomes become fully active for a brief period, and this is followed by random inactivation of one of them in each cell. However, in mice and some other mammals, in the extraembryonic tissues, which are responsible for transferring nutrients to the fetus, selection seems to have favored the stabilization of the inactive state of the father's X chromosome: it becomes thoroughly silenced in these tissues. This might have evolved because, as you suggest, the extraembryonic tissues are where an extorting paternal X chromosome would do most of its extorting. In fact we can extend your idea: it might be that the gene-poor Y chromosome is also a victim of a maternal strategy of inactivation! Maybe in the past the father imprinted his sex-determining Y chromosome so that it extorted nutrients from the mother, and she used a similar drastic strategy of inactivating genes on the extorting Y chromosome. If the mother inactivated paternally imprinted "greedy genes" on the Y chromosome, and they remained inactive from one generation to the next, it might be one of the reasons why, over evolutionary time, the Y chromosome degenerated and is now so small and carries so few genes.

I.M.: I wonder how you could find evidence supporting these hypotheses. Maybe comparing the growth patterns of abnormal embryos that have only a single X chromosome (either a maternal or paternal one) would give us a clue. And maybe making crosses between species with different numbers of genes on their Y chromosome (if there are such species) and looking at the size of their offspring would help. But I shall resist the temptation to speculate further, and turn to another question about the Haig hypothesis, this time a more philosophical one. It seems to me that the logic of Haig's idea fits very nicely with the selfish gene point of view—it's selfish genes in fathers against selfish genes in mothers. It really doesn't fit very well with your biological philosophy, does it?

M.E.: You are confusing the selfish gene view with the conflict-oriented view. We certainly do not deny that there are lots of conflicts in the world—conflicts between mates, between parents and offspring, between predators and prey, between hosts and parasites, and so on. It is clear that adaptations are shaped by these conflicts as well as by the other interactions that organisms have with their environments. That much is obvious from *any* point of view. The selfish gene point of view is that the gene, rather than any other unit, is the beneficiary in any selection process. Talking about conflict between imprinted genes may sound similar to this, but the two ideas are distinct. Logically, thinking in terms of evolutionary conflict has nothing to do with whether the single gene (an allele) or any other entity is the unit of selection. However, it is true that conflict-oriented theorists do tend to think in terms of selfish genes. As you know, we don't take a selfish gene view of the world: we prefer to think about the selection of heritable phenotypic traits, rather than genes. In most cases, we believe, the single gene is not a unit of selection and evolution, because its selectable effects are network-dependent, and on the average selectively neutral. From our point of view, Haig's conflict hypothesis is fine—we have no problem accepting it. However, if we needed to flesh out the hypothesis at the cellular and molecular level, we would have to consider the developmental networks that underlie the marking process, and think about how these networks evolved, rather than concentrate on individual genes.

I.M.: I have to reluctantly agree that the connection between the conflict and selfish gene viewpoints is not logically necessary, although the two do seem to overlap. But let me move on, and ask one last question about imprinting. It's about the term "imprinting" itself. I find myself imagining imprinting as a positive process in which a sex-specific mark that will make a gene active or inactive is stamped on it by the parent. But if I understand the process at all, imprinting refers merely to a difference in the marks inherited from the two parents, which causes the two alleles or chromosomes in the offspring to function differently. Surely it doesn't imply anything about genes being actively modified, or about being modified for activity rather than inactivity. I am not sure whether this is just my problem, or a general one, but maybe a different word, one that does not have the connotation of imposing an explicit shape on something, would be better.

M.E.: This is not just your problem. The usage is sometimes very confusing for biologists too. "Imprinting" really does not tell us anything about the process itself, except that the result is an epigenetic difference between two homologous regions of chromosomes. Unfortunately there is no way in which the term can be changed now, because it has become part of the jargon of biology. We do need to be careful, however. Judging from what is written, the "active" connotation can sometimes be a problem for biologists.

I.M.: Let's leave imprinting now and move on to genetic assimilation. You seem to think that genetic assimilation has been enormously important in evolutionary adaptation. If I understand your reasoning, this follows from your view that it is traits and developmental networks, rather than single genes, that are the units of selection and evolution. You maintain that most of the time what is selected is not alternative alleles of a single gene, which you insist are usually more or less neutral in their effects, but alternative developmental variants, which incorporate several genetic differences. From this it follows that most evolution occurs in conditions that reveal phenotypic differences in development, and it involves genetic assimilation. Is this correct? Is that what underlies your view that genetic assimilation is so important?

M.E.: Yes, we think that genetic assimilation and selection in stressful conditions have been central to adaptive evolution. For genetic assimilation to occur, however, the environment has to have a dual role: it must both affect development and have selective effects. This is not always the case. Sometimes a change in the environment does not induce differences in the way organisms develop and behave, or if it does induce differences, they are selectively neutral or even harmful. The environment is then just selecting—determining who will be most successful at contributing offspring to the next generations. Genetic assimilation is irrelevant in such cases. But very frequently—probably in most cases—the environment does have a dual role, leading to an adaptive adjustment to development as well as determining who can survive and reproduce. The genetic tuning of the environmentally induced response is therefore likely to be very important. We must stress though that it is not *always* necessary to have several different genetic variations brought together in order to produce a difference that is visible to selection. Sometimes a change in a single gene can have consistent, beneficial, phenotypic effects. More usually, however, variation in a single gene is unseen, and only shows up in the "right genetic context" (in the presence of certain other alleles) or in the "right environment" (usually a somewhat unusual environment). Generally speaking, we think that particular combinations of several alleles are needed for visible and selectable phenotypic differences.

I.M.: But such combinations of genes cannot be passed on! The sexual process that constructs a successful combination will also destroy it!

M.E.: Initially the chances that offspring will resemble their parents with respect to the new useful character will not be very high. But if selection is persistent, the population will gradually become enriched with the "right" genes and hence the "right" combinations, until eventually there may be some genetic assimilation, perhaps even full assimilation, with 100 percent probability of passing on the trait. When several genes are involved, evolution may at first be slow, but because the phenotypic variation produced by the various gene combinations can be quite substantial, there will be selection both for the trait and for its canalization—for its developmental stability. Sometimes selection for stability in a pathway may be as important as selection for a particular phenotype. For example, poisonous insects often have a distinct warning pattern—a vivid combination of red, yellow, and black markings which, after a few bad experiences, their predators learn to associate with their foul taste. Once they have learned the pattern of markings, the predators avoid the insects. In such cases, the exact pattern of colored markings that the species adopts is relatively unimportant, so long as all share the same pattern. Selection is therefore primarily for stability, rather than for a particular "best" phenotype. In most cases, however, genetic assimilation will affect both the stability and the product of a developmental pathway.

I.M.: I suppose that when new species evolve, the stabilities of the features that have something to do with reproduction are particularly important. Animals have to recognize their own species, and the male and female bits can't vary too much if they are going to copulate. But I was wondering about something else—about the domestic dog, with whom I am a lot more familiar than I am with silver foxes! How does the evolution of that humanly constructed species, which is what I believe it is, fit with your ideas? Do you think that here, too, epigenetic changes led the way for epimutations and genetic assimilation, and eventually it all culminated in the domestic dog?

M.E.: It is quite possible and likely, but it isn't easy to unravel the evolution of the dog. There is little doubt that selection for "the dog" was complex, and initially it was probably not very planned. Presumably at first man wasn't actively involved at all: it could be simply that wolf packs were hanging around human camps, scavenging for food remains, and those that were least fearful and most friendly got the most food and survived. There is no doubt that in dogs, as in silver foxes, domestication led to changes in the females' reproductive cycle. In dogs it also led to an absence of paternal care—domestic dogs are the only canids in which fathers do not help to look after their young. Nevertheless, although hormonal changes undoubtedly occurred at some stage, it is very difficult to reconstruct the epigenetic part of the pathway of dog evolution, because we cannot tell when and whether hormonally mediated epigenetic changes unmasked genetic variations, and if they did which were selected. We can feel rather more sure about the behavioral part of the pathway, where dogs-to-be that could readily learn to behave in a way that suited humans survived, and became better at it. In this case we are thinking about genetic assimilation of behavioral traits, through selection of the ability to learn to behave in a certain way. What was once the nurture of the wolf became part of the nature of the dog. We must leave this topic, however, because interactions between the behavioral and genetic systems are the subject matter of the next chapter.

One of the take-home messages of the previous chapter was that in evolution the role of "the environment" is quite subtle. Traditionally, it is seen as the agent of selection, determining which variants survive and reproduce. Yet, because it influences development, it also has a role in determining which variants are there to be selected. The consequences of this dual role are that environmental effects on development may guide the selection of genetic variants. In the previous chapter we looked at this in relation to environmentally induced developmental modification of an organism's form, describing how, through natural selection, an induced change can eventually become a permanent part of the phenotype. We now want to extend the same idea, the idea of genetic assimilation, to behavior, and show how natural selection can convert what was originally a learned response to the environment into behavior that is innate.

We will also be considering another factor that complicates the way in which we have to think about the role of the environment in evolution. It is that the organism itself is often responsible for selecting the environment in which it lives and for constructing some aspects of it. If you release English rabbits and hares into the countryside, the rabbits will head for the hedge and the hares will opt for the open field. They themselves decide where they will live and reproduce. Both types of animal will also modify their environment. This is very obvious with rabbits, whose feeding and burrowing habits often transform the landscape. Farmers have their own ways of describing such activity ("ruin" and "destroy" are two of the milder terms they use), but biologists call it "niche construction." All organisms do a bit of niche construction (we gave some examples in chapter 5), but its effects on evolution are particularly significant for animals that inherit a niche in the form of the artifacts, behaviors, and cultures of their elders. Changes in habits and traditions can result in these animals creating a very different social and physical environment for themselves and their descendants. It is therefore wrong to think of them as just passive objects of environmental selection. This is especially true for humans, whose elaborate cultural constructions form such a large part of their environment. What we transmit through our behavioral and symbolic systems obviously has profound effects on the selection of the information that we transmit through our genes.

We will leave the complexities of the effects of human culture on genetic evolution until later, and start this chapter with a relatively simple problem—the evolution of instincts. Instincts are complex behaviors that occur either without having to be learned at all, or with very little learning. They are clearly adaptive, and many would develop through learning even if they were not inborn, so how and why did they become permanent parts of the animals' makeup? Was it simply a chance combination of rare and random mutations that made many small mammals show fear and avoidance responses upon first hearing hissing, snakelike noises? What type of selection resulted in hand-reared spotted hyenas, who have never had anything to do with lions or their mother's responses to lions, reacting with fear when they first experience a lion's smell? How did natural selection lead to newly hatched seagull chicks responding to a long object with a red dot (which vaguely resembles a parent's beak) by pecking it?

Genes, Learning, and Instincts

The evolution of instincts fascinated and puzzled the early evolutionists. The Lamarckian explanation—that a learned behavior could gradually but directly become an inherited one—was obvious and satisfactory for many people, but it wouldn't do for the neo-Darwinians. They had to explain the evolution of instincts in terms of natural selection. One of the earliest attempts to do so was made by the Scottish biologist Douglas Spalding. You will recall that in chapter 6 we used an adaptation of his story about Robinson Crusoe's parrots to illustrate the differences between a symbolic and nonsymbolic communication system. In fact Spalding's original story had a very different goal—it was intended to provide an evolutionary explanation of instincts. The original story reads as follows:

Suppose a Robinson Crusoe to take, soon after his landing, a couple of parrots, and to teach them to say in very good English, "How do you do, sir?"—that the young of these birds are also taught by Mr. Crusoe and their parents to say, "How do you do, sir?"—and that Mr. Crusoe, having little else to do, sets to work to prove the doctrine of Inherited Association by direct experiment. He continues his teaching, and every year breeds from the birds of the last and previous years that say "How do you do, sir?" most frequently and with the best accent. After a sufficient number of generations his young parrots, continually hearing their parents and a hundred other birds saying "How do you do, sir?" begin to repeat these words so soon that an experiment is needed to decide whether it is by instinct or imitation; and perhaps it is part of both. Eventually, however, the instinct is established. And though now Mr. Crusoe dies, and leaves no record of his work, the instinct will not die, not for a long time at least; and if the parrots themselves have acquired a taste for good English the best speakers will be sexually selected, and the instinct will certainly endure to astonish and perplex mankind, though in truth we may as well wonder at the crowing of the cock or the song of the skylark. (Spalding, 1873, p. 11)

In this story, the learned utterance, which was first taught by Mr. Crusoe and then transmitted from parents to young, eventually became independent of learning and cultural transmission (figure 8.1). Mr. Crusoe had selected for the best learners—for parrots that needed to hear the utterance fewer times than had their ancestors in order to learn it. Eventually so little learning was needed that the behavior was considered practically inborn. It became an instinct.

Notice that Spalding cleverly used another Darwinian mechanism, sexual selection, to explain why the parrots went on using English after Mr. Crusoe's death. The parrots continued to speak English, said Spalding, because they had acquired a taste for the language, and good speakers were more likely to be chosen as mates. A parrot could enhance its reproductive success by impressing the opposite sex with its language skills.

Figure 8.1
Mr. Crusoe's trained parrots.

Sexual selection was an idea that Darwin had proposed two years earlier, in *The Descent of Man*. It was his way of accounting for the evolution of seemingly ridiculous characters such as the peacock's tail. Such an adornment, he said, could not have been established through natural selection, because it certainly didn't help its owner to survive. But if there was sexual selection—if females preferred the males with the most beautiful tails— those males would have more young, and beautiful tails would become more common. It was an ingenious idea, and Spalding made use of it, but it was not widely accepted at the time. After some initial arguments, sexual selection was generally forgotten until it was revived and reformulated about a hundred years later. Since its reincarnation it has been applied widely. It is now the basis of explanations of the supposedly innate differences in the talents, values, and attitudes of men and women that we looked at in chapter 6.

In Spalding's thought experiment the selection of Mr. Crusoe's parrots was initially artificial. However, it is not difficult to see how, even without human involvement, behavior that was at first learned could become innate through natural or sexual selection. Consider a population of songbirds in which the young have to learn their song from adults. Imagine that a new type of predator arrives in the area, so both young and experienced male birds are forced to sing less than usual to avoid being detected and attacked. Thanks to the predators, the young will hear the adults' species-specific song less often than formerly, and have less chance to practice it. Consequently, if females continue to prefer good singers as mates, there will be strong selection for rapid and accurate song learning. Those young males who learn the song quickly will win more mates and produce more offspring, and some of these offspring may inherit their song-learning talent. If this situation persists, with both predation pressure and sexual selection remaining strong for many generations, it will result in the birds needing to hear very few bouts of singing (or none at all) to learn their species-specific song. The song will have become almost entirely innate.

You can use a similar argument to explain the evolution of fear responses. If young mammals have to learn from their own experience or from the experience of their parents how to avoid a new predator, and learning is time-consuming and exposes them to danger, the fastest learners will be the most likely to survive, and natural selection may culminate in an "instinctive" fear-and-avoidance response.

Spalding was not the only nineteenth-century evolutionist to suggest a Darwinian mechanism that could transform an initially learned response or habit into an instinctive one. In 1896, the American paleontologist

Henry Fairfield Osborn and two psychologists, Conway Lloyd Morgan in England and James Mark Baldwin in America, each independently came up with a somewhat similar idea about how natural selection could convert acquired characters into inherited ones. At the time, the battle between the neo-Lamarckians and neo-Darwinians was at its height, and their suggestion seemed to be a simple way of reconciling the views of the two opposing camps. Baldwin described the idea as "a new factor in evolution" and referred to it as "organic selection." Now, somewhat unjustly and inappropriately, the evolutionary mechanism that each of the three scientists hit upon is usually known as the "Baldwin effect." What each suggested was that when animals are faced with a new challenge, individuals first adapt to it by learning. If the new challenge—the selection pressure—is ongoing, this individual learning allows the population to survive long enough for congruent new hereditary changes to appear and make learning unnecessary. In this way neo-Darwinism, which focused on hereditary determinants and selection, was wedded with the views of the neo-Lamarckians, who focused on learning and individual adaptability, to produce a theory that explained the inheritance of acquired characters.

Baldwin, Osborn, and Lloyd Morgan put forward their ideas just before the dawn of Mendelian genetics. Understandably, they were somewhat vague about the nature of the hereditary variations that would replace the individual learned responses, but they were very clear that learned behavior would guide the selection of "congenital variations in the same direction as [the] adaptive modifications." They also recognized that this was a gradual and cumulative process. In many respects, therefore, what we now call the Baldwin effect resembles Waddington's genetic assimilation, although Waddington always maintained that there was a conceptual difference.

The assessment of whether or not genetic assimilation is the same as the Baldwin effect can be left to the historians of biology. What we want to pursue here is an idea that was important in the thinking of Lloyd Morgan and Baldwin, the two psychologists—the idea that where behavioral changes lead, inherited (genetic) changes follow. For obvious reasons we will use Waddington's framework of genetic assimilation rather than the Baldwin effect, since we want to show how a learned response can be transformed into an instinctive one through selection acting on combinations of preexisting alleles.

We can start by thinking about animals learning how to handle a new type of food, or how to dig a burrow to hide from a new type of predator. Since learning the new activity is both risky and costly in terms of time and energy, those individuals who can carry it out without having to invest too much in learning will survive and reproduce more successfully than others. Before the new challenge appeared, variation in the ability to learn this particular action was selectively unimportant—there was no advantage in being a fast or efficient learner, because no such learning was necessary. But once there is an urgent need to learn, the previously hidden differences between individuals are exposed, and this variation can be organized into more effective genotypes through repeated sexual shuffling and selection. Gradually, the ability to learn the activity improves. The behavior becomes more canalized. Eventually, after many generations of selection, some individuals may respond so quickly that the learned response is in effect instinctive.

Most assimilation processes will not end up with a completely internalized, instinctive response. Assimilation is more likely to be only partial: some learning will still be needed, but it will have become much more rapid and efficient. Whether assimilation is complete or partial, what has happened is that through selection the mind has been molded so that an adaptive behavioral response becomes more likely.

Expanding the Repertoire: The Assimilate-Stretch Principle

So far, this account of the evolutionary interplay of genes and learning has been rather one-sided. In fact, we have shown that selection for more effective learning undermines learning! The consistent pressure to learn, and to learn fast, leads to the behavior being controlled more and more by the genetic inheritance system, and less and less by learning. We arrived at this point because we chose to describe evolution in a stable environment, with consistent, multigenerational selection in the same direction, for faster and faster learning about predators, or food, or social responses, or some other aspect of life. Suppose instead that the selective environment is not stable, but changes and fluctuates. This might be so if, for example, a fruit-eating species arrives in a region where the fruit supply varies from place to place, changes with the seasons, and there are many species competing for it. In such a situation we would expect an increased reliance on individual and social learning to evolve from more instinctive responses. However, even when there is a stable environment and selection for fast learning leads to more instinctive responses, it still need not lead to simplified behavior. It could lead to behavioral sophistication.

To see how this can happen, imagine a species of bird in which males are capable of learning a sequence of four consecutive actions—four movements in a display dance, for example. Assume that these four movements are somehow learned from experienced males, and that learning more than these four is very difficult for the bird, because the learning capacity of the species is limited. Females find the male dance very attractive, and choose the best dancers as mates, so there is consistent selection for the males to learn the dance quickly, efficiently, and reliably. The result of this intense selection is that one of the four steps is genetically assimilated: it no longer has to be learned—it becomes innate. Males now need to learn only three movements, so it will be easier for them to learn their dance. But something else has happened: part of their unchanged dance-learning capacity has been "liberated." Potentially, they can still learn four movements, but they now have to learn only three. Consequently, if females prefer the most beautiful or interesting dances, the males may introduce an additional movement (originally learned by trial and error) into the dance. There will now be five movements, one of which is innate and four learned. If elaborate dances continue to be attractive to the female, then, through the selection that she imposes, another previously learned movement may be genetically assimilated. This will again free up learning capacity, so males may add yet another learned movement; there will be two innate and four learned movements. Gradually, by genetically assimilating formerly learned movements, the sequence will get longer and longer, although the amount that has to be learned remains the same. Avital and Jablonka called this general process, which is shown in figure 8.2, the assimilate-stretch principle: part of a behavioral sequence that formerly depended heavily on learning is genetically assimilated, and this allows a new learned element to be added. They suggested that the assimilate-stretch principle might underlie the evolution of many complex patterns of behavior.

There is another interesting effect of genetic assimilation: it may lead to the evolution of categorization, which will change the animal's perception of its environment. Think about a population of monkeys that is threatened by a new aerial predator, monkey-eating eagles. Obviously, individuals need to learn to recognize and avoid the eagles. Those monkeys who are best able to identify and memorize the shape of the eagle, its mode of flight, and so on have a better chance of surviving. The new predator therefore exposes hidden genetic variation in these abilities, and after many monkey generations of selection, the population will consist of individuals with a genetic makeup that enables them to identify and avoid the predator much more efficiently than before. But let's assume that the avoidance of monkey-eating eagles has been only partially genetically assimilated, because selection was not intense or long-lasting enough.

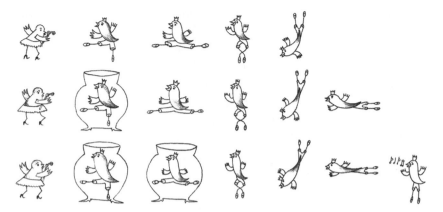

Figure 8.2
Evolution through the assimilate–stretch principle. In the top row, a male bird performs a four-movement dance for the female on the left; each movement has to be learned. The middle row shows the addition of a new learned movement, following genetic assimilation of the first of the original four movements. The bottom row shows a six-movement dance, with two of the original movements having been assimilated. For simplicity the figure shows assimilation at the beginning of the sequence, and new movements being added at the end, but the addition of new learned movements and assimilation of existing ones could occur anywhere in the sequence.

Whereas a fully assimilated avoidance response would have been specific to the monkey-eating eagle, partial assimilation has led to a response to all aerial predators whose shape and pattern of flight vaguely resemble those of the monkey-eating eagle. What this means is that the monkeys have, in effect, formed a new conceptual category, that of "aerial predator." They will now perceive some aspects of their world in terms of this new category.

Cultural Niche Construction

Up to this point, we have been concerned with the way in which the genetic system can take over behavior that animals formerly had to learn. How they learned was not important for our arguments: it didn't matter whether they learned from each other or through their own trials and errors, so long as there was a consistent need to learn the particular activity in every generation. We now want to move on to the more complicated situation in which learned behaviors or the consequences of an animal's activities are transmitted from generation to generation. When they are, changes in the genetic and behavioral systems inevitably interact. The reciprocal influences of the two heredity systems can be seen most easily in human populations, but they are not limited to humans or even to social animals. Whenever the activities of one generation shape the conditions of life for the next, there will be feedback between the inherited genes and the inherited niche. The niche that is inherited may be an altered aspect of the chemical or physical environment, such as the changed soil that results from the activities of earthworms, or the warren system excavated by rabbits, but in the cases we want to look at it is the culture that humans construct for themselves.

One of the first persons to emphasize the influence of "social heredity," as he called it, on the selection of "biological" (genetic) qualities was James Mark Baldwin, one of the originators of the idea we now know as the Baldwin effect. He recognized that cultural factors often determine how likely it is that people with various mental and physical qualities will survive and reproduce. The idea was not taken very far, however, and until recently evolutionary biologists paid little attention to the way in which genetic and cultural changes impinge on each other. Most people have preferred to think of human evolution in terms of either culture or genes. One of those who hasn't is the American anthropologist William Durham, who for some years now has been discussing the coevolution of genes and culture.

Durham has analyzed some fascinating examples of the way in which changes in human lifestyles have influenced the frequencies of some of their genes. One of these involves the genetic changes associated with dairying. Somewhere around six thousand years ago, following the domestication of cattle, humans began to use milk and milk products such as cheese and yoghurt as food. Using fresh milk as food is not straightforward, however. People in the Western world, who have been told that "milk is the perfect food" and have been brainwashed with slogans such as "drinka pinta milka day," are usually surprised to learn that most of the world's adults get little or no nutrional benefit from drinking milk. When this was discovered in the mid-1960s, people began to realize that for many countries sending food for famine relief in the form of dried milk didn't help most of the intended beneficiaries. Indeed, it could actually do them harm. The problem is that in order for milk to provide the simple sugars that can easily be absorbed into the bloodstream, the lactose in milk has to be broken down in the small intestine. This requires the enzyme lactase. But in all mammals, including most humans, the ability to digest lactose declines after weaning—there is only a small amount of lactase present in the adult intestine. Consequently, if adults drink fresh milk, they get little benefit from it, and may even end up with indigestion and diarrhea as the gut bacteria work on the undigested lactose. Usually people have no problems digesting milk products such as cheese and yoghurt, because these foods contain little lactose: microorganisms have broken down most of it during their manufacture. It is only fresh milk and fresh milk products that are not digested properly.

People who are able to drink milk with no ill effects can do so because they have a variant allele of the lactase gene. This allele affects the gene's regulation, with the result that lactase activity persists into adulthood. The effect is dominant, so adults possessing a single copy of the lactase-persistence allele are "absorbers," who can get the full nutritional benefits from milk. What is interesting is that the distribution of the allele is very nonrandom: absorbers are particularly common in populations of northern Europeans (and their overseas descendants), and in certain communities in the Middle East and Africa, but in most populations they are the minority.

Whenever such odd distributions of genetic variants are found, biologists start looking for an evolutionary reason for it. Sometimes they can find no reason other than chance, but in this case analysis of the data led to the conclusion that a high frequency of lactose absorbers is causally related to cultural practices associated with dairying. It was suggested that the domestication of cattle had altered the selective environment in which humans lived, and in some populations made the ability to break down lactose when adult an advantage. Consequently, through natural selection, the frequency of the allele that enabled people to do this increased.

On the basis of both ethnographic and genetic evidence, Durham concluded that cultural evolution leading to milk drinking and an increase in the frequency of the lactase-persistence allele occurred several times, not necessarily always for the same reason. Among the pastoral nomads of the Middle East and Africa, hunger and perhaps thirst were probably very common, and animals that had originally been domesticated for meat offered a potentially valuable source of food and drink in the form of fresh milk. Possessing the lactase-persistence allele was therefore beneficial in these nomadic populations, because it enabled a person to get the full nutritional value from milk. Someone with the allele was more likely to thrive and have children than someone without it, so absorbers became more common.

One might expect from this argument that absorbers would be common in *all* dairying populations, but they are not. The reason is probably cultural. In many dairy-farming communities, such as those around the Mediterranean, milk is used as a food, but largely in the form of cheese, yoghurt, or similar products. These foodstuffs have far less lactose, and can be digested more readily. Consequently, having the lactase-persistence allele is of no advantage. Indeed, it may be a disadvantage, because there is some evidence (not very strong) that lactose absorbers are more prone to cataracts and some other medical conditions. Why Mediterranean people should use processed milk products rather than drink fresh milk is tied up with their cultural history—with factors such as how often and when they and their herds moved, and how dependent they were on domestic animals and milk. The allele for lactase persistence was probably present in most of these populations, but it just wasn't very important, so it never became common.

One region where the lactase-persistence allele did become common, even though there was a tradition of mixed farming rather than just dairying, was central and northern Europe. In Scandinavian countries, over 90 percent of the population are absorbers. According to Durham, the reason for this may be that drinking milk is beneficial not only because it is an excellent energy source but also because lactose, like vitamin D, facilitates the uptake of calcium from the intestine. In sunny regions, people normally have enough vitamin D, because sunlight converts precursor molecules in the skin into the vitamin. As you move further north, however, there are increasingly long periods with little sunlight, and people also tend to keep their bodies well covered against the cold. For them, vitamin D is sometimes in short supply. This leads to poor absorption of calcium, and a consequent tendency to develop rickets and osteomalacia. Drinking milk helps to prevent these problems, because lactose promotes the uptake of calcium, which is plentiful in milk. So by enabling milk to be digested, the lactase-persistence allele reduces bone diseases in its carriers, and consequently has spread in northern populations.

The story of the evolutionary interplay of the lactase gene and dairying culture does not end there, however. Durham and his colleagues have shown that the importance of cows in the local myths and folklore of Indo-European cultures increases with latitude. In southern cultures, the myths are about bulls, sacrifice, and slaughter; in the later, more northern cultures, there is more emphasis on cows, milk, and nurturing. In the north, cows were described as the first animals of creation, and they were not

Figure 8.3
The adoration of the cow.

sacrificed, but lived to produce milk, which was drunk by giants and gods. Milk was the source of their great strength and ability to nurture the world. These myths clearly reflect the importance of fresh milk to the populations, and probably had an educational value far greater than that of "drinka pinta milka day"! By encouraging milk drinking, the myths further enhanced selection for lactose absorbers, so the cultural and genetic changes became mutually reinforcing.

The dairying story is a good example of the coevolution of genes and culture, with changes in both favoring dairy practices and milk consumption. There are other examples of coevolution in which the interaction is not as harmonious. One that Durham describes is the outcome of the adoption of slash-and-burn agriculture in certain parts of Africa. Deforestation produced open sunlit areas, with freshwater ponds. With the ponds came mosquitoes, and with the mosquitoes came malaria. As a result, the frequency of the sickle cell allele of the hemoglobin gene increased. We described the molecular biology of this allele in chapter 2, but not all of its effects. Being homozygous for the sickle cell allele—having two copies of it—results in severe anemia, and usually in early death. But having a single copy (being heterozygous, a "carrier") protects individuals against malaria, because the malarial parasite doesn't do very well in their red blood cells. Since people who were carriers survived better in the deforested, mosquito-infected areas created by the new agricultural practices, the sickle cell allele became more common. The unfortunate consequence was that more people inherited the allele from both parents, and developed the devastating anemia. In this way the genetic change that followed the changes in agriculture was extremely detrimental, although without it the communities might not have survived at all.

Both domestication and deforestation are good examples of the way in which persistent environmental changes resulting from cultural evolution can alter the relative advantages and disadvantages of carrying certain alleles. We have to confess, however, that examples of coevolution that are as convincing as these are few and far between, probably because very few people are making detailed studies of this type. But there are plenty of hints of other associations. It is often suggested that the differences in the incidence of a particular genetic disease are related to the cultures of the groups concerned. Take Tay-Sachs disease, for instance. It is a recessive disorder that appears in the first few months of life and results in the death of the affected children before they are four years old. This devastating disease is much more common among Ashkenazi Jews than in almost any other group. Some scientists argue that this is just chance, but others believe it could be an indirect result of Jewish culture and history. There is suggestive evidence that carriers of the Tay-Sachs alleles (there are several different alleles) are less likely than other people to develop tuberculosis. For reasons associated with cultural intolerance, Jews were often forced to live in slums and ghettos, where TB was rife. So, the argument goes, because carriers of the Tay-Sachs alleles survived the ghettos better than noncarriers, the alleles became more common in Jewish populations. There isn't a great deal of evidence for this historical-cultural effect on the frequency of the Tay-Sachs allele, although it is certainly plausible. Today things are working the other way around—the allele is affecting culture. It is changing the ways marriage partners are chosen: because the disease is so common and distressing, many Jewish communities now provide premarital counseling and testing services aimed at reducing the number of afflicted children.

The dynamics of the interactions between cultural and genetic changes is complex and difficult to unravel, and we have made no attempt to describe them for either the milk or malaria stories. Common sense and a lot of indirect evidence suggest that learned, socially transmitted behaviors occupy the driver's seat of coevolutionary change, simply because adaptation can occur much more rapidly through behavior than through genetic change. New learned habits are likely to be the first adaptive change, and these will then construct the environment in which genetic variations are selected. There is a problem here, however: the idea makes sense if the cultural change in the conditions of life is persistent and stable, but not if there are rapid and frequent changes. If culture is continually altering the perceptual, cognitive, and practical aspects of the niche it constructs, how can genetic evolution keep pace with it? This is one of the problems that make understanding the coevolution of genes and culture in humans so difficult and so challenging. No one doubts that during human evolution cultural change has been and is enormously important and at times rapid, but how and how much it has affected genetic evolution is far from clear. Consider language, which today we see as probably the most important factor in human cultural evolution. How is the evolution of the capacity for language related to the cultural evolution that promoted it and that it promotes? And how do genes tie in?

What Is Language?

Language is something that everybody knows how to use, some with great eloquence and beauty, yet it is notoriously difficult to define. Obviously language is a powerful symbolic system of communication and representation. But what kind of symbolic system is it? Is language a matter of words and their meaning? Or is it to do with the rules of grammar? Or is it about its use in practical situations? Or is it all of these things? We need to understand what the language faculty is before we can discuss how it evolved.

There are two highly contrasting types of answer to the question of what language is. The first answer is that of the American linguist Noam Chomsky and his followers. According to the Chomskians, the essence of language is in its formal structure—in its grammar. What Chomsky has in mind is not the grammar of a particular language such as English or Hebrew, but universal grammar (UG), which is common to all languages. Because it is universal, UG can be uncovered by the rational analysis of any of the world's languages. Although each language has its own rules, according to Chomsky there are universal principles or superrules that guide the formation of these rules. The American linguist Steven Pinker has likened UG to the common body plan found in a group such as the vertebrates, which all have a segmented backbone, four jointed limbs, a tail, a skull, and so on. Although birds, whales, frogs, and humans look so different, and the hind limbs of the whale and the tail of humans and adult frogs are not obvious, these animals nevertheless all have an identical basic architecture. In the same way, languages may seem very different, but they too have a common basic structure. There is a set of superrules that ensure that, whatever the language, phrases are constructed in a way that allows them to be interpreted. For example, whatever the language, if a combination of the words "cruel," "bites," "man," and "dog" is to mean anything, they have to be organized in a certain grammatical conformation—in English, for example, in a particular word order.

Through their analyses of sentence structure, linguists have concluded that one of the things that is at the core of our unique language capacity is an abstract computational system that allows recursion: we have the ability to generate an infinite variety of expressions by embedding one language element within another. We can construct and understand sentences such as, "Then came the Holy One and killed the angel of death, who killed the slaughterer, who killed the ox, that drank the water, that quenched the fire, that burnt the stick, that beat the dog, that bit the cat, that ate the kid, that my father bought for two zuzim." That particular sentence, which sounds much better in Aramaic, is the end of a song that is sung during the Passover meal. In spite of its complexity, even children understand and enjoy it. It has the same structure and appeal as the English nursery rhyme "This Is the House That Jack Built," which ends

This is the farmer sowing his corn
That kept the cock that crowed in the morn
That waked the priest all shaven and shorn
That married the man all tattered and torn
That kissed the maiden all forlorn
That milked the cow with the crumpled horn
That tossed the dog
That worried the cat
That killed the rat
That ate the malt
That lay in the house that Jack built.

According to Chomsky and his school, children grasp such complex structures because the basic structure of UG is already set up in their brains at birth. It is part of our genetic heritage. We have an innate understanding of recursion and various other devices and rules about what you can and cannot do with different classes of words and phrases. In other words, we have what Chomsky calls a "language organ"—a mental module for language. Each language (including the sign languages developed by deaf people) is a particular implementation of UG. UG has a series of alternative possibilities (*parameters*) built into it, and the language the child experiences determines which of the possibilities are used. For example, word order is important in English, and during an English child's development the language module will be triggered to "fixed word order" (so "cruel dog bites man" means something different from "man bites cruel dog"); in other linguistic environments, where word order matters less, "free word order" will be triggered, along with some other parameters about modifying words with tags—learned suffixes, prefixes, or changes in the word that determine what role "dog," "man," and "cruel" have in the sentence. This is how Chomsky himself described the language organ and its use:

We can think of the initial state of the faculty of language as a fixed network connected to a switch box; the network is constituted of the principles of language, while the switches are the options to be determined by experience. When the switches are set one way, we have Swahili; when they are set another way, we have Japanese. Each possible human language is identified as a particular setting of the switches—a setting of parameters, in technical terminology. (Chomsky, 2000, p. 8)

We've illustrated Chomsky's description in figure 8.4.

Children learn with remarkable ease not only the words of their language but also the language-specific rules about modifying words and using tags to determine their role in phrases and sentences. They begin to generalize and apply these rules very quickly, without, of course, formally learning the local rules of grammar or the exceptions to the rules. Parents of young children who learn a second language often witness them applying a grammatical rule (of which the children are totally unaware) to the wrong language. A four-year-old Hebrew speaker who went to England discovered and became fascinated by the slugs that were a feature of his life in that wet country. "Slug" was one of the first English words he learned. "Look, slugim!," he cried excitedly when he saw these new and wonderful creatures marching along the damp pavements. He was applying Hebrew language rules to an English word: Hebrew is a language with gender, and most singular male nouns end in a consonant; the plural of male nouns has the ending *-im*, and female nouns *-ot*. So, naturally, the plural of "slug" is "slugim."

Early linguistic experience adjusts the unchanging principles of the UG according to the alternative (but fully innate) parameters that are built into

Figure 8.4
Chomsky's switch box. Three children are exposed to different languages—English, Hebrew, and Polish. All the children have the same networks (the same universal grammar), but exposure to the different languages has set the parameter switches in their brains in different ways. As a result, the children use the grammatical rules of their own language. In the sentence spoken by the English child, the role of each word is determined by the word order, and neither the verb nor the nouns are gendered. The sentence spoken by the Israeli child has the same meaning, but the grammatical structure is "friend(male)-my(suffix) kissed(male) nose-my(suffix)." In Polish, the same meaning is conveyed by a sentence in which the word order is less important than in English, and words indicating case are used: "friend(male) to-me kissed(male) in nose."

the system. According to the Chomskian view, we have to learn aspects of a language such as vocabulary, when and how to say things, and so on, but the grammar—what really makes this system of communication into a language—is not learned. UG is what *allows* linguistic learning. It follows from this view that general intelligence and linguistic capacity are separate entities. It is recognized that the various aspects of our evolved cognition interact, just as morphological organs like the heart and kidney interact, but nevertheless the various "cognitive organs" are taken to be as distinct and independent of each other as the kidney and the heart. Something else that follows from the Chomskians' view of language is that whatever chimpanzees can do in the way of communicating with symbols when living in a linguistically structured environment, it is not language, because it lacks grammar.

From an evolutionary point of view, there is a problem with the Chomskian approach. If the language organ is so intricate and complex, then as Darwinians we have to assume that it evolved largely through the cumulative effects of natural selection. But if so, what was selected? What function was adaptive? If UG has to preexist in order for language to be acquired, then it is difficult to see how it can have arisen through function-driven step-by-step evolution. Chomsky himself thinks that the special component that endows language with its uniqueness—the computational system that links sounds and meaning—did not evolve through selection for improved communication. Until recently, Chomsky insisted that UG can tell us nothing at all about its own origins or function, and tended to avoid the question of the evolution of the language faculty. When he tackled it at all, he described it, almost jokingly, as a saltatory event: some genetic change, in itself possibly trivial, produced a perfect linguistic genius, all at once, by bringing together the various cognitive faculties of the hominid brain to create a new, extremely intricate and specialized language organ. Needless to say this view is rather difficult for a Darwinian to accept, and recently Chomsky has changed his approach somewhat and produced a more specific theory. He developed the new evolutionary scenario with two evolutionary-minded biologists, Marc Hauser, who studies the ontogeny and evolution of communication, and Tecumseh Fitch, who studies animal vocal and auditory systems.

The new version of Chomsky's theory describes something the authors call the "faculty of language in the broad sense" (FLB), which is made up of three interacting subsystems. One is the sensorimotor system that is responsible for producing and receiving linguistic signals—it's the speaking and hearing bit; the second is the conceptual-intentional system that underlies the ability to categorize, organize, and understand social and ecological cues—it's the thinking bit. The third subsystem, which links the first two, centers around the computational system. This subsystem maps the internal representations that are generated by the conceptual-intentional system into the sound or signs that are produced by the sensorimotor system. Hauser, Chomsky, and Fitch call this computational system, which embodies recursion, the "faculty of language in the narrow sense" (FLN). The biases and constraints that FLN imposes on language, which restrict the set of languages that can be learned, are more or less equivalent to UG.

Hauser, Chomsky, and Fitch believe that our sensorimotor and conceptual-intentional subsystems are based on mechanisms we share with nonhuman animals. These subsystems evolved in the usual Darwinian manner, with hominids gradually gaining important and possibly even crucial adaptations such as being able to produce clear and discrete sounds, an improved social intelligence and a theory of mind, and the ability to imitate sounds. What was really new, however, was the association of these two subsystems with FLN. This is what resulted in language. FLN is assumed to have evolved for some other reason, such as number quantification, navigation, or some other ability requiring recursion, and was not initially part of a communication system. Once the FLN interacted with the other two subsystems, it enabled generativity and an almost perfect mapping between speech and meaning. In other words, the full human linguistic communication system emerged.

There is no reason to doubt that combining several different preexisting faculties can lead to important and surprising evolutionary novelties. This has happened many times during the evolution of life, such as, for example, when feathers, whose origin was in protective scales, got involved in temperature regulation and locomotion. However, it is difficult to accept that an exquisite adaptive specialization like flight or language is the result of emergence alone, without subsequent elaboration by natural selection. It is much more reasonable to adopt the traditional *adaptive* Darwinian explanation, which is that recruiting an existing system (such as feathers or the computational capacity of FLN) into a new functional framework (locomotion or communication) is followed by its gradual evolutionary refinement and adjustment within this new framework. One would expect the properties of FLN to become more adapted to the conceptual system, which would mean they would not be abstract and meaning-blind, as Chomsky's UG theory says they are. Even those Chomskian linguists who believe that UG evolved gradually through natural selection (and there are some) still take it for granted that the components of UG are blind to meaning. As we will argue later, many aspects of the structure of language *are* adapted to their functions, and this obviously affects how the origins and evolution of language should be seen.

We will now leave the ideas of the Chomskian school and look at the other main group of answers to the question of what language is. The view of those known as functionalists is diametrically opposed to that of the Chomskians. Rather than seeing language as a special faculty of the mind, they regard it as a product of general cognitive processes and mechanisms. According to their views, there is nothing unique about learning language.

It develops alongside and in essentially the same way as other, nonlin-guistic, cognitive functions. It is the general constraints imposed by having a certain type of body, certain senses, and a certain type of brain that shape the acquisition of language, just as they shape any other mental faculty. Grammatical rules stem not from an innate universal grammar, but from such things as the physical properties of the speech channel, the con-straints on memory, a limited attention span, and the way *all* information is processed. Since it is clear that humans (i.e., linguistic apes) and chimpanzees (i.e., protolinguistic apes) have similar bodies and brains, but the human brain is much larger, the mature language faculty of humans is seen as an emergent property of their larger brains. The evolu-tion of language is therefore simply an aspect of the evolution of general intelligence. Language, according to this view, has not evolved as such; it emerged when the brain of highly intelligent, social, communicating hominids reached a critical size. There is no need to postulate a dedicated "language organ." For functionalists, the bonobo chimpanzee's ability to understand spoken English just as well as a two-and-a-half-year-old human may be of significance when trying to understand the evolution of language.

At first sight the functionalist approach to language does make evolu-tionary sense. It suggests that, as with many of the other structures and functions in animals, the mechanisms underlying language development are general, and only the outcome is specific. Because it assumes that lan-guage has evolved by natural selection, its structure should therefore be related to its function(s) in the same way as the structure of the immune system or an eye is related to its function. The problem is that there are at least two features of language that are not compatible with the purely func-tionalist view. The first is that language is a very constrained system of communication. If you find this statement surprising, think of all the things that *cannot* be expressed very well in words, but can be expressed exquisitely through pictures, or through music, or through dance, or through grimaces, and so on. (Can you really describe your father's smile in words?) Then think of the many things that can be expressed almost *only* through language, such as "How do I get to San Francisco without going on the freeway?" If you do this, you soon realize that language is a very specialized system of communication, not a general-purpose tool. Any evolutionary explanation has to account for this. Just as when thinking about the frog's visual system it is not enough to say it has evolved for seeing, because we need to know why that particular system has evolved in the way that it did, it is also not enough to say that language evolved for communication. We have to explain how the linguistic system that humans possess has evolved to be the idiosyncratic and constrained communication system that it is.

A second feature of language that is not easy to explain in simple functionalist terms is the speed and ease with which a fully mature language capacity is acquired by children. In fact, even when children have no proper exposure to language, they nevertheless rapidly develop a system of linguistic communication. One of the best examples of this is the way a new, mature, fully grammatical sign language developed from scratch in a community of deaf children in Nicaragua. After the Sandinistas came to power in 1979, they built a school for deaf children, and brought in children who had previously been communicating with their family through nonlinguistic mimetic gestures. Attempts to teach the children to lip-read failed dismally, but the children soon invented, by themselves, a sign language. The first version was crude and not very grammatical, but as new deaf children came to the school and were exposed to the crude version, a fully grammatical language developed. It took only about ten years. This suggests that, rather than being just a tool of general intelligence, the human brain has evolved an organization that makes it specifically biased toward the rapid acquisition and invention of language.

It seems to us that neither the basic functionalist view, nor the Chomskian view of an FLN that somehow maps between sound or signs and meaning, provides an adequate framework for explaining the evolution of language, or the peculiarities of its structure and acquisition. Recently, the Israeli linguist Daniel Dor has developed a view of language which we believe is more compatible with general evolutionary ideas. It is a view that relates the structure of language to its special communication function, which is why it appeals to evolutionary biologists.

One of the things that Dor and other linguists have found is that the grammatical structure of phrases and sentences is associated with the types of concepts the words in sentences embody. It seems that when we use language, we automatically but unconsciously classify events and things into various categories, and treat these categories differently when we construct phrases and sentences. For example, the grammatical patterns we use depend on whether the participants in an event are active or inactive; on whether an action leads to a change in state of the object or it does not; on whether events are factual or hypothesized; on whether things are countable (like bottles and people) or not countable (like beer and fog); on when events happened (past, present, future); and so on. What is fascinating is that although there are endless ways of classifying things, events, properties, and so on, the categories that are reflected in differences in grammatical patterns are only a small subset of all those that we could use. So, although in all languages the distinction between active and inactive participants in an event is reflected in some aspects of grammar, as is the difference between factual and hypothesized events, and the difference between an action that leads to a change in state and one that does not, other categorical distinctions are not. In no language is the difference between the categories "friend" and "foe" marked by differences in grammar; similarly, the difference between the categories "boring events" and "interesting events" is not reflected in grammatical differences in any language. Other categories of things or events are grammatically marked in some languages, but not in all.

What Dor concludes from this is that language is structurally designed to communicate some things much better than others. Its design enables it to deal well with messages that are grounded in a rather constrained set of categories having to do with events and situations, their time and place, and the participants in them, all of which are reflected in grammatical structures. There is a core set of categories that are identifiable in all languages, although the way that they are indicated grammatically varies from language to language. In addition, different languages may structurally distinguish some categories that are not distinguished in others. Dor's view of language therefore encompasses both the universality and diversity that characterize language.

How Language Changed the Genes

After this rather long introduction about different people's views on what language is, we can get back to our original theme which was the evolutionary interplay of the genetic and cultural inheritance systems. Daniel Dor and Eva Jablonka realized that characterizing language in the way we have just described reframes the question of its evolution. It is no longer about how the Chomskian rules of UG came into being, or even about how the recursive mechanism of FLN became part of the human communication system. And it isn't about how a general-purpose language tool emerged as a by-product of the evolution of a large brain and improved general intelligence. Instead it is about accounting for the evolution of a functionally specialized and constrained communication system, in which a universally shared core of semantic categories is mapped into structural regularities in speech, and a further cluster of semantic categories are structurally marked in some, but not all, languages.

Dor and Jablonka see the evolution of language as the outcome of continuous interactions between the cultural and genetic inheritance systems, with both niche construction and genetic assimilation being important. We will outline the picture they paint by starting with a group of early hominids who have various ways of communicating—through gestures, grimaces, body language, and some restricted symbolic vocalizations. Their linguistic system is very simple, however, consisting of single-word utterances and short unordered strings of words. These people can certainly think and feel a lot more than they say: they have a good understanding of social relations, and can attribute intentions and wishes to other members of the group. As group-living animals, they have the need to communicate with each other and to share information, so their limited linguistic system is important to them (particularly when they have no visual contact with each other), and they use it frequently and comfortably. It becomes increasingly important as the group's culture develops, and they acquire more and more information that needs to be learned and communicated.

Imagine now that one or a few persons come up with a linguistic innovation. It might be a new word, or a grammatical structure such as a convention about word order that makes it clear who did what to whom, or a tag that when added to words indicates, for example, possession, or more-than-one. The innovation may arise by chance during the play of youngsters, or perhaps it is invented as a result of changed ecological conditions; maybe it occurs because the group size has increased and social relations have changed, or perhaps it is acquired through social interactions with another hominid group. Whatever its source, the innovation is still entirely cultural, involving no genetic change.

Most innovations, even potentially useful ones, never get incorporated into the group's language. Even if they are invented, innovations that relate to things like emotions or manual instructions rarely last long, because such things can be communicated far better through facial expressions, body language, or mime. The types of new vocalizations or structures that are eventually adopted are those that are good tools for communication, and are easy to learn, remember, and use. New words or structures that can be used in many situations usually survive better than those with a restricted use. For example, ways of indicating causal relations (the "because" words and structures) are rapidly adopted and widely applied. Even words and structures that are not particularly easy to understand and use sometimes become incorporated into the groups' language if they reduce ambiguity or enable information to be transmitted concisely.

Through use, the original inventions are improved and streamlined, and other new inventions that build on them are adopted. Gradually, as words and structures accumulate, the amount that has to be learned increases. Clearly, the expanding language is changing the social niche that the people occupy. They have to adapt to it. We can assume that the ability to use language is probably becoming more important for individual and group survival, because, for example, language is used when planning communal actions such as hunting, or information about poisonous or medicinal plants is communicated through speech. Better mastery of language may also affect a person's social and sexual status if good speakers play a more central role in group activities, and are therefore regarded as desirable mates. So, for all sorts of reasons, better language learners and users are at an advantage.

Notice that, so far, all of the language evolution that we have described has occurred through cultural changes. We now want to look at the impact of these changes on the genetic system, since it is reasonable to assume that the ability to learn to understand and use language is influenced by genes. Some individuals will have a genetic constitution that makes them better at acquiring and using the culturally expanded linguistic system, and through the selective advantage this confers, the proportion of these individuals in the population will rise. They will be people with better general intelligence, better memory, better voluntary motor control of vocalizations, and a more sophisticated social awareness; they will all learn to use language quickly and well. But more than this will be involved. Selection for the ability to acquire and use language will expose variation in people's capacity to remember *words* (rather than general memory), in their ability to recognize social intent expressed *through words* (rather than through other communication systems), and in their ability to relate the conceptual distinctions that are fundamental to thinking to the grammatical structures of phrases. For example, for reasons unconnected with language, our group of people recognize the difference between animate and inanimate, between present and future, and between male and female, but they vary in how well they can distinguish these and other categories and in their ability to link the categories to words and the way words are strung together to make grammatical structures. Those that are able to do these things well will be at an advantage when learning language, and will thrive and multiply because of it.

What we are describing here is partial genetic assimilation of the language faculty. The faster learners of the various facets of the sophisticated, culturally constructed linguistic system are at an advantage, and in time the population comes to consist of faster and better learners of the culturally evolved words and language structures. Just as with the evolution of the dance that we described earlier, the fact that there is partial assimilation of some linguistic features that formerly took a lot of learning frees up the linguistic learning capacity. More linguistic evolution through cultural innovation can now follow, more partial assimilation can occur, followed by more linguistic innovations, and so on. The assimilate-stretch principle is at work here.

Not all aspects of language are likely to be genetically assimilated—only those that are used repeatedly and consistently, and survive changes in people's conditions of life and social habits. Elements such as particular words or specific grammatical markers (say adding an *s* for plurals) change so rapidly that there is no way they can be assimilated—there simply isn't time. Cultural evolution moves faster than genetic evolution. It is far more likely that the conventions that correspond to the stable aspects of life— to the fundamental distinctions between categories of things or events— will be assimilated. But even the way core categories such as male/female, one/more-than-one, now/not now, animate/inanimate are distinguished is unlikely to be completely assimilated, because the ongoing process of cultural evolution puts a high premium on flexibility. The markers for these distinguishable categories will never become innate knowledge, but through partial genetic assimilation the rules will become very easy to learn. Something else will also happen: as genetic assimilation occurs, it will tend to channel and constrain future language evolution. It must now comply with the increasingly genetic component of its acquisition and use, not just the general constraints of perception and general intelligence.

Let us sum up. Dor and Jablonka's view is that language evolution has involved interactions between the cultural and genetic systems. It has resulted in a channel of communication that is more attuned to some categories of things, states, and events than others. Some of these categories are recognizable in all languages, and they are indicated by various grammatical structures and tags that reflect core distinctions between the categories. Linguistic evolution was culturally driven, but it constantly interacted with other cultural and genetic evolutionary processes that were going on at the same time, such as social and technological evolution, the evolution of the vocal apparatus, and the evolution of voluntary motor control of sound production, to name but a few. Cultural evolution led to the expansion of the environment as it was perceived by humans, and as a result individuals were faced with more information than they could learn and communicate. Through natural selection, some of the culturally

Figure 8.5
Language evolution: from chimplike utterances to a Japanese poem.

acquired features of language were genetically assimilated, so they needed less learning. The assimilate–stretch principle was very important in this: as old conventions were assimilated, new linguistic conventions could be learned. The process of linguistic evolution was thus an interactive, spiraling process, in which cultural evolution guided and directed genetic evolution by constructing a cultural niche that was constantly changing, yet kept some aspects stable. It is those stable aspects that have been partially genetically assimilated and resulted in languages manifesting a blend of universality and variability.

Dialogue

I.M.: I understand that your main intention in this chapter was to show how symbolic and nonsymbolic cultural evolution can drive genetic evolution. Cultural evolution constructs the environment in which genes are selected. Metaphorically speaking, culture is the horse that drags the genetic cart. I am prepared to accept that cultural evolution can be thought of as a Lamarckian process, if you insist on this terminology, but what I do not understand is what the Baldwin effect has to do with Lamarckism.

Why was it seen as a way of reconciling Lamarckism and Darwinism? It seems to me that it is a perfectly conventional Darwinian process.

M.E.: Yes, it is a Darwinian process. The Baldwin effect was thought to reconcile Lamarckism and Darwinism because it explained how, through selection, a learned or otherwise acquired character can become innate, or "inherited." Of course neither Lamarck nor Darwin had any need for such reconciliation. Lamarck took the inheritance of adaptations for granted, and Darwin thought that both selection and use and disuse are required to explain evolutionary change. It was in the context of the debate between the *neo*-Darwinians and the *neo*-Lamarckians that the Baldwin effect was seen as a reconciliation.

I.M.: The Baldwin effect and genetic assimilation seem to me to be very good explanations for the existence of mental modules! I'm not entirely clear about your position on this issue. You attacked the idea of modules quite a lot in chapter 6, but now you seem to be more well disposed toward them. You obviously believe that some have evolved, rather than just emerged, because you described inborn behaviors such as that of the hyena cubs who are afraid of the smell of never-encountered lions. You implied that this sort of behavior evolved though genetic assimilation, which makes sense. So what exactly is your position on this cognitive module issue? You have confused me.

M.E.: We accept that genetic assimilation leads to some modular organization of behavior. But we stressed that assimilation is in most cases incomplete—it is the extreme, rare, and spectacular cases that show complete assimilation. We can also easily envisage the opposite process, with selection for a wider rather than a narrower response. Initially selection might be for a better memory for edible food items (you might say a "food-selection module" is formed), but there might then be an advantage in generalizing this memory to other things, such as predators or competitors. We really do not have a "position" in the sense that we adhere to an exclusively modular or exclusively general-intelligence image of the mind. We believe that there are some domain-specific adaptations and some adaptations are domain general—that there is a spectrum. It's a boringly middle position, but this is what our understanding of biology leads us to believe.

I.M.: Is there any experimental evidence that behavior can be genetically assimilated?

M.E.: To the best of our knowledge, no experiments have been done. They could be done with *Drosophila*, because fruit flies can learn and remember. The assimilation of behavior has been modeled, however, and the models show nicely how a behavior that is initially learned by trial and error can be converted into an innate one for which no learning is required. There is also plenty of indirect evidence for the genetic assimilation of behavior, and it is the simplest way of explaining a lot of evolutionary changes. Assimilation makes it unnecessary to assume that there is a direct feedback from acquired or learned characters to the hereditary material, as naive Lamarckians postulated. It also makes it unnecessary to assume that random mutations and selection, unmediated by the organism's behavior, produce adaptive "instincts."

I.M.: I can see that canalization and genetic assimilation are central to your arguments. But what about selection for plasticity—for increasing the ability of individuals to respond to different conditions? Surely this is as important as increased canalization. Why don't you stress this aspect of evolution?

M.E.: You can use the same line of reasoning to explain the evolution of increased plasticity as that which we used to explain canalization. The starting point is an environment that is changeable or constantly has new components introduced into it—new predators or new potential food sources, for example. Such environmental conditions will unmask variations in the capacity of individuals to make adjustments to changed conditions—variations that are not apparent in unchanging environments. Those individuals that are able to respond to varying conditions are more likely to survive and reproduce, so genes enabling this will increase in frequency. Selection for this type of plasticity has probably been enormously important in evolution, but it presents no special theoretical problems for evolutionary biologists. It makes intuitive sense—there is no enigma such as there is with moving from flexible "learned" or "acquired" characters to "inherited" ones. Nevertheless, both genetic assimilation and selection for increased flexibility involve the same basic principles—the unmasking and selection of previously hidden variation. West-Eberhard, in particular, has stressed the generality of such processes in evolution, and has suggested the term "genetic accommodation" for the genetic stabilization through selection of new phenotypic responses.

I.M.: Let me get back to the evolution of culture, which seems to have involved something like this genetic accommodation idea. You maintain that cultural changes have influenced the selection of our genes and the evolution of our language, but I wonder whether this is really so. Was there really any significant "cultural evolution" in the early evolutionary history of *Homo*? Just because we see a lot of progressive cultural change today, it doesn't mean that this was so 2 or 3 million years ago. And if it wasn't, your arguments about the evolution of language collapse.

M.E.: In chapter 5 we described a lot of animal traditions. Traditions can be seen in every aspect of animal life—in modes of foraging, criteria for selecting mates, ways of avoiding predators, decisions about where to live, practices of parental care, the use of communicative signs, and so on. Wherever you look, you see them. They are particularly evident in higher primates, our evolutionary cousins. We mentioned an analysis of long-term studies of nine African populations of common chimpanzees, which showed that there are thirty-nine different cultural traditions, five of which have something to do with communication. If chimpanzees have evolved traditions, surely it is reasonable to assume that early *Homo* did too. We are not making very many assumptions about hominid cultural evolution, especially in its early stages. The pace of cultural evolution may have been rather irregular, and different aspects of culture may have evolved at different rates, but we do think that to deny that cumulative cultural change—"cultural evolution"—has occurred throughout the evolution of our lineage would be absurd.

I.M.: I suppose that is what is behind your reluctance to accept Chomsky's emergent perfect language organ. You insist on something more gradual, and I sympathize with your attitude up to a point. The problem with any "sudden-emergence" view of language is that the first person with the mutation, the lonely genius, would need someone to communicate with.

M.E.: You can argue, as Hauser, Chomsky, and Fitch did, that you start with an FLN or something similar that has nothing to do with communication—something that was selected because it organized the ability to navigate better, for instance. The "linguistic mutation" that allowed the FLN to interact with the conceptual-intentional and the sensorimotor subsystems could become more common simply because the offspring of the genius inherited her gene. This fortunate family and lineage would be able to convert whatever primitive communication signs their less well-endowed neighbors used into real language, and would have an enormous advantage on many fronts.

I.M.: You still need a kind of genetic miracle—the great mutation, the great emergence, the Big Bang. Is this assumption really necessary? You said that some Chomskians do think that language evolved. If you accept the basic Chomskian assumption that UG or FLN is an abstract computational system, you can imagine that its rules accumulated gradually, in the context of the evolution of communication. If someone invents a completely arbitrary rule that helps to avoid ambiguity in communication in general, and if others learn and adopt it, surely this will help communication enormously, and the rule and its users will be selected. Even seemingly crazy rules may be better than no rules at all! The most important thing is that a shared useful convention is established. The nature of the convention, whether it is arbitrary or not, is much less important. So there would be selection for conventions, then full genetic assimilation, then the invention of another meaning-free convention, then genetic assimilation again, and so on. Why is this not enough? You'll end up with a whole set of abstract, meaning-blind, but very effective rules.

M.E.: Something like this has been suggested by the American linguist Steven Pinker and his colleague Paul Bloom, a psychologist. But if you really think about which conventions or rules people invent and accept, you will realize that it is the conventions that help people think and communicate about their everyday experiences and needs that have the greatest chance of being both invented and accepted. The process of invention is not some kind of random neural firing, with luck alone allowing you to hit the target. Think about two rules that allow better mutual understanding: one is a meaning-blind rule, the other is meaning-related. Suppose the problem is recycling glass bottles, and there are two possible rules. The first is "put blue glass in the first bin, green glass in the second, brown glass in the third, and clear glass in the fourth"; the second rule is "put blue glass in the blue bin, green glass in the green bin, brown glass in the brown bin, and clear glass in the white bin." Which rule has a better chance of being invented and getting established? We think the latter. Surely the empirical finding that a lot of grammatical structures are *not* meaning-blind and are based on the way people think about the world is to be expected. Even if a few structures that were arbitrary did become established just because people started using them and they were understood, to believe that *all* grammatical markings are meaning-blind is stretching it.

I.M.: But it seems that an awful lot of important categories of things and events are *not* associated with special grammatical rules. You said that the difference between friend and foe, and between boring events and interesting events, is not associated with particular grammatical rules in any language. I'm not surprised that the boring/interesting difference is not recognized—I doubt our early ancestors had time to be bored, so it wouldn't have figured much in their thoughts. But surely the category "foe" was important. So why didn't it become grammatically marked in some way?

M.E.: As you know very well, asking why something *did not* evolve is not a very proper question! But if you want our speculation, the "foe" category wasn't grammatically marked because it was always indicated in a different way—by the tone and pitch of the voice. Grammatical marking would therefore be redundant.

I.M.: You know, it seems to me there isn't such a big difference between your position and the Chomskian one. You both accept that there are preexisting biases that make language learning easier than it would otherwise be. You just don't want to call it a language organ, although I am not sure why not. You differ, it seems to me, only in the way you characterize the bias.

M.E.: There is a superficial similarity, in that we agree that the human nervous system is biased to develop in a "linguistic way." However, Chomsky postulated that there is a distinct "language organ." We cannot accept that, although like Chomsky we don't agree with the extreme functionalist views. We can accept a *limited* emergence position—that the linguistic system got started as the result of the evolution of and coming together of a more powerful memory, better voluntary motor control of the vocal system, more sophisticated categorization of social relations and events, more varied vocal communication (including vocal imitation), and so on. We can even imagine how such a combination could have allowed the organization of signs into simple phrases. But we believe that this was merely the basis of language, and that the more special linguistic regularities could not have emerged in this way. They would need to evolve by cumulative natural selection—mainly cultural selection, but some genetic too. The other main difference between our position and Chomsky's is that he postulates a formal, nonfunctional UG, whereas the theory that we prefer, that of Dor and Jablonka, is based on a specifically functional characterization of grammar.

I.M.: Can you then accept the idea that there is a general computational mechanism in the brain—an FLN or something like it—that allows recursion, is generative, and can link different domains? I would have thought that a general mechanism that is used for other functions as well as language would fit nicely with your ideas, particularly since you think that language is just a very special type of symbolic communication.

M.E.: We are not sure that this distinct, abstract FLN isn't a mirage. We see no obvious reason why recursion, which seems to be the essence of FLN, should not have evolved in the context of social communication (particularly gossip), rather than somewhere else. Sit on any city bus and you will overhear something like "Did Jack tell you what Victoria told Mark that Jill had heard James say . . . ?" Gossip and recursion go together! Another problem with FLN is that it doesn't seem to provide any explanation for the baroque complexities of the inherited parameters that are a feature of Chomsky's UG. Even if we accept that something like FLN is a component of language, we believe that it must have evolved in association with the conceptual and sensorimotor systems. We cannot accept that it just happened to fit and organize the other systems into a new and perfect linguistic supersystem. It had to adapt to them, especially to the conceptual system. If it did, the rules cannot be entirely abstract and meaning-blind.

I.M.: The rules-as-emerging-from-interactions view seems to put you closer to the functionalists. Yet you seem to contradict yourself about this. On the one hand, in chapter 6, you convinced me that a combination of existing features can lead to a new and very specific talent like literacy. On the other hand, you do not accept the functionalists' explanation of the evolution of language as something that is based on general-purpose cognitive tools, with the specificity of language being the consequence of the peculiarities of the evolution of a large human brain, the anatomy of the vocal apparatus, the constraints of speech production, and general learning mechanisms. I can't see what is wrong with the functionalists' reasoning.

M.E.: If you can derive the unique grammatical regularities of language, its speed of acquisition and use, or even the de novo invention of language, solely from considerations of general intelligence, brain development, motor development, and cultural evolution, we will be delighted. It will be most parsimonious and elegant. No one has yet done it. We accept the functionalists' general arguments, but we think that they do not explain sufficiently well why we can acquire and use language so easily early in life. We think that to account for these it is necessary to postulate that rules relating to core universal concepts have been partially genetically assimilated.

I.M.: Why not assume that the basic genetic evolution of language was determined by general intelligence and by the conceptual "guidelines" that our perception, motor and social behavior dictated, while the complex UG rules were driven by cultural evolution?

M.E.: We cannot rule that out, but we find it difficult to believe, first, because of the speed of early linguistic learning; second, because of the *limitations* of linguistic communication; and third, because of the inevitability of gene-culture coevolution when the process is lengthy and directional, as language evolution is likely to have been.

I.M.: But couldn't a linguistic rule that is common to all languages simply be the result of it having arisen independently in each of them through cultural evolution?

M.E.: It is possible that *some* of the grammatical regularities that are present in all languages are the result of convergent cultural evolution. The common, *already evolved* structure of language probably imposes very strong constraints on new innovations. So all languages could have homed in on similar, new, culturally invented structures. But this would occur only after the language system had already evolved the basic rules, which would then guide the formation of all new ones.

I.M.: You know, you left out quite a lot from your evolutionary scenario. You didn't say much about what made linguistic behavior so advantageous, or about how language first emerged. Surely both are rather important for your story.

M.E.: That is true. We hinted that social evolution was very important in the process of language evolution, and that you would need to think about several parallel advantages, because language is, and probably was from its inception, multifunctional. A lot of interesting work is being done on the type of selection that may have been involved, and on the stages of language evolution, but we didn't try to deal with every aspect of the problem, just with gene–culture interactions. We have avoided the question of evolutionary origins not only with respect to language but also for all the other inheritance systems we have described. We are going to tackle the problems of origins in the next chapter.

9   Lamarckism Evolving: The Evolution of the Educated Guess

We hope that readers who have reached this penultimate chapter are by now convinced that DNA is not the be all and end all of heredity. Information is transferred from one generation to the next by many interacting inheritance systems. Moreover, contrary to current dogma, the variation on which natural selection acts is not always random in origin or blind to function: new heritable variation can arise in response to the conditions of life. Variation is often *targeted*, in the sense that it preferentially affects functions or activities that can make organisms better adapted to the environment in which they live. Variation is also *constructed*, in the sense that, whatever their origin, which variants are inherited and what final form they assume depend on various "filtering" and "editing" processes that occur before and during transmission.

Some biologists have great difficulty in accepting this "Lamarckian" aspect of evolution. To them it smacks of teleology, seeming to suggest that variations arise for a purpose. It appears as if the hand of God is being introduced into evolution by the backdoor. But of course there is nothing supernatural or mysterious about what happens—it is simply a consequence of the properties of the various inheritance systems and the way they respond to internal and external influences. We know, however, that we have left something out—something that may make it look as if we are still retaining naive teleology. We have assumed all along that the inheritance systems through which potentially adaptive variations are generated and transmitted *already exist*. Could it be that in assuming this we have slyly introduced some mysterious intelligence into evolution right at the beginning? If we are to get rid of any suspicion of invoking the hand of God, we need to explain how such intelligent systems came into being in the first place.

Before we start looking at the origins of the systems that introduce instructive elements into evolution, we want to highlight an old and well-established piece of evolutionary biology, which is central to a lot of our arguments. It is that many new adaptations begin as by-products or modifications of characters that were originally selected for very different functions. For example, the vertebrate jaw had its origins in the skeletal elements that supported the gills (respiratory organs) in primitive jawless fish. The front gill support was gradually co-opted for new, feeding-related functions, and became the jaws of later fish. Modification did not stop there: in later vertebrates, changes in the manner of feeding led to altered jaw articulations, and freed up some of the bones at the hind end of the jaw. These were put to a new use—to carry vibrations. They eventually ended up as the three tiny bones in our middle ear. So what started as breathing aids became feeding aids, and the feeding aids later evolved into hearing aids. A structure that was originally selected for one function evolved to have a very different one.

Often when an existing structure is recruited for a new job, its old function is not lost, so it ends up with several functions. Mammalian hair is a good example: it probably evolved originally for insulation and temperature regulation, and for many mammals this has remained its most important job, but in some it also has a role in courtship displays and camouflage. In addition, some hair now has a sensory function. The "whiskers" on a mammal's snout are hairs that have evolved to become exquisitely sensitive to touch. We have shown this and other examples of evolutionary modification and diversification of function in figure 9.1, and in the following sections we describe how comparable changes have contributed to the evolution of the various heredity systems.

The Origin and Genetics of Interpretive Mutations

Much has been written about the origin of life. It was once seen as a metaphysical problem, but it has now acquired scientific status, and scientists working on prebiotic evolution are closing in on plausible and testable theories about how living things came into existence. We are not going to dwell on these because the origin of life is really outside the scope of this book. We want to start at the point where, through a complex process of chemical and biochemical evolution in which natural selection probably played a big role, DNA had become the hereditary material. Natural selection continued to affect its organization and activities, and at some stage it led to the mechanisms that generate what in chapter 3 we called "interpretive mutations," which introduced a Lamarckian dimension into evolution. The question is, how did these mechanisms evolve? How did it

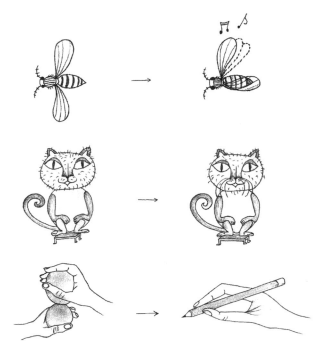

Figure 9.1
Evolutionary changes in function. At the top, fruit flies vibrate their wings, structures that were originally selected for flight, to produce a courtship song. In the middle, mammalian hair, originally selected for the regulation of body temperature, has a sensory function in the whiskers. At the bottom, the human hand, selected for manipulating objects, now plays a role in symbolic communication.

come about that genetic changes sometimes occur preferentially at a particular site in the genome or at a time when they are likely to have useful effects?

When we described the systems producing interpretive mutations, we said that there are mixed views about the massive genome-wide increase in mutation frequency that is sometimes seen in cells experiencing a severe stress. Some people doubt that it is an adaptation at all. They maintain that it is pathological—just a reflection of the fact that everything in the organism is collapsing, including the systems that maintain DNA. The spate of mutations is simply the result of the cell's failure to repair its DNA properly, they say. Other people take a very different view, arguing that the ability to produce this burst of mutation is an adaptive response, which has been fashioned by natural selection.

There is one particular cellular system, known as the SOS response system, that is particularly interesting and relevant to the two opposing views about stress-induced mutation. As its name suggests, the SOS system cuts in when things get really desperate, and a cell cannot reproduce. If, in spite of all the maintenance and repair that goes on, DNA damage is so severe that it arrests the process of replication, a protein that recognizes the problem is activated. This activated protein destroys the molecules that are associated with the control regions of SOS genes and keep them inactive. Once these repressor molecules are removed, the SOS genes are turned on. The proteins they encode enable a bypass and patch-up job to be done on the DNA, so that replication can be completed. However, the resulting DNA copies are very inaccurate. Therefore, although the SOS system allows replication to continue, it acts as an inducible global mutator system, causing numerous changes in daughter DNA molecules.

Many of the genes that have a role in the SOS response have other cellular functions, so we can only guess at the origin of the system and how natural selection has molded it. However, its present activity has led the French microbiologist Miroslav Radman to call some of the genes involved "mutases." He believes very firmly that, whatever the origins of the SOS system, it has now evolved to become a system to control mutability, increasing the mutation frequency in times of stress. It was selected because lineages in which chance genetic changes linked the error-prone DNA repair systems to cellular changes caused by stress survived better than others did. Although their enhanced mutation rate meant that in stressful conditions most cells died, a few survived because they were lucky enough to produce mutations that enabled them to cope.

Obviously, since most mutations are likely to make things worse rather than better, it would be much more efficient if the extra mutations induced by stressful conditions were restricted to those genes that, if changed, could rescue the cell. And we know that there are situations where cells do produce mutations not only at the right time but also in the right place. In chapter 3 we described Barbara Wright's work, which showed that the bacterium *E. coli* can sometimes make very appropriate mutational guesses. She found that when starved of a particular amino acid, the bacteria increased the mutation rate in the very gene that might, if mutated, enable the cell to make the amino acid missing from its food. This is certainly an adaptive response, and it is not too difficult to think how it may have evolved. Selecting genetic changes that link the existing mechanisms that turn genes on and off to the error-prone mechanisms that repair DNA would do the trick. In times of stress, the inducible system that turns a gene on would then also turn on the production of mutations in that gene.

It is even easier to imagine how another of our categories of interpretive mutations evolved by natural selection. This is the one that is often found in pathogenic microorganisms, where there is a consistently high rate of genetic change in certain restricted regions of the genome. In these "hot spots," mutation is going on all the time at a rate hundreds of times faster than elsewhere. This is not the disaster that it would be at other sites in the genome, because the genes in hot spots code for products that need to change frequently. Think, for example, about a pathogenic organism that is constantly at war with its host's immune system. The immune system recognizes the pathogen by its protein coat. The pathogen can evade detection for a while if it changes its coat, but the host's immune system will soon catch up, and recognize the new coat. Yet another coat change is needed. The pathogen has to constantly keep one step ahead of its host's defenses. If its coat protein genes are in mutational hot spots, then there is a good chance that it will be able to do so.

Mutational hot spots exist because certain types of DNA sequence are much more prone to replication and repair faults than others. For example, short repeated sequences in and around genes tend to cause problems, because replication enzymes "slip," missing out or duplicating some repeats. They make the region error-prone. If there is an advantage in frequent genetic change, then short repeats or other sequences that make the DNA liable to be replicated or repaired inaccurately will be selected. These sequence features originate by chance, but they will be retained and accumulate if by making genes more mutable they enable the lineage to survive in ever-changing conditions.

From this rather sketchy account it should be clear that there is really no great mystery about the origins of the systems that generate interpretive mutations. Some probably began as by-products of emergency DNA repair systems, which sometimes became linked to the systems controlling gene activity, whereas others had their origin in common and random DNA changes such as the introduction of small repeated sequences. That is all that was necessary to form the initial, accidental, and crude versions of the interpretive mutational systems, with their seemingly "purposeful" anticipatory effects. Of course, their subsequent evolutionary elaboration required further selection, but there is no reason to doubt that such selection would have taken place.

The Origin of Epigenetic Inheritance Systems and the Genetics of Epigenetics

In chapter 4 we described several very different epigenetic inheritance systems. What they had in common was that all of them can transmit non-DNA information from mother cells to daughter cells, and they all provide ways through which induced changes can be inherited. These EISs play a key role in the development of multicellular organisms, where the ability of cells to pass on information about their functional states is indispensable, yet they are also found in protozoa. This suggests that even though they seem to be tailored for multicellular life, EISs are really very old. If so, to understand their origins and subsequent evolution, we have to ask about their role in ancient unicells. Of what advantage was it to these cells to be able to pass on their epigenetic state to daughter cells?

One possible answer is that EISs were selected because they enabled early cells to survive conditions that were constantly changing. For a lineage in such an environment, switching among several alternative heritable states was probably an advantage. While cells in one state survived through one set of conditions, those in other states did better in different circumstances. The genetic system can transmit alternative states, but there are several reasons why epigenetic variants are often more appropriate. The first is that the rate at which they are produced is usually far greater than for mutations; if conditions change frequently, this can be a significant advantage. The second and related reason is that epigenetic variants are often readily reversible, whereas mutations usually are not. The third reason is that the production and reversion of epigenetic variants may be functionally linked to the changing circumstances; this is rarely true for mutations.

Constantly and erratically changing conditions may therefore have been part of the driving force for the evolution of epigenetic inheritance, but we believe that the more predictable aspects of the environment were of even greater significance. Almost all organisms live in environments that cycle in a fairly regular way: there are daily cycles of light and darkness (usually accompanied by temperature changes); there are cycles associated with tides; there are seasonal cycles, and so on. If organisms are to survive, they have to adapt to these regularly changing conditions. They do so in different ways. Long-lived organisms, which endure through many cycles of change, usually adapt physiologically or behaviorally. For example, as winter approaches, many plants shed their leaves, some animals hibernate, mammals often grow thicker fur, and birds sometimes migrate; in spring everything changes back again. In contrast to this physiological and behavioral adaptation, very short-lived organisms, which go through many generations during each phase of the cycle, may adapt genetically: through natural selection, genetic variants that are appropriate for one phase are replaced by those appropriate for the other phase. Between these short-lived and long-lived organisms are the types that interest us, those which go through several (but not many) generations during each phase of the cycle. Imagine, for example, a cycle with two phases, ten hours hot and ten hours cold, and a unicellular organism that divides every hour. What is the best strategy for this organism in these conditions, where each phase of the cycle is ten times longer than the generation time?

The answer, which seems intuitively obvious but is backed up by mathematical models (intuition is not always a safe guide in evolutionary thinking), is that the best thing for the organism to do is to switch state once every ten generations. To do so, it needs a memory system that will retain and pass on a newly acquired state for ten cell divisions (figure 9.2). If it has such a system, offspring will usually inherit the currently adaptive state, so they will not need to invest time and energy in finding the appropriate response to their present conditions. They will get the information "free" from the parent cell. Natural selection will therefore tailor the stability and heritability of the different functional states to the length of the phases in the cycle. Because the organisms go through only a few generations in each phase, it is the epigenetic systems, with their high rates of variation and reversion, that natural selection is most likely to hone into efficient memory systems. Mechanisms that generate precise directed DNA changes might also evolve in such conditions, and indeed may be the basis of cell memory in some unicellular organisms. Nevertheless, the adaptive flexibility of epigenetic regulatory systems would have made them prime targets for modifying into the memory systems of unicells living in regularly cycling conditions.

There are clearly several reasons why epigenetic inheritance systems would have been targets of selection in early unicellular organisms, but so far we have avoided a critical part of their evolution. We haven't looked at how they all got started. This is what we intend to do in the remainder of this section. Because the four types of EIS that we described in chapter 4 are so different, we will deal with each of them separately, asking how the system originated and how it was later modified.

Self-Sustaining Loops
Guessing at the evolutionary origins of this type of EIS is not difficult, because transmitting self-sustaining loops is an almost inevitable con-

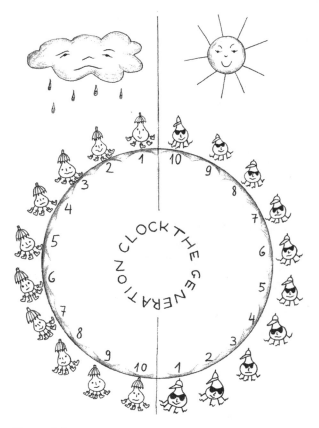

Figure 9.2
The generation clock. When living in an environment that switches regularly every ten hours, a unicell that divides every hour evolves a memory that enables it to retain its adapted state for ten generations.

sequence of cell growth and multiplication. We described a simple positive feedback loop in chapter 4. Basically, a system is self-sustaining if, once a gene is turned on by some internal or external inducer, its product interacts (not necessarily directly) with the gene itself and maintains its activity. With such a system, the gene remains switched on for as long as enough of its product is present. If for some reason the concentration of product falls, the gene switches off and stays off.

When a cell in which the gene is active divides, its two daughters share the molecules of the gene product that are present, and, provided each receives enough, the product will maintain the activity of the gene in both daughter cells. The transmission of the gene's active (or inactive) state is thus an automatic consequence of the nature of the feedback loop and cell division: inheritance is simply a by-product of the system that maintains gene activity within the cell. If inheriting the mother cell's state of gene activity is an advantage, then anything that makes it more reliable will be incorporated. For example, natural selection may favor genetic changes that make the gene product more stable, or that make the interaction between the gene's regulatory region and the product more reliable, or that increase the amount of the gene product in the cell and thus ensure that both daughter cells get enough to maintain activity.

Most self-sustaining loops are part of large and complex cellular networks involving many genes, each engaged in several regulatory interactions. In such networks, the number of genes and how the gene products interact with other genes and with each other determine the behavior of the system. Using simulations based on theoretical models of regulatory networks with some very simple properties, the American biologist Stuart Kauffman has shown that networks that are made of many interacting genes, each regulated by a few others, can assume several different states, each of which is very stable and is transmitted automatically following cell division. Dynamic stability is, in Kauffman's words, "order for free," because it is an inevitable consequence of the regulatory structure of the network. The system can be fine-tuned by natural selection, which adjusts the components and their connections, but most new mutations do not substantially alter the states that emerge from the web of interactions.

Structural Inheritance
The origins and the original advantages of the second type of EIS, the one that reproduces three-dimensional structures and transmits them to daughter cells, are also easy to envisage. We know that as cells grow and multiply, complex molecular structures, such as those that form the internal skeleton of cells or their membranes, assemble and disassemble. The process of construction is probably quicker and more reliable if existing structures are used as templates or guides for the assembly of new ones. An existing structure can be thought of as a kind of cast, which attracts free components to it. If such "guided assembly" improves cell function and maintenance, or stabilizes interactions between cells, then natural selection will act on genetic changes that affect the chemical and topological properties of the components, altering their affinity for preexisting structures and for each other. In early cell evolution, this type of selection would not only have improved cell maintenance, it would also have been important for the growth of structures following cell division. In other

words, structural inheritance, like the inheritance of self-sustaining loops, was probably an automatic by-product of systems that were selected to maintain cellular structures and functions. Subsequently, templated assembly was stabilized and improved through the selection of genetic variations that affected the structural properties of the component molecules.

Some of the work from Susan Lindquist's lab has shown how quite simple genetic changes can enhance or reduce the chances that a protein will adopt a structure that has self-templating properties. In chapter 7 we described the Lindquist group's studies of a yeast protein with two alternative conformations, the normal one and a rather rare prion form [PSI⁺], which affects the fidelity of protein synthesis. As part of this study, Liu and Lindquist constructed strains with mutations that altered the number of copies of a short string of amino acids that is normally present in the protein in five not quite identical copies. They found that whereas the normal protein with its five imperfect repeats has a low probability of becoming a prion, the addition of two more copies made the prion conformation, with its self-templating properties, much more likely. In contrast, deleting four of the repeats led to a loss of the protein's prion-forming tendency. Something similar has been found with the gene for the protein that forms mammalian prions: alleles that increase the number of copies of a string of amino acids in the protein also increase the likelihood of developing the prion disease. This does not mean that repeated amino acid sequences inevitably turn a normal protein into a potential prion, but these examples do indicate the kind of genetic change that can provide selectable variation affecting the conformation and templating properties of proteins.

The nature of the variation that was selected during the evolution of more complex heritable structures, such as the cortical organization of *Paramecium*, is still largely a mystery, but Tom Cavalier-Smith has developed some intriguing ideas about the evolution of complex cell membranes. Given appropriate conditions and the right kind of lipid molecules, simple membranes can form spontaneously, but the protein-containing membranes found in today's organisms cannot. Very early in biological evolution, membranes ceased to be capable of self-assembly, and became structural inheritance systems, in which existing proteins and lipids guide the insertion of similar molecules to form more of the same type of membrane. According to Cavalier-Smith, these "genetic membranes," as he calls them, have been of major importance throughout evolution, from the origin of the first protocells to the emergence of present-day eukaryotic cells. Like most biologists today, Cavalier-Smith accepts the formerly scorned nineteenth-century idea that various types of bacteria were incorporated and integrated into the ancestors of eukaryotic cells, eventually to become cell organelles that are now essential for survival. From his work with single-cell organisms, he has concluded that the membranes of these former bacteria have retained their basic structure and composition, even though the genomes of the original bacterial cells no longer exist. Like other genetic membranes, the ancient bacterial membranes were and are enormously stable. Cavalier-Smith believes that the additions and losses of different types of heritable membranes were probably rare events in the history of life, but when they did occur, they were of fundamental importance.

Chromatin Marking

Chromatin undoubtedly plays a major role in the storage and transmission of cellular information, but guessing how it evolved to become a system that responds to environmental and developmental cues and transmits information about the response to daughter cells is not easy. Part of the problem is that we still know very little about how the histones and other chromatin proteins interact with each other and with methylated and nonmethylated DNA. What we do know makes it obvious that many different selection pressures must have been involved in the evolution of chromatin. Its structure would have been affected by selection for packing long DNA molecules into small nuclei, for mechanisms that accurately distribute replicated DNA to daughter cells, for the ability to modulate gene expression, for protecting DNA, and so on. With so many different factors influencing the evolution of chromatin, working out how heritable marks fit into the overall picture is difficult.

One type of chromatin marking which has been the subject of a lot of evolutionary speculation is methylation, the first system to be described. A case has been made for this EIS being a by-product of a system that evolved primarily to defend cells against foreign and rogue DNA sequences. Present-day fungi, plants, and animals all have a methylation-based defense system, and, although it is not exactly the same in all groups, this wide distribution suggests that it is very ancient, predating the time when the major kingdoms diverged. Today, DNA methylation seems to act as part of a genomic immune system, in which cells detect foreign DNA sequences and render them harmless. The system is needed because when viruses invade a cell they hijack the host cell's resources and multiply, and copies of viral-type DNA sometimes become integrated into the host's genome. If this happens, then provided several copies of the foreign DNA are present, the host cell is able to recognize that it doesn't belong, and methylates it. The same treatment is given to rogue DNA sequences such as transposable elements, which multiply and spread within the genome rather like cancer cells within a body. They too are methylated and kept inactive. Exactly how the methylation machinery recognizes duplicated DNA sequences is not known, but such sequences can produce unusual paired conformations, and it may be these that turn them into targets for methylation. Once methylated, the DNA attracts specific types of protein that bind to it and prevent its transcription. Foreign and rogue sequences are thus silenced, and, unless something goes wrong, they can do no more damage. Obviously, since keeping unwanted DNA sequences in check is essential if cell lineages are to survive, in the early evolution of such a defense system any genetic change that made a protein function as a methylase, helping to rapidly re-establish the methylated state after DNA replication, would be selectively favored.

If DNA methylation was indeed selected originally for its role in genome defense, it could then have been recruited for the regulation of normal gene expression and cell memory (figure 9.3). Repeated sequences, perhaps derived from foreign DNA, which were situated in or around a gene whose

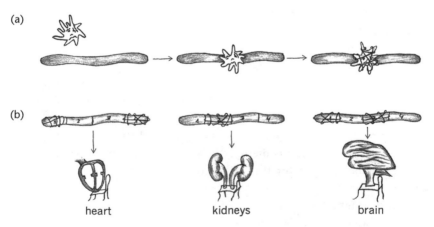

Figure 9.3
Methylation in defense and regulation. In (*a*) a foreign genetic element (e.g., a virus) invades the cell (left), and its DNA gets integrated into the host's genome (middle), but the cell's evolved defense system (right) silences the integrated DNA by methylating it (the element is enchained). In (*b*) the methylation system is used to silence normal genes in various tissues where their products are not wanted.

product was not always required, could have become signals for methylation and inactivation. The heritability of the methylated state, which ensures that once DNA is inactive it remains that way in daughter cells, would enhance its value for gene regulation. Gradually the system would evolve further, with other DNA sequences becoming targets for methylation, and more proteins that are able to recognize methylated DNA taking part in the control of silencing.

The problem with the idea that the methylation system originated as a cellular defense mechanism is that it leaves a lot of observations unexplained. It does not account for the observation that the methylation silencing machinery seems to come into play only when something else, such as chromatin remodeling through histone modification or the association of regulatory proteins, has already shut down gene activity. It looks as if methylation silencing merely stabilizes and maintains an already established state, rather than being involved in initiating it. Another problem for the genome defense idea is that germ-line DNA often has reduced levels of methylation, yet the germ line is the very place where defense is most needed, because movement of transposable elements in germ-line cells could increase the burden of mutations in future generations. The genome defense idea also offers no clues about why there are such different levels of methylation in different species. For example, there seem to be no methylated bases at all in the nematode worm *Caenorhabditis elegans*, and in the fruit fly *Drosophila* there are so few that they remained undetected for years. For these and other reasons, we doubt that methylation marking originated primarily as a defense system, or that silencing foreign and subversive DNA is its primary function in most organisms today. We think it is more likely that its role in stabilizing gene regulation preceded or accompanied its use in genome defense, and that it became involved in defense only after protein marking systems had been established.

Protein marks were probably originally transmitted by accident, through remnants of DNA-associated protein complexes remaining attached to the old DNA strands during replication. Having hitchhiked their way to daughter DNA molecules, they nucleated the formation of new aggregates of proteins, thereby helping to re-establish the same chromatin structure as in the parent chromosome. If the inheritance of such rudimentary protein marks was beneficial, selection would have favored changes in the proteins that enhanced their transmission. Selection would also have acted on DNA: repeated DNA sequences, which offered many similar binding sites for a control protein, may have made chromatin inheritance more reliable, as would a specialized system of DNA marking such as methylation. Enzymes that methylate various DNA bases are present in bacteria, so they probably existed prior to the origin of eukaryotes, and could have been modified and incorporated into the chromatin marking system.

The evolution of all types of heritable chromatin marks must have been closely associated with variations in the DNA sequences that carry them. Since only a subset of marks is heritable, we have to ask what type of DNA sequences are capable of carrying the "stubborn" marks that are transmitted to subsequent generations. In present-day organisms, the stubbornness of marks is associated with DNA features such as the density of CG nucleotides and the number of repeats of various short sequences. Arrays of tandem repeats and clusters of CG sites now carry many of the marks that survive the restructuring of chromatin during mitotic and meiotic cell division and the development of the early embryo. In early evolution, variations in such sequences, which can arise in many ways, including through errors in replication and through the activities of transposons, would have provided the raw material for the selection of heritable states of gene activity that depend on differences in chromatin structure.

RNA Interference

The last of our EISs, RNA interference (RNAi), seems to be ubiquitous, so it probably has very ancient origins. We have already referred to the possibility that it evolved as a genomic immune system (see chapter 4), but we'll outline the idea again here. With this system, double-stranded RNA structures are cut into fragments, which then mobilize enzyme systems that destroy all copies of the original RNA transcript and others with the same sequence. Sometimes they also cause the methylation and inactivation of the gene from which this RNA is transcribed. In a way that is not understood, in some organisms elements of the mobilized silencing system can move from cell to cell, so that once initiated, silencing spreads. Since RNA viruses and transposons often produce double-stranded RNA during replication, it has been argued that the RNA interference system might have evolved to contain their activities. This defense function then became involved in the long-range and long-term regulation of gene expression.

Although this evolutionary scenario is plausible, we suspect that even more ancient regulatory functions preceded the defense function of small RNAs. Many biologists believe that in early evolution, long before DNA took over the leading role as information carrier, life centered on RNA. In this "RNA world," RNA was both a carrier of information and an enzyme that affected molecular reactions. It is not difficult to imagine that in these circumstances, where there were networks of RNA-mediated interactions, natural selection could have led to some RNA molecules responding to changes in conditions in a way that inhibited the activities of other molecules with a similar sequence. They might have modified the structure of these molecules by base-pairing with them, for example. Later in evolutionary history, as the division of labor between nucleic acids (eventually DNA) as information carriers and proteins as the main enzymes and regulatory molecules increased, vestiges of earlier RNA control systems may have remained. These could have become modified to fit the new information system and defend it against foreign RNA and DNA sequences.

What we have just said is very vague and speculative, and based on no evidence whatsoever. The reality is that biologists know so little about how the RNAi system works that it is premature to speculate at all about its origins. However, as with the chromatin-marking EISs, we are reluctant to accept the view that RNAi evolved primarily for genome defense, simply because we see no good reason why it should not have been selected as an epigenetic control system that contributed to cell heredity right from the beginning. We firmly believe that even in very primitive single-cell organisms the capacity to transmit existing cell states was often an advantage, and systems that enabled this would have been selected. We suspect that one of the reasons why there has been so much emphasis on the idea that DNA methylation and RNAi evolved as defense systems stems from the context in which they have been studied. Much of what we know about them, especially about RNAi, has been discovered because they have "defended" plant and animal cells against experimenters trying to modify genes and gene expression. The DNA or RNA the scientists insert is frequently silenced. Since RNAi and DNA methylation so successfully impede the efforts of human genetic engineers to manipulate the cellular machinery, it has been easy to assume that they evolved to limit the activities of the earlier cell manipulators—rogue DNA sequences and viruses. But even if defense is now an important role of these systems, it is not necessarily the original one or the primary one in all present-day organisms.

The Origins of Animal Traditions: Selection for Social Attention and Social Learning

We are in less murky waters when we leave epigenetic systems and think about the origins of behavioral inheritance, because with behavior it is easier to make use of one of evolutionary biologists' favorite tools, comparisons among existing species. Some animals transmit information through their behavior, whereas others seem not to, so we have a chance of discovering the circumstances in which such behavior is an advantage. We have to be a little careful, of course, because often we do not know whether animals are transmitting information behaviorally: what is information for them may not seem like information to us, and vice versa.

We really have no idea what sort of information and how much information is transmitted through behavior-influencing substances present in eggs, feces, mammalian milk, and so on, because so few studies have been made of these routes of information transfer. The transmission of behavior-influencing substances may be more common than we think, because natural selection would surely favor young receiving "advanced information" about the world they are soon to experience. It is easy to imagine how such a system could evolve. Suppose that an incidental consequence of a mother eating a particular food is that traces of it occur in her feces. If offspring benefit from finding and eating the same food, natural selection will favor changes in the mother's physiology and behavior that make traces of the food substance in her feces more certain, and changes in the young that make them more likely to respond to these food traces. Through natural selection, what was originally an accidental side effect can be shaped into a more assured route of information transmission.

In chapter 5 we discussed various other types of socially mediated learning through which information is transferred between individuals. The term "socially mediated learning" (often called simply "social learning") is a very broad one, and this breadth is significant. All that is needed for learning to be described as "socially mediated" is that an experienced individual—an animal that knows and does something, or has a particular preference—influences another, naive individual in a way that makes the latter develop and practice a similar behavior or have a similar preference. Commonly all that happens is that through its behavior an experienced animal draws the attention of the naive one to some aspect of the surroundings which it hadn't taken much notice of previously. The naive animal then learns the same behavior through its own trials and errors.

As far as we know, there is nothing fundamentally different in what goes on in the brain during social learning and non–social learning, and one doesn't exclude the other. It is just that with social learning other animals are the relevant part of the environment in which learning takes place. Even when a pattern of behavior is learned socially, it is usually fine-tuned by non–social learning as the individual fits it into its own idiosyncratic behavioral repertoire. Learning goes on all the time, and behavior is constantly being adjusted, especially in young animals.

From what we have just said, it probably sounds as if socially mediated learning is merely an inevitable by-product of youngsters learning in social conditions. It is, but this doesn't really explain anything about the origin and evolution of this type of behavior, because it presupposes that young animals pay attention to the activities of nearby experienced individuals. Why should they? Surely a youngster could learn just as well through its own trials and errors. Who wasn't told as a child that they would get on a lot better if they concentrated on what they are doing themselves instead of watching what other people are up to? What sense is there in youngsters paying attention to the activities of others? Is there any advantage in doing so?

There is, of course. There are two obvious problems with purely asocial learning. First, when an animal learns completely on its own, it is likely to make mistakes, and some of these can be costly: it costs a lot to experiment with a new type of food if it proves to be poisonous. It can cost even more to learn to be wary of predators by having an encounter with one of them (figure 9.4). It is much better, in evolutionary terms, to acquire this information through observing how experienced adults behave, simply because you are more likely to survive. The second problem with asocial learning is that there is a lot of useful information that an animal may not discover for itself. Will an inexperienced tit standing by a closed milk bottle realize that it is a source of nutritious cream? In general, when animals learn exclusively asocially, each one has to invent the wheel for itself. But when they are social and observe experienced wheel users, they get a lot of clues about wheels. Being attentive to parents and other individuals is therefore likely to evolve whenever social learning consistently and significantly reduces the cost of learning on one's own, or when it enables animals to get useful information that they would otherwise be unlikely to acquire.

Social learning is common in vertebrates, particularly in birds and mammals, and it has probably evolved independently many times. Wherever there is a social organization that involves young animals living with experienced older individuals, some of the habits of the older generation are likely to be passed on to the younger one. Sometimes this can go on for many generations. If it does, the habits may evolve culturally as new variations appear and the behavior is adjusted. Whether or not this happens depends on how stable the constructed ecological and social conditions are, how easily the behavior is learned and remembered, how useful it is, the interactions that develop between the effects of the behavior and other types of behavior, the opportunities for it to be

(a)

(b)

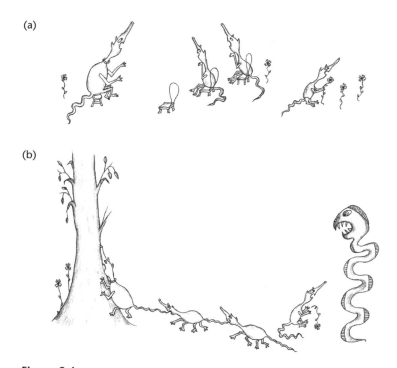

Figure 9.4
The benefits of social attention. In (*a*) two youngsters are paying close attention to their mother, while one ignores her. The outcome of the youngsters' differing behaviors is shown in (*b*), where the two attentive youngsters follow their mother away from danger; the fate of the inattentive one is uncertain.

transmitted, and so on. But, given learning and social groups consisting of closely associating experienced and naive individuals, animal traditions are almost inevitable.

What Is Needed for the Evolution of Communication through Symbols?

There are many animal traditions with different and independent origins, so biologists can compare them and try to come up with useful generalizations about how they evolved, but only once, as far as we know, has a culture with a fully fledged symbolic system emerged. This happened in our own lineage and nowhere else. Therefore we cannot compare symbolic systems in different groups to arrive at generalizations. To reconstruct the evolutionary path that led to our unique type of information-transmitting system, we have to rely on what we know about ourselves and other highly social and intelligent animals, particularly our close relatives, the great apes.

We should really start our account of the possible origin of symbolic thinking and communication by considering the evolution of human cognition and consciousness, but we are not going to venture into this vast area of complexity and uncertainty. Instead we will take a simpler option, and outline parts of an experimental study which has, in our opinion, provided some of the most important insights into what is needed for the beginnings of symbolic communication. The work that we shall be describing is that of Sue Savage-Rumbaugh and her colleagues at the Georgia State University Language Research Center in Atlanta, who have been studying language-instructed chimpanzees.

There are two species of these close relatives of ours, the common chimpanzee and the pygmy chimpanzee, or bonobo. We will focus on the work done with the latter species. The natural home of bonobos is in the forests of the Congo, where they live in large mixed-sex groups. Compared with most populations of the common chimpanzee and human beings, they seem to be very nonaggressive animals. Group wars and infanticide, behaviors that are sometimes found in the common chimpanzee as well as in our own species, seem not to occur among bonobos. Their society seems more egalitarian, lacks male dominance (if anything, females are slightly more privileged), and both heterosexual and homosexual sexual acts are practiced frequently and with great creativity. Sexual activity apparently has a very important role in bonobo society, helping to cement friendships and diffuse conflicts.

Sue Savage-Rumbaugh has found that when young laboratory-reared bonobos are exposed to human culture, human speech, and human-made lexical symbols, they acquire the ability to understand and to communicate purposefully using symbols. In other words, they learn to use language. Notice that we said when the young are *exposed* to the human symbolic system, not when they are *trained*. This is significant, because Sue Savage-Rumbaugh did not teach them in a highly structured way, using food rewards for good performance, which is how animals are usually trained. By accident she discovered that young bonobos, like young children, acquire a fairly sophisticated ability to understand spoken English and to communicate with symbols merely by being exposed to intense interactive communication with speaking adults.

The first bonobo to manifest a spontaneous ability to comprehend and communicate with symbols was an infant male called Kanzi. He acquired his basic skills when he was between six and a half months and two and a half years old, while the scientists were trying to teach his adoptive mother to communicate with a board of lexigrams. These are abstract visual images that stand for words, which the bonobos and their human caretakers touch in order to say something. This is important because, if they are to develop the ability to use language, young chimps need to express themselves actively, when and if they want to, and the anatomy of their vocal tract is not suitable for communication through human-type sounds. People around Kanzi spoke to him in normal English, just as one does to a human infant, and they would sometimes point to the corresponding lexigrams, but they didn't try to teach him in any systematic way because they were concentrating on his mother. Remarkably, without his human caretakers realizing it, Kanzi picked up the beginnings of language. His caretakers discovered this only after he was separated from his mother, but once they realized it, they were able to build on the language he had absorbed, until eventually he (and later his sisters) developed the linguistic proficiency of a human infant.

The correctness of his reactions to various complex linguistic requests proved that Kanzi really could understand spoken English. In carefully controlled tests, his comprehension was found to be similar (in fact slightly superior) to that of a two-and-a-half year old human child. When the human and bonobo youngsters made mistakes, their mistakes were very similar. It is true that Kanzi's ability to "talk" (using lexigrams with gestures and vocalizations) was less impressive than his comprehension of spoken English, but the same is true of human infants, who at first can understand language far better than they can speak it. We also have to bear in mind that large boards with hundreds of lexigrams on them are cumbersome, and even humans have difficulty using them, so Kanzi was handicapped.

By human standards, Kanzi's language skills seem limited, but he certainly used symbols and applied rules such as those of word order to understand complex sentences. We doubt that he did so because he had a special Chomskian type of language organ, or FLN, or any kind of mental language module that evolved in his ancestors. As far as we know, bonobos in their natural habitat do not communicate with linguistic symbols. (We have to be a little cautious in saying that, because rather little is known about bonobos in the wild, and if people go on destroying their habitat it is unlikely that very much more will ever be known.) There is convincing evidence that bonobos understand the emotions and intentions of others, and can anticipate them and sometimes manipulate them. There is also evidence that they use sticks and branches to mark locations and direc-

Figure 9.5
Learning to read.

tions for foraging, and there are some intriguing hints that they use social rules to manage their group life. So the initial conditions for the evolution of symbolic communication all seem to be in place. Nevertheless, it looks as if a symbolic-linguistic system has not developed in bonobos living in natural conditions. Given their ability to learn such a system when presented with it in captivity, what is missing? What started the evolution of our hominid symbolic system?

Two related sets of conditions seem to have pushed our ancestors along the route to language. The first was an altered ecological and social environment, which provided a strong and persistent motivation for better communication. Most anthropologists agree that leaving the forests had a snowball effect on the lifestyle and social organization of early hominids. This gave the impetus for developing new ways of communicating. The second and related set of conditions has to do with anatomy and physiology. What seems to be missing in bonobos is a ready way of producing communication signs that is under effective voluntary control. There are anatomical constraints on vocalization, and since movement from place to place requires the use of their forelimbs, fine control of hand and finger movements is unlikely to evolve. It was probably the increased motor control over hand movements and vocalizations, and the ability to imitate both gestures and vocal sounds, that were crucial in hominid language evolution. Once upright posture appeared in the lineage leading to humans, the hands were somewhat freed. Through selection for better foraging and toolmaking, fine control of the hands improved, and hand gestures could be used for communication. Another result of the upright posture was the descent of the larynx and other changes in the vocal tract, so sounds became better articulated. The voluntary production of clearly distinct sounds, especially consonants, meant that the voice could become specialized for communication.

A social system that allowed the cultural transmission, through imitation, of voluntarily produced sounds and hand gestures could have produced a preliminary linguistic community, but how far could the symbolic system have evolved through cultural evolution alone? This is where the work with the bonobos is important, because it gives us an idea of the extent to which such a system can be stretched once the community has the appropriate symbols. Since there were no lexigrams or users of spoken English to provide symbols in early hominid evolution, we have to assume that as controlled vocal or gestural communication got under way, a simple system of signs and rules evolved through cultural transmission. Sooner or later, communities would have reached a stage comparable to that of Kanzi, using a simple language that combined vocal and gestural signs in a rule-guided manner. However, for the system to go beyond what vocally endowed bonobos could achieve—to go beyond the linguistic system of Kanzi or a two-and-a-half-year-old human child—cultural and genetic evolution had to go much further. As we argued in the previous chapter, our view is that various features of the emerging language system that were initially culturally transmitted were later genetically assimilated. Constant interactions between the genetic and cultural systems were needed to produce the fully fledged and idiosyncratic linguistic system. Nevertheless, before that happened, hominids could have traveled quite a long way along the route to a symbolic system of information transfer simply as a result of the evolution of better general intelligence, better general voluntary control over movements and sounds, better memory, and *a lot* of cultural evolution.

Transitions on the Evolution Mountain

From the accounts we have given in this chapter, it should be clear that although much still remains to be learned about the origins of the various inheritance systems, there is no need to invoke the hand of God in the birth of any of them. In fact, as with evolutionary origins in general, we could have begun most of our accounts with "In the beginning—was a by-product of. . . ." This by-product was in most cases later transformed by natural selection into something very different from the adaptation with which it was originally associated.

Although the origins of the nongenetic inheritance systems were unexceptional, some of the effects they had were far-reaching, so we will end this chapter by taking a general overview of how they have affected the history of life. The panoramic view presented in figure 9.6 suggests that some of the great evolutionary transitions—from unicells to multicellular organisms, from individuals to cohesive social groups, from social groups to cultural communities—were all built on new types of information transmission. As new ways of transmitting information were added, and new types of organisms with different capacities for evolution emerged, the role and importance of existing inheritance systems changed.

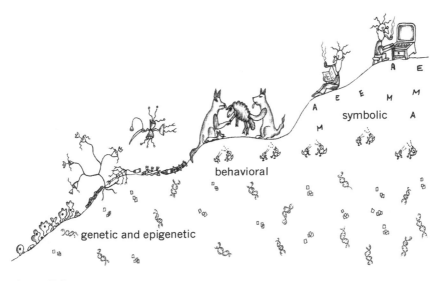

Figure 9.6
The evolution mountain: moving up the mountain, there are new types of inheritance systems underlying new types of organism.

Near the bottom of the evolution mountain are simple unicellular organisms that have both genetic and epigenetic inheritance systems. The nature of the primitive soup from which life emerged and the structure of the first real cells are unknown, but at some stage individual genes became linked together to form chromosomes, and the DNA information system with its genetic code and translation machinery was established. Later, eukaryotic cells evolved—cells in which chromosomes were enclosed within a nucleus and cell division was through mitosis and meiosis. All of these transitions were associated with changes in the organization, transmission, and interpretation of the information associated with DNA. But, in parallel, epigenetic systems, which depend on the production of the genetically encoded proteins that are their building blocks, were also modified.

As epigenetic systems became more elaborate, they became more effective information-transmitting systems and, as we argued in chapter 4, they enabled the evolution of multicellular organisms with many cell types. Epigenetic and genetic inheritance systems (including interpretive mutations) continued to play the major role in the evolution of plants, fungi, and simple animals, as well as unicellular organisms. However, once more complex animals with a central nervous system had evolved, behavior and behaviorally transmitted information became important. Through behavioral transmission, animals had the potential to adapt in ways that were impossible or unlikely through transgenerational epigenetic inheritance or gene mutations. With animals' increasing reliance on socially learned information came complex social structures and relations, and group traditions. Eventually, in the primate lineage, symbolic communication emerged and led to the explosive cultural changes we see in humans, where symbols have taken the leading role in evolution. As has happened throughout evolutionary history, a higher-level inheritance system now guides evolution through the lower-level systems, including the genetic system.

By taking a view of evolution that focuses on the transmission of information, we are continuing a recent trend in evolutionary thinking. Interest in the nature, storage, and transmission of biological information grew out of the debate about units and levels of selection and evolution, which we mentioned in chapter 1. The heart of the problem is that groups are made up of individuals, individuals are built from cells, cells contain chromosomes, chromosomes have genes, and selection can occur at any or all of these levels. Yet the higher-level entities are obviously integrated units of function, which can reproduce as a unit, even though they are made up of component parts similar to those that in the evolutionary past were themselves independently reproducing entities. So how did the higher-level entity evolve? Why doesn't selection among lower-level entities disrupt the functioning of the higher-level entities? For instance, why doesn't competition among selfish cells destroy the ability of an individual animal or plant to function as a whole?

The most influential of the books that discuss these issues is John Maynard Smith and Eörs Szathmáry's *The Major Transitions in Evolution*. In it the authors offer a comprehensive analysis of the evolution of new levels of complexity, identifying eight major transitions: (1) from replicating molecules to populations of molecules in compartments (protocells); (2) from independent genes to chromosomes; (3) from RNA as both an information carrier and enzyme to DNA as the carrier of information and proteins as the enzymes; (4) from prokaryotes to eukaryotes; (5) from asexual clones to sexual populations; (6) from single-cell eukaryotes to multicellular organisms with differentiated cells (plants, fungi, and animals); (7) from solitary individuals to colonies and social groups; and (8) from primate societies to human societies with language. Maynard Smith and Szathmáry suggest that all of these transitions were associated with changes in the way that information is stored, transmitted, or interpreted. They show how higher-level entities can evolve through selection acting on lower-level units because the latter can benefit more by cooperating than by competing. With most transitions—the exceptions are (3), (5) and (8)—a unit that once reproduced independently became part of an integrated system that formed a new unit of reproduction. The independent gene, for example, became part of a chromosome; for the single cell, the new unit was a multicellular organism. According to Maynard Smith and Szathmáry, once the old entity became part of a larger reproducing unit it could no longer survive and reproduce independently, because along with the emergence of the higher-level entity came mechanisms that ensured its stability and prevented it from disintegrating into its component parts.

Our approach to evolution is similar to Maynard Smith and Szathmáry's, in that it is centered on the transmission of information. It differs, however, in that we focus on new or modified types of hereditary information, and see them as crucial factors in the evolution of new levels of organization, whereas Maynard Smith and Szathmáry see all evolution between the emergence of the first cells and the acquisition of language by hominids in terms of changes in the genetic system. Naturally, they recognize that EISs are important for the development of multicellular organisms, but they do not see them as distinct information-transmitting systems that can affect evolution directly. Similarly, information that is transmitted from generation to generation behaviorally has no direct role in their evolutionary scenarios. EISs and behavioral transmission are seen as outcomes rather than as direct agents of evolution. With the exception of variation provided by the linguistic system, which drives cultural evolution, Maynard Smith and Szathmáry assume that genetic differences provide all the heritable variation on which natural selection acts. This restricted view of heredity means that there is no room in Maynard Smith and Szathmáry's evolutionary ideas for instructive processes, other than in human societies. This, we believe, is a mistake.

We are convinced that for all major changes in evolution one has to think about at least two dimensions of heredity: the genetic and epigenetic. With many animals, a third dimension, that involving behaviorally transmitted information, is also relevant, and for humans the symbolic systems adds a fourth dimension. All four ways of transmitting information introduce, to different degrees and in different ways, instructive mechanisms into evolution. All shape evolutionary change. Yet, so far, the existence of the instructive aspect has had little impact on evolutionary thinking. This must soon change. As molecular biology uncovers more and more about epigenetic and genetic inheritance, and as behavioral studies show how much information is passed on to others by nongenetic means, evolutionary biologists will have to abandon their present concept of heredity, which was fashioned in the early days of genetics, nearly a century ago. If Darwinian theory is to remain in touch with what is already known about heredity and evolution, efforts must be made to incorporate multiple inheritance systems and the educated guesses they produce.

Dialogue

I.M.: I have a general problem with some of the things you said in this chapter, but before we get to that let me ask you something specific about DNA methylation. You said that it is a good cell memory system, and defends cells against genomic parasites. You also said that it was probably present in early single-cell organisms. If so, why isn't it found in all organisms today? According to you, there is very little DNA methylation in fruit flies, and none at all in the nematode *C. elegans*. Wouldn't these animals also benefit from having a good cell memory system and the ability to silence genomic parasites?

M.E.: Yes, but for them methylation would probably be an unnecessary luxury. Methylation has a high cost—it increases mutation. Methylated cytosines are very prone to change spontaneously into thymines, and changing from a C to a T in some critical DNA sequences could be disastrous. So, unless a good cell memory is essential, it is better to do without methylation marking. For long-lived animals like us, where cells are constantly dividing in order to replace those that are damaged or worn out, a reliable way of transmitting epigenetic information is indispensable. But for small animals with short life spans and little cell turnover—animals like *C. elegans* and fruit flies—long-term cell memory is unnecessary. For them, the mutational hazards of methylation outweigh the benefits, and they have abandoned or reduced their use of this EIS. Selection has favored alternative memory systems, such as protein marking and steady-state systems.

I.M.: That makes sense. So let me ask you now about this general problem that keeps bothering me. It is whether or not organisms have *evolved* the ability to make evolutionary changes. You said in this chapter that some people think that producing a burst of mutations when conditions get tough for bacteria is an evolved response, and you've mentioned various other seemingly "clever" ways in which organisms generate variations that could promote evolutionary change just when it's needed. Do you accept that these are adaptations for the capacity to evolve, or do you think that in spite of their influence on evolution they are really just incidental by-products of other things?

M.E.: This is something that people get very hot under the collar about. Some see the idea of selection for evolvability—for the ability to generate relevant selectable phenotypic variation—as a real heresy. It seems to them to bring a designer back into evolutionary thinking. Natural selection has no foresight, they insist. Systems that benefit organisms in the future cannot evolve unless group selection occurs. And as you will have gathered, invoking group selection is something people are very sensitive about. Evolution for evolvability implies selection among lineages rather than among individuals. The lineages that survive are those that respond to life-threatening conditions by generating heritable variations, some of which turn out to be useful. Of course, even if such selection between lineages does occur, it doesn't mean that the variation-generating systems initially evolved for this end.

I.M.: I can't see what is so objectionable about lineage selection. It seems to me a lineage can be thought of as an individual extended in time, certainly if it's an asexual lineage, as many bacterial lineages are. Surely anything that improves its chances of survival will, by definition, be selected. Think of all the ways in which you said severe conditions could increase the amount of selectable variation. You mentioned variations generated through the SOS system, through increased activity of jumping genes, and variations revealed through the effects of Hsp90 and those strange yeast prions. Do you really think that these are all just by-products? Surely at least some of these systems were selected because they promote evolutionary change.

M.E.: We don't want to commit ourselves on this, because there is not enough evidence. It is too easy to assume that because a particular aspect of an organism's biology promotes evolution, it evolved for this reason. Genomic imprinting, for example, is a very effective way of ensuring that mammals retain sexual reproduction. It prevents parthenogenesis (virgin birth), because if you need imprints on both maternal and paternal genomes for normal development, you can't get rid of sex. Since lineages with sex remain on the evolutionary scene longer than parthenogenetic lines (sex generates selectable variation, giving more opportunities for change), you could argue that imprinting evolved to avoid a reversion to asexuality, which would probably lead to extinction. However, we think that it is extremely unlikely that this is an evolved function of imprinting. Nevertheless, it is an important incidental effect, and one that preserves evolvability. Our feeling is that most of these systems that seem to promote evolutionary change did not originate as adaptations to enable this. Nevertheless, we do accept that some may now be maintained partly because they generate stress-induced variation. For example, as we said in chapter 7, we think that in modular organisms, such as plants, the stress-induced movement of transposable elements is often beneficial and may have been maintained and sophisticated through natural selection. In general, we think there is no simple yes or no answer to your question. In some groups stress-induced variations accelerate evolutionary change and may have been maintained by selection for that end; in others they have remained as nonselected by-products of other adaptations.

I.M.: But surely direct evidence for or against the selection of systems that produce stress-induced variation could be obtained. Are there no experiments?

M.E.: Yes, there are a few that are very informative. For example, there is some fascinating work by a group of French scientists who looked at the mutability of hundreds of different *E. coli* strains that had been isolated from various sites all over the world. Some were taken from the air, some from the soil, some from water, some from animals' feces, and so on. They found that all of these strains responded to stresses like starvation by increasing their mutation rates. What was really interesting, however, was that the size of the increase in stressful conditions differed among the strains. And it wasn't just random variation—how much mutator activity was induced was clearly correlated with the environmental niches from which the strains came. It was more in the strains isolated from the feces of carnivores than in those isolated from omnivore feces, for example. The results strongly suggest that the mechanisms underlying stress-induced mutagenesis have been adjusted by natural selection. In other words, this is an evolved system.

I.M.: What about the Hsp90 chaperone and the yeast prion? You said in chapter 7 that they can unmask genetic variation in stressful conditions. Is there any experimental evidence showing that they have been selected to make evolution faster? They, too, can increase evolvability, but is it just a side effect?

M.E.: It is very difficult to know. In whatever way the evolvability-enhancing functions originated (we suspect they were probably side effects), once in place they allow adaptation to stressful and fluctuating conditions. Independently generated, formerly cryptic genetic variations produce visible and therefore selectable phenotypic effects. We think that lineage selection may have played some role in maintaining the unmasking function, but we really cannot be sure. One way of finding out might be to create strains that produce a lot of functional Hsp90 in stressful conditions (perhaps by giving them additional, stress-inducible, Hsp90 genes), and see how they compete with normal strains when under stress. Similarly, taking strains that can form the prion [PSI$^+$] structure and other strains that cannot, and seeing how they compete over many generations in stressful, fluctuating environments, might be informative. If the strains with the capacity to unmask variation do better than those lacking it, it would give some support to the idea that this may have been important in evolution because it promotes genetic change in conditions of stress. Needless to say, it would not give a general definite answer to the question of the evolution of evolvability.

I.M.: The ability to generate heritable variation when organisms find themselves in adverse conditions must surely have been important throughout evolution. If stress both induces variation and reveals variation, couldn't it be responsible for a lot of rapid adaptive evolution and even produce new species?

M.E.: We believe that stress has been very important, although direct experimental evidence is lacking. We can get some clues, perhaps, from the type of work Belyaev did on domestication. Belyaev, you will recall, was the Russian biologist who studied silver foxes. He believed that the remarkably rapid changes in our domestic animals occurred because the stress of domestication led to new variation that could be selected.

I.M.: Wait a minute! Surely taming and breeding wild animals is so stressful for them, and involves such intense selection, that it is too extreme to be relevant to anything that happens in nature!

M.E.: We think not. On the contrary, we believe that persistent, stressful, changed conditions are exactly those that may have initiated the origin of many new species. If a few individuals arrive in a new region where they have to live in conditions to which they are not adapted, they will be behaviorally and physiologically extremely stressed. Stress may induce epigenetic and genetic variations that enhance the rate of genetic change and adaptation to the new environment. If the population remains isolated, an incidental by-product of these adaptations could be the formation of a reproductive barrier between it and the mother species. This could be the outcome of differences that prevent mating taking place, perhaps involving changes in the reproductive cycle, such as were seen in Belyaev's silver foxes and have occurred in dogs. Alternatively, if individuals do mate with the mother species, genetic and epigenetic differences affecting chromatin and chromosome structure (which were also seen in Belyaev's silver foxes) could result in offspring that either do not develop properly, or are sterile because meiosis and germ cell production are abnormal. New species can probably arise in many different ways, but the stresses that induce heritable epigenetic and genetic variations may often have been a major factor in initiating their formation. As we see it, stress-induced variation has often been significant in adaptive evolution and speciation.

I.M.: It's a pity you have no evidence, although I know that you will tell me that no one has good direct evidence about how new species originate. Just one last question before I leave the origin of EISs and the fascinating subject of evolvability. What we see as a result of EISs is the continuity over several cell generations of different phenotypes which are all based on the same genotype. So why not think about the transmission of these variations as a manifestation of phenotypic plasticity that is extended over time? Wouldn't this way of looking at things help you to understand the origin and evolution of EISs?

M.E.: Yes. Sometimes it is a helpful way of thinking, not only about EISs but also about the other nongenetic inheritance systems. All can be viewed as temporally extended effects of development that were selected because they improved the survival and reproduction of the individual's descendants. That is how people usually see maternal effects—the consequences of development that are carried over into the next generation. There is no reason why you shouldn't extend this and have grandparental effects, great-grandparental effects, and so on. All could be seen as temporally extended plasticity. But there are two reasons why this should not be the only way of looking at these ancestral effects. First, if you see phenotypic inheritance solely in terms of gene-based developmental plasticity that is extended in time, you are likely to overlook the autonomy of nongenetic inheritance once it is place, and the effects it has on the evolution of other inheritance systems. In other words, you might not pay much attention to the type of evolution shown by Jaynus creatures and tarbutniks. A second reason for not allowing the genetic perspective to dominate is that if you think about early evolution—about the origins of life and what went on before genes as we now know them existed—you realize that phenotypic continuity based on self-sustaining chemical cycles and structural templating must have preceded (and probably formed the basis) of the genetic system. We avoided going into this, but the studies and models that people have made suggest that the early evolution of life involved interactions between self-perpetuating systems that were not what would now be called genetic.

I.M.: The first of your reasons was good enough for me! I want to turn now to something I didn't understand about social systems. You said that the evolution of social learning and traditions is related to the evolution of social attention—to the evolution of paying attention to what others are doing. But social attention is itself part of the evolution of sociality, is it not? How can you separate the two?

M.E.: You can't. It's like most things in biology, a cycle (or rather, a spiral) of causes and effects, with more social cohesion leading to more social attention, and vice versa. We were not trying to map an evolutionary path for the social learning that leads to traditions. It will be different in different groups, because the route to enhanced social cohesion is always idiosyncratic. We highlighted social attention because it is crucial for social learning, but we were not claiming that it always initiates the evolution of social systems.

I.M.: My own attention was very much taken with this ape Kanzi. I know nothing about the experiments, but if he couldn't speak very well, and if we can judge his comprehension only by his behavioral responses to various linguistic requests, how can we be sure that he understood the signs in the same way that a human does? He may have been using a different kind of understanding—a nonlinguistic, nonsymbolic kind. Maybe although he and the human child responded in the same way, they did so for different cognitive reasons.

M.E.: This is a valid point, and a common criticism of the interpretation of the results that Sue Savage-Rumbaugh and her colleagues got with Kanzi. But look at some examples of the kind of spoken English sentence to which Kanzi usually responded correctly:

Go get a coke for Rose.
Tickle Rose with the bunny.
Go get the doggie that's in the refrigerator.
Can you make the bunny eat the sweet potato?
Take the carrots outdoors.
Go outdoors and find the carrots.
Pour coke in the lemonade.
Pour lemonade in the coke.

These are just eight of more than 600 sentences used to test Kanzi and a human child. As we said, Kanzi performed slightly better than the two-and-a-half-year-old. The mistakes of both youngsters were similar, commonly due to misunderstanding a particular word. All we can say is that if something looks like a duck, quacks like a duck, and walks like a duck, as far as we are concerned it is a duck unless proved otherwise. Kanzi seemed to understand English word order, and could decipher the different meanings of the same words when they were combined with other words in a sentence. He even made up new word combinations and certainly understood combinations he had never heard before, since most of the 600-plus sentences to which he responded correctly were novel combinations for him. He referred to future acts, to imagined events. If this is not language comprehension and use, the critics need to come up with a more convincing explanation and suggest appropriate and fair tests. A general argument from the a priori position that apes cannot understand language, and therefore whatever they do it is not linguistic, will not do in science. We think the burden of proof is now on the critics.

I.M.: So can you tell me how it has come about that bonobos, and presumably other great apes, have an intelligence that allows them to understand symbols? This is something they never do in the wild. Maybe the big evolutionary jump is not between symbol-using humans and non-symbolic apes, but between the great apes with their potential to understand and communicate with symbols and the rest of the animal world that cannot.

M.E.: As we suggested, the very sophisticated social and ecological intelligence of great apes, in combination with an appropriate system for producing controlled communication signs and cultural evolution, could probably lead to simple symbolic communication. But it is not trivial to evolve a cultural system of communication signs that are voluntarily produced, memorable, distinct, and easy to use. If a system of signs is available, we know from Kanzi that the signs can be learned naturally, without a great deal of formal structured tutoring, presumably because of the great plasticity of the youngster's developing brain. So we think that the major hurdles on the path to symbolic communication (and they are great ones) are the evolution of the production system and the conditions that promote cultural evolution. What is needed is a system of production in which there is voluntary control of gestures and vocalizations, and also, crucially, cultural evolution that shapes these gestural and vocal signs into essential, conventionalized, complex social tools that can be used for many purposes. How bonobos and other great apes (and possibly also dolphins and whales) evolved the remarkable ability to learn and to comprehend symbols when presented with them is a difficult question. It seems that the large brain of a young social animal with a strong motivation to communicate has far greater potential than was at one time believed. Thinking more generally, we would say that all animals have a surplus of possibilities, because every structure has many potential uses. Humans can fly to the moon, and bonobos can understand basic English.

I.M.: This is too vague, but I suppose you will say that the potential for symbolic communication will be different in different animals, depending on the fine details of their brains, social systems and relations, ecological opportunities, and so on. So we'll leave it for now and look at your evolution mountain, which evokes a host of questions, especially about that large part at the bottom, before multicellular organisms evolved. You said that epigenetic inheritance was involved in evolution right from the start, but what did it have to do with those early transitions that led to present-day cells? I know that early evolution is not one of your main topics, but give me an example!

M.E.: We will give you the most obvious example. Think about the evolution of chromosomes. Maynard Smith and Szathmáry have given a very convincing explanation of why selection favored independent genes becoming linked together, but that doesn't go far toward explaining the evolution of the chromosomes we find in cells today. If you go on linking genes together, you ultimately end up with very long DNA molecules, which have to be packaged into cells in a way that allows their orderly replication and leaves the genes available for transcription. You need appropriate support, protection, and anchoring systems for the DNA molecules, and during evolution various proteins were presumably co-opted and modified for these functions. Critically, we believe, the packaging had to be compatible with the ability to quickly re-establish states of gene activity after cell division, so there had to be some kind of reliable, reproducible, chromatin-marking systems. Nothing much is known about this aspect of chromosome evolution, yet it is obviously important if we are to understand how chromosomes evolved. In eukaryotes, whose chromosomes have histones and a whole battery of special chromatin proteins, it is fairly obvious that DNA sequences and EISs must have evolved together.

I.M.: I have another, rather different, mountain-related question. Your mountain gives a definite impression that evolution is progressive over time: more inheritance systems underlie the organisms, more sophisticated educated guesses can be made, and the creatures become more complex. Your picture of a language spiral in the previous chapter presents a similar view—it is an image showing a linear progression. I thought that was a rather outdated view. Isn't the notion of evolutionary progress considered to be very improper in biological circles these days?

M.E.: Only if progress is seen as the goal of evolution. We do not subscribe to this progressionist view. The story we have been following, that of the evolution of new information systems and their relation to the emergence of new types of individuals, is one particular trajectory of evolutionary change among many others. We followed certain paths because we find them more interesting than others. However, there are many alternatives, some going in the opposite direction—from the more functionally complex (e.g., multicellular organisms) to something simpler (single cells), and others branch off in all sorts of directions. But, as many people have noted, if you start an evolutionary pathway from the simplest possible beginning, it is almost inevitable that complexity will increase in some lineages. Nevertheless, where, when, and how it does so, and what type of complexity it will be is far from being determined. We can imagine a parallel world in which evolution occurs without the appearance of all the information systems with which we are familiar. In our Jaynus world, for example, organisms did not have a central nervous system, so they did not have the behavioral system of information transmission. Our imaginary tarbutniks started their virtual existence without a system of social learning, and it is very easy to imagine a world without a symbolic system—symbols are, after all, relative newcomers on our planet. We do not feel that we need to apologize for taking an interest in the evolution of a cumulative increase in complexity and in the number and sophistication of information-transmitting systems.

I.M.: A last general question then: Whatever their origins, the evolution of all your nongenetic systems—the Lamarckian systems, if you must call them that—is also about the evolution of evolution, isn't it?

M.E.: Yes. Ever since they evolved, the additional inheritance systems, with their Lamarckian properties, have been shaping evolution, creating new ways of evolving, enhancing the rate of evolution, and sometimes giving it definite new directions. The inheritance systems did not originate for that end, but these were their effects. Accepting the position we have outlined means rethinking the definition of heredity and changing the way one approaches evolutionary problems. The four-dimensional view has many and varied repercussions. We'll deal with some of them by using a last big question-and-answer session in the next, final, chapter. It will be helpful if you start by summarizing what you think is the main message of our book. We shall take this as our starting point.

I.M.: You have taken me on quite a journey through the wonderlands of heredity, and I am certainly not going to go through it all again. Instead I want to look at the core of your argument—at the claims that you made at the beginning of the book, in the Prologue. Your first claim was that there is more to heredity than genes. I doubt that anyone will dispute this, even with respect to epigenetic variations (of which most nonbiologists have never heard). Your next claim was that some of the inherited differences between individuals are not just random accidents. Here again, it's difficult to disagree: new heritable variations can obviously originate in many different ways. Certainly some are the results of accidents, but a lot occur because organisms have evolved systems that bias when, where, and what type of variations occur. Biologists may quibble about the randomness or otherwise of gene mutations, but I think that no one will deny that many epigenetic and cultural differences are nonaccidental. To use your own language, there are many different instructive processes that lead to educated guesses. This almost inevitably means that the next claim that you made—that acquired information can be inherited—must be valid. You told me that evolutionary biologists have a problem acknowledging that something that has been "acquired" can be inherited, because it is associated with Lamarckism, but I find it very difficult to see how anyone can seriously argue that induced or learned epigenetic and cultural variations cannot be passed on. It seems to me that the general case against the inheritance of acquired information should be dropped: as far as the *inheritance* of acquired information is concerned, your case is convincing.

It is the move to what this means for evolution that really matters and is more contentious. It means—and this was your last major claim in the Prologue—that evolutionary change can result from instruction as well as selection. This makes your version of evolutionary theory rather different from the prevalent one. I can show how it differs if I do what you did in chapter 1, where you highlighted the main features of some of the historical transformations of Darwin's theory. For your postModern (or is it post-postModern?) Synthesis, I see the major features as follows:

• Heredity is through genes and other transmissible biochemical and behavioral entities.
• Heritable variation—genetic, epigenetic, behavioral, and symbolic—is the consequence both of accidents and of instructive processes during development.
• Selection occurs between entities that develop variant heritable traits that affect reproductive success. Such selection can occur within cells, between cells, between organisms, and between groups of organisms.

Clearly, because you give weight to the epigenetic, behavioral, and symbolic dimensions of heredity, evolutionary change does not have to wait for genetic changes. They can follow. Phenotypic modifications will usually come first.

This version of Darwinism certainly has consequences for how we should view patterns and processes in evolution. You indicated several of them. Maybe the most obvious is that by introducing multiple heredity systems and nonrandom variations it is possible to give additional or alternative accounts of evolutionary changes such as the transition to multicellularity and the initiation of speciation. But of more general interest are the implications your version of Darwinism has for the dynamics of evolutionary change. It implies that evolution can be very rapid, because often an induced change will occur repeatedly and in many individuals simultaneously; there is also a good chance that such a change will be of adaptive significance, since it stems from already-evolved plasticity. Even without selection, evolved plasticity will bias the direction of evolution, simply because induced variations are nonrandom. However, as I see it, one of the most important implications of the version of Darwinism that you have espoused is probably that when the conditions of life change drastically, it may induce large amounts of all sorts of heritable variations. The genome, the epigenome, and the cultural system (when present) may all be restructured, with the result that there can be rapid evolutionary changes in many aspects of the phenotype.

This summary of your views isn't exhaustive, I know, but I hope that it is a fair précis of the main claims that you have made. Will it do?

M.E.: Yes, although we have to resist a pedantic urge to elaborate.

I.M.: Good. So let me now turn to some general matters. There are three aspects of your four-dimensional view that particularly interest me. I am a practical person, so first I would like to hear what kind of differences this whole approach makes to medicine, to agriculture, and to the environment—to things that matter to people. You mentioned some of the implications for the world outside biology as you went along, but I would like you to pull them together for me. Then, since I also enjoy philosophizing, I would like you to tell me what difference your viewpoint makes to biological thinking and issues in the philosophy of biology. Finally, I would like to know what ethical differences your approach makes. This is something you should not try to evade. But before you start with the practical implications, I would like to know what kind of objections you most often encounter when you talk about your views with your biological colleagues. I am sure that there are recurring questions.

M.E.: Yes, there are. There are four common questions. But before we come to these, we need to describe the wider general problem. It is not always well articulated, and it is only partially covered by the questions we get asked, but it is the most fundamental problem that biologists have with what we say. As you have already indicated, most have no problem accepting that there are four types of inheritance systems rather than one; they are also prepared to accept that not all inherited variants are random in origin, and not all are transmitted through blind copying processes. The sticking point is the Lamarckian implications of this—the difference that it makes to the way in which we have to interpret evolutionary processes. Many biologists are reluctant to recognize that if development and learning impinge on the generation and transmission of variations, we need to know how this works in order to understand the causes of evolutionary change. For as long as it was possible to assume that all heritable variations stem from what are essentially mistakes or can be treated as such, the origin of the variations that shaped evolutionary adaptations could be dismissed as irrelevant: it was selection alone that was central to the study of adaptation. However, once you take a broader view of heredity and accept that not all changes in transmissible information are the result of accidents, this exclusive focus on selection is no longer legitimate. To understand adaptation, you also have to study the instructive processes involved in the generation and transmission of variation, and the way in which they interact with selection. Some biologists find this difficult to accept.

I.M.: Is this difficulty related to the kinds of questions you get asked?

M.E.: Indirectly. The first question, the one that comes up after almost every single talk we give, is a lengthy version of "But it's still all genes, isn't it?." The argument is that since it is genes that make proteins and control, or at least affect, the formation of proteins and their interactions (and all subsequent higher levels of organization), heredity and evolution can be boiled down to genes. This question is the reason why we keep repeating, like a mantra, that we are talking about heritable variations that do not depend on variations in DNA sequence. Some people behave like a rubber band—they understand our arguments as we discuss them, but soon after they repeat the question yet again, asking, "Why doesn't it all boil down to genes?." That is why we told the story of the Jaynus creatures and the tarbutniks. "Freeze" a genotype, or even an epigenotype, and you can still get interesting heredity and evolution. So if you were bored when we repeated our mantra time and time again, we apologize, but this is why we did it. We were afraid of the rubber band syndrome.

I.M.: Maybe as a nonbiologist I am immune to this disorder! I think a more interesting objection to your 4D view—one which is almost the mirror image of the one that you have just dealt with—would be that you are separating things that cannot be separated. It is nonsense to separate the genetic, epigenetic, and behavioral aspects of development. Genes are absolutely necessary, always. Epigenetic aspects are just as fundamental—after all, you are talking about the control of gene expression. And behavior, when it occurs, is essential too. You are reifying the distinctions that you have made.

M.E.: Of course you cannot separate levels of organization in the living, functioning organism. But you can distinguish between heritable variations at different levels of organization. You can therefore distinguish between different dimensions of heredity and evolution. When you are looking at the developing and interacting organism, when you ask questions about the ongoing processes of development, then of course you cannot separate them.

I.M.: What are the other objections you encounter?

M.E.: The second one is rather specific to epigenetic inheritance systems. Many people argue that epigenetic inheritance is unimportant for understanding heredity and evolution, and that anyway heritable epigenetic variations do not have large morphological effects.

I.M.: These are two separate points, are they not? I was satisfied with the magnitude of the epigenetic effects. The *Linaria* example—that monstrous flower variant—was clearly dramatic and rather morphological in its nature, and those poor obese yellow mice were also quite convincing. You said in chapter 4 that there are other examples of epigenetic variants that are passed on, but I really don't need more examples. What I need is evidence that the selection of epigenetic variants really has been important in evolution.

M.E.: We have no direct evidence, but we are sure that it is worth looking for it. The critics think there is no point in doing so. That's the difference. From basic biological principles and by extrapolation from what has already been found, we believe that it is highly likely that there is much more epigenetic inheritance than has so far been identified. People have hardly started looking for it. We predict that many more cases will be found, especially in plants and simple organisms, but not only there. Look at those yellow mice—they are probably just the tip of an iceberg. We are convinced that now that the DNA sequences of the genomes of so many organisms are known, and people are beginning to look at methylation and other aspects of chromatin structure, they will find a lot of variation. Many accidental and induced epimutations will probably have only short-term effects, but some will be heritable, although with different degrees of stability. The point is that epigenetic variants exist, and are known to underlie the inheritance of phenotypic traits, including some that do not show typical Mendelian patterns of inheritance. They therefore need to be studied. If there is heredity in the epigenetic dimension, then there is evolution too.

I.M.: May I guess what the third objection is? Don't people complain that your view is too messy? That understanding evolution is much easier when reduced to genes, and that, in fact, a lot has been achieved with gene-based models? That simplification leads to models that can actually be studied, while complexity leads to profound paralysis?

M.E.: No, this is not the third most common objection. We've heard that one too, but it is not common. Nevertheless, we'll answer it, since you asked. Simplification is fine, but it has to be of the right kind. It is no good if it misleads you or limits you too much. You can make a chair using just one tool, a knife, but it's much easier if you also use a saw, a hammer, and a chisel. It may be more complicated to learn to use four tools rather than one, but you can make chairs a lot more easily, and a lot better; moreover, you can construct things that you can't make with a knife. At one time people tried to explain the working of an animal in terms of the way it was put together anatomically, using the principles of mechanics, but you will agree, we hope, that when biochemistry was added, it helped. In some ways it has made the endeavor more complicated, because we have to think about more things at the same time, but adding a biochemical dimension has also made things simpler, because activities that were very difficult to explain in terms of simple mechanics are much easier to understand with the help of biochemistry. The same is true of heredity and evolution. For some time people used gene-based explanations, and it worked for the cases they chose to study. But the more that heredity has been studied, the more difficult it has become to explain everything in terms of genes. So it's time to add nongenetic systems, because we shall do better with these additional ways of thinking about heredity.

I.M.: So what is the third most common objection?

M.E.: The third one runs something like this: "Had you shown that Lamarckism really exists, that would be interesting. But you haven't. As you admit, there is no evidence of precise directed mutations, and your interpretive mutations are just an expansion of neo-Darwinian ideas. Nobody has ever seriously questioned the occurrence of Lamarckian-like behavioral and cultural evolution, but this is not really Lamarckian, because the acquired changes in information are not transferred directly to the genes."

I.M.: So?

M.E.: We would like to believe that after all that we've said there is no need to answer this one. First, biology and heredity do not start and end with genes. Second, the very notion of what is and what is not Lamarckian is problematical. As we said in chapter 1, just as no one today believes in Darwin's Darwinism, no one today believes in Lamarck's Lamarckism. These terms have acquired very general meanings. You can avoid this terminology, but it comes at a price, because you detach yourself from a tradition of thought and from some general but important things that these vague general terms refer to. The specific version of Lamarckism that some critics deem worthy of being called "Lamarckism" is one that assumes direct adaptive feedback from the soma to the DNA in the germ line. This was the position of August Weismann (although he didn't know about DNA, and would have said "determinants" instead), and today this is the position of the philosopher David Hull. According to this view, interpretive mutations are not instances of Lamarckian inheritance, because the induced changes are in DNA; and the transmission of acquired epigenetic variations, variant food preferences, and new patterns of behavior are not instances of Lamarckian inheritance either, because they do not involve the transfer of information from the phenotype to the genotype (or memotype).

Most Lamarckians, past and present, would disagree with this assessment. They recognize that a lot of variation, be it genetic, epigenetic, or behavioral-cultural, is targeted, constructed, and, in the case of humans,

also future-oriented. When such variation is transmitted between generations, they call it Lamarckian inheritance. In other words, whenever heredity is "soft," in the sense that what is inherited is molded by environmental
and developmental factors, it is considered "Lamarckian." It seems to us
that the anti-Lamarckians confuse several claims. Lamarckians *do not* make
any claims about information going from protein to DNA, from product to
plan, from a cake to the recipe. We agree with Dawkins and Hull that there
are good reasons for thinking that such reverse construction may be very
difficult, and in most cases unlikely. Lamarckians also do not make any
assumptions about the nature of the hereditary material. What they *do*
claim is that heritable adaptive changes are the outcome not only of natural
selection but also of internal (evolved) systems that generate "intelligent
guesses" in response to the conditions of life. Lamarckians argue that such
informed sources of variation have important consequences for the study
of heredity and evolution. Their problem is that Lamarckism has such a bad
name, and has been used as a stigma for so long, that a lot of people simply
assume that any form of Lamarckism is nonsense. As a result, anything
that makes scientific sense is, by definition, not Lamarckism! People are
often confused about both the historical and the theoretical aspects of
Lamarckism. We hope this book will do something to change that.

I.M.: I am not at all sure that it will. Language is very powerful. I think
that you may be doing yourselves a great disfavor by branding your position as "Lamarckian" or even "Darwinian-Lamarckian." If what you want
to do is convince people that your point of view is valid and fruitful, why
do you insist on using a term that, as you obviously recognize, makes
people think that you are a pair of confused, muddle-headed idiots? Surely
it will make them refuse to listen to you. Why not avoid "Lamarckism"
altogether, since what you are doing anyway is making an idiosyncratic
blend of theories and data, which have little to do with Lamarck's original theories and even less to do with some of the versions of Lamarckism
that existed in the early twentieth century? Besides, you may be confusing people. For a lot of people Darwinism and Lamarckism are incompatible. You are either one thing or the other. And here you come along and
say that you are both good Darwinians (and you insist on that!) and at the
same time you are also good Lamarckians. Don't you think that you would
lose nothing by simply dropping the term? I cannot think of any harm it
would do, and I can think of a lot of good in terms of persuasion. Rhetoric
is not unimportant, as your favorite philosopher knew well.

M.E.: Would you also have us drop "Darwinism" because it has so
many connotations and has lost some of its original meaning? Today's

Darwinians recognize that to explain evolution they have to think about a lot more than natural selection. People like Steve Gould and Gabby Dover, for example, are seen as good Darwinians, but both have emphasized the importance of aspects of evolution that do not depend directly on natural selection. Darwinism today is not synonymous with the theory of natural selection. So we believe that, in spite of the problems, which we recognize, it is time for people to be more open-minded about Lamarckism too. Doing so might aid understanding in the slightly longer term, because this research tradition is very interesting, and although it certainly includes masses of rubbish (as does any other research tradition, including Darwinism), we can learn from it. Pretending that our position is ahistorical or focusing only on the Darwinian source of influence on what we think would be misleading.

I.M.: I disagree. I think you are tying yourselves to a terminology that is detrimental to your cause. But if you choose to bang your heads against a brick wall, that is your own problem. Let's move on. What is the fourth most common objection or question you hear?

M.E.: This is what we call the comparative question. It has general and specific versions. The general question is, What is the relative importance of the various inheritance systems? The specific versions take the form, What is the relative importance of culture versus genes for human beings?, or, How important are epigenetic variants compared to genetic variants in plants?

I.M.: Can you answer it? Are there any independent measures?

M.E.: In a sense it is an absurd question. The importance of a given heredity system depends on the trait, the time scale considered, the genetic and social population structure, the ecological circumstances, and so on. It's like asking what is more important, the environment or genes. It's simply a nonsensical question in this general form. This doesn't mean that one cannot develop some measure to estimate the effect of different inheritance systems for well-defined cases. The American mathematical biologist Marc Feldman and his colleagues have devised ways in which the effects of cultural inheritance can be taken into account. Similar considerations led Eva Kisdi and Eva Jablonka to suggest how the effects of epigenetic inheritance could be measured.

I.M.: I understand that there is a measurement called "heritability," which tells you how much of the variation between people is caused by genetic differences. I have read that for a lot of personality traits, like the novelty-seeking behavior that you discussed in chapter 2, heritability is around 40 to 50 percent. This surprises me, because if, as you led me to believe, the interactions between genes are so complex and dependent on upbringing and lifestyle, one would expect the genetic reasons for differences between people to be much less. Do you think that the estimates of the genetic component are exaggerated because they don't take into account other sources of transmissible information?

M.E.: Yes, this is part of what we are claiming. But before we try to answer this question in more detail, we must clarify what heritability is, because it is an extremely misunderstood term. Heritability has a precise technical meaning in biology: it is a measure of the proportion of the visible, phenotypic variation in a particular trait, at a particular time, in a particular population, living in a particular environment, which arises from genetic differences between individuals. It is a population measure, *not* a measure of the relative role of genes and environmental factors in individual development. It was developed for use in agriculture—for studying crop yields, milk production, and such things—but now, regrettably, it is often used for human traits. One way in which it is estimated is by looking at how closely relatives resemble each other. You compare identical twins, nonidentical twins, parents and children, cousins, and so on. Because you know the genetic correlations between various relatives from the principles of Mendelian genetics, from the phenotypic correlations you can estimate what proportion of the total variation is attributable to genetic variation.

What has to be appreciated, however, is that heritability is not fixed: with the same genotypes, a trait may have a low heritability in a variable environment (where a lot of the variation is attributable to environmental factors), and a high heritability in a stable environment. If the heritability of a trait is low, it does not mean that there are no genetic differences between individuals—it may just mean that the trait is very well canalized. Another thing that is important to understand is that when a trait has a high heritability because there is a close correlation between relatives, it does not mean people are the same or even similar! The scores for a measured personality trait in parents and children might be closely correlated, yet very different because each child is 10 points higher on the scale than the average of its parents. But let's return to these quoted estimates of 40 to 50 percent heritability.

In most cases, all heritable variation has been assumed to be genetic. Yet if a component for cultural inheritance of the type that Feldman suggested is introduced into the calculations, it could decrease heritability significantly. We assume that the same would be true if epigenetic inheritance were to be included. You also have to take into account behavioral inheritance—for example, self-perpetuating prenatal effects like those we mentioned in chapter 5, and the transmissible effects of parental care. In some cases it may be very difficult to distinguish between the various sources that contribute to "heritability," and the whole attempt to have separate estimates might be invalid. Heritability estimates for humans are based on so many simplifying assumptions that their usefulness is always doubtful, but by failing to isolate nongenetic inheritance, we believe they greatly exaggerate the genetic contribution to variation. The bottom line is that the geneticist's estimates of heritability can be misleading because there is more to heredity than genes.

I.M.: But estimates of heritability, even if limited, are still of practical importance in agriculture, and maybe they would be even more useful if you could factor in the epigenetic component. So how about dealing now with some of the practical implications of your approach. I can see how the symbolic and genetic systems are now meeting and changing our world through the cultural practices of genetic engineering, but if you go beyond this rather extraordinary meeting point, what difference does the 4D view make to our lives?

M.E.: We can only scratch the surface of an answer. The little that we know already suggests some obvious and quite dramatic implications. Let's start with medicine, and with interpretive mutations and epigenetic variations. If microorganisms have systems that increase the rate of mutation in stressful conditions, then these systems may have to be targeted when we fight microbial diseases; otherwise we could lose the arms race. It may mean, for instance, that when designing antibacterial treatments, in addition to the drugs that kill the bacteria, we have to include something that inhibits or destroys their mutation-generating systems, so that any survivors will not be able to mobilize these systems and launch a drug-resistant strain.

As for EISs, we have already mentioned several areas of medicine where their importance is beginning to be recognized. First, many cancers are known to be associated with cell-heritable epigenetic modifications such as alterations in methylation patterns and other aspects of chromatin organization. Knowing this may enable better risk assessment and diagnosis, and, since epigenetic changes are potentially reversible, for some cancers it may mean better methods of prevention and treatment. Second, epimutations may be involved in some hereditary diseases. We know that peculiar patterns of inheritance can be associated with defects in imprinting, and human geneticists are on the lookout for these, but it is unlikely that heritable epigenetic defects are limited to imprinted genes. The yellow mice that we mentioned in chapter 4 had an inherited tendency to develop diabetes and cancer because of the methylation mark on a transposon-derived sequence. We are ready to bet that similar inherited disease–influencing epigenetic marks will be found in humans because there are thousands of transposon-derived sequences scattered all over the human genome, and some of these are likely to carry variable epigenetic marks influencing gene activity. Nevertheless, most epimutations will probably affect only a single individual, because they occur in somatic cells. They may be one of the reasons why identical twins are occasionally profoundly dissimilar with respect to some aspects of their phenotype.

A third and obvious area of importance is in the study of prion diseases like Creutzfeldt-Jakob disease and kuru. A fourth is in environmental medicine and epidemiology, because there is growing evidence that factors such as starvation and treatments with some drugs (possibly including thalidomide) can have transgenerational effects. It is already known that a person's nutrition may affect the health not only of their children but also of their grandchildren. This is true for men as well as for women. Animal studies suggest that stress and hormone treatments can also have effects lasting several generations. Epidemiological research programs and medical practice will have to accommodate information like this, so that we know how to avoid passing on the effects of our sins or misfortunes to future generations. A fifth area of importance is associated with the new work on RNA interference. We may be able to silence disease-related genes by introducing artificial small interfering RNAs, which will recognize and inactivate the genes' mRNAs, and perhaps turn off the genes themselves. In theory this could help us to cure many viral diseases and treat cancers, at least in their initial stages. These are very early days as far as this technology is concerned, but we are cautiously optimistic.

I.M.: What you say about the transgenerational effects of nutrition is frightening. If it's really true, it means that social injustice may have its roots in the gametes and wombs of people's ancestors. People with stressed and malnourished ancestors are born disadvantaged. Are scientists working on these things?

M.E.: Oh, yes. Scientists are now compiling and studying data about the long-term and transgenerational effects of various stressful conditions. Fortunately some data already exist, having been collected for other purposes. That's partly how we know that the children of women who were undernourished during pregnancy are more likely to develop cardiovascular disease and diabetes, and have a host of other problems. Long-term, multigenerational effects are certainly being looked at, but not yet beyond the second generation. The studies using mice and rats suggest that we should be looking at later generations, although at present this is difficult. The persistence of ancestral epigenetic states means that methods of compensating for the misfortunes of ancestors may be needed to ensure that the present generation does not start with an epigenetic disadvantage. Merely ensuring that individuals who carry detrimental ancestral marks develop in a normal environment may not be enough; it may be necessary to provide people with special diets or other treatments that will counteract their epigenetic heritage. That is why the type of work done with the yellow mice, where, you will recall, it was found that nutrition affected the epigenetically marked yellow allele's expression, is so important. It's not just that the results are interesting for biologists: they also point to the kind of thing that needs to be looked for in epidemiological studies.

There is also a lot of research into other medical aspects of epigenetic inheritance. People are working hard on the epigenetics of cancer; prion diseases are studied far more than they were before BSE; and as you can imagine, there is intense research on RNAi, because the pharmaceutical companies are interested. In addition, biotech companies are now mapping the DNA methylation profile of human tissues, hoping to develop ways of diagnosing abnormal methylation patterns that could be indicative of diseases, and to find drugs that will alter disease-related epigenetic marks. So yes, the work has certainly started, and it is likely to grow. An era of epigenetic engineering is just beginning.

I.M.: And what about agriculture? Do we need a 4D approach there too?

M.E.: Of course. The importance of epigenetic inheritance in agriculture is already widely acknowledged, because it has caused so many problems in genetic engineering aimed at crop improvement. That's how a lot of heritable epigenetic effects were discovered: newly inserted foreign genes were silenced through DNA methylation or RNAi, so the scientists were frustrated in their efforts to have useful foreign genes stably expressed. However, they have found ways around the problem. On the positive side, since some epigenetic variations can be induced by environmental means, it may be possible to develop agricultural practices that exploit these inducing effects and thus develop "epigenetically engineered" improved crops. The siRNAs will be great tools too, because by using them scientists may be able to silence whatever genes they want to by introducing artificially made siRNAs.

Another area where you have to look beyond genetics is cloning, which has always been important in agriculture. Plants have been reproduced from cuttings since time immemorial, and more recently they've been produced from single cells. But as you know, cloning animals has been far less successful. Dolly the sheep led the way and showed that cloning from adult cells is possible, but the process is still very inefficient, and many cloned animals are abnormal. In some cases the abnormalities are known to be associated with methylation differences. So far no one has come up with a way of resetting epigenetic marks so that development is normal. When they do (and we think they soon will), it will have enormous implications, and not just for agriculture. Scientists have already taken the first steps toward cloning humans, although, thank goodness, only with the aim of providing a supply of embryonic cells for therapeutic purposes—for replacing a person's defective or damaged tissues. We know that there are still many problems to overcome, particularly that of imprinting, and there are probably others that we don't know about, but it may not be impossible to go all the way with human cloning.

I.M.: Aren't you horrified by the prospect? What about the ethical problems?

M.E.: There is already a lot of public debate, and that's healthy. We must make sure our legislators listen to it. Pressure from individuals and companies wishing to profit from the technology may have to be resisted. At present, even if ethically acceptable (and we doubt that it is), attempts at reproductive cloning—using a person's somatic cells to try to produce a similar individual—would be scientifically irresponsible. There are too many biological problems still unsolved. However, we think that in the long term, cloning for therapeutic purposes has great potential, although we realize that people have qualms about it. It's worth remembering, perhaps, that at one time people regarded the idea of organ transplants with distaste, but now most of us take them for granted. The same acceptance will probably come for at least some uses of cloning. And cloning technology will probably be far better than organ transplants.

I.M.: I am impressed with your optimism! What about the other dimension of heredity, behavioral inheritance? Does that have any uses?

M.E.: One possibility we have already mentioned is imprinting planteating insects on particular weeds, and thus trying to control those weeds. It may not be feasible, but scientists certainly need to know about the ways in which insect food preferences are influenced by their mother's food, and they are working on the subject. Transmitted food preferences may be one of the reasons why insects shift so rapidly from their original host to a new crop plant.

I.M.: You are touching on ecology now, so tell me, what are the more explicit implications of your view for this sphere? What does it mean for conservation and biodiversity issues, for example?

M.E.: We have repeated many times that to some extent organisms construct their own environments. We were concerned mainly with the way the behavioral and symbolic systems construct ecological and social niches, but EISs and developmental legacies have a role too. Let's recapitulate a little, starting with cultural evolution. With both humans and animals, some habits are self-sustaining and self-perpetuating because they construct the sociocultural environment that allows their own transmission. It is most obvious with language and the song dialects of birds. These must be learned and practiced in order for them to be learned by others. If, either by accident or through terrible human cruelty and perversion, children reach their teens without any exposure to language, then they never learn language properly. You need normal development in a normal linguistic community to maintain linguistic communication. If *all* practice of a language is impossible, the language disappears—it goes extinct. The same is true of birds' song dialects.

To maintain traditions and culture, you need continuous social and cultural niche construction, which may involve interactions with other species. The chimps' nut-cracking tradition is unlikely to survive if there are no nut trees. Similarly, if we introduce a new competitor, the feeding habits that birds learned from their parents may change. This in turn may affect other species, both competitors and predators. There is always a complicated web of interactions among species, which together construct and sustain the common niche. You see this very dramatically in places like a tropical rain forest, where the extent of the interactions really hits you. Something that to the nonspecialist looks from a distance like "a tree," on closer inspection turns out to be a mass of plants of all descriptions, among which there are all sorts of insects, birds, frogs, and so on. There are hundreds of species in close visible interactions, and it is often impossible to guess where one plant ends and another begins. And this is just what you see with the naked eye—what happens on the microscopic scale is almost unimaginable. We don't know what type of interactions these are, but it would be surprising to us if the maintenance of such communities did not involve a lot of epigenetic inheritance, with plants inducing heritable changes in each other.

I.M.: And when you destroy one species, you may destroy the community? Is that it?

M.E.: Every organism is a community, with parasites and mutualists. The American biologist Lynn Margulis has been emphasizing this point for years. But the extinction of one species may not destroy the web of interactions in a rain forest "tree"—it may be quite robust, just like genetic webs. This is not the point. The point is that we *do* destroy whole communities; we go in for wholesale destruction. When we destroy rain forests, we destroy huge, complex, very beautiful, ecological webs. And we destroy them forever.

I.M.: Maybe we should freeze seeds, embryos, and the DNA of plants and animals, and keep them for use in a better, more ecologically sane future, if there is one.

M.E.: It wouldn't work. You would have to reconstruct the community, and often these communities are very old, with historical memories that are stored in their epigenetic and behavioral systems. These are part of their "identity," part of their stability. You cannot freeze these memories: they have to be maintained and transmitted through use, so you cannot reconstruct the communities from their component parts. The history is gone, and with it the specific community.

I.M.: So when you destroy a community, you destroy all the epigenetic and behavioral variations of the members. But how can you be sure that these communities cannot reconstruct themselves?

M.E.: We know too little to be sure. But even if the same genes and alleles exist in other combinations in other communities, the epigenetic marks and the historical heritage that they carry are gone. It is rather like destroying a culture or a language, and consoling ourselves with the thought that since the human DNA still exists, that culture or language can be reconstructed. It's absurd! DNA is not enough. Phenotypic (not just genotypic) continuity is essential. We do not know what can be reconstructed and what cannot, but there's sure to be a lot of information that is lost forever. There are some well-known examples from human history. One is Easter Island, in the South Pacific, where humans exploited the lush tropical paradise so completely and irreversibly that over the course of a few centuries the island's plants and animals, and its fascinating human culture, all became extinct. There are many more such cases. We are ignorant, yet incredibly arrogant. We are destroying whole ecosystems on an unprecedented scale, and we cannot even foresee the consequences.

I.M.: I could have preached the same sermon without knowing anything about different inheritance systems!

M.E.: Maybe, but if you do consider them, you can see even more clearly how irreversible and how disastrous our deeds are. We destroy much more variation and diversity than we imagine, and it is often on the nongenetic aspects of this diversity, especially its historical dimension, that the stability of communities may depend. You see it clearly when you try to reintroduce endangered species of social animals into the wild. Just taking a group of animals and dumping them in their natural habitat, hoping that their genes will tell them what to do, is usually useless. They die. In order for reintroduction programs to succeed, you have to teach the animals, who have lived for a long time in unnatural conditions in zoos or wildlife parks, how to behave, what to eat, which predators to avoid, and so on. Then it may work. But in many cases we may not know what to "teach" animals or plants, and how to regenerate the web of interactions.

I.M.: About this huge interacting web of organisms and environments—how far are you ready to take this image? Would you agree with James Lovelock that our whole planet, Gaia, is one great unit, a living, self-sustaining, self-perpetuating entity?

M.E.: There is little doubt that there are self-sustaining, complex, and as yet poorly understood webs of interactions among organisms and their environments, and that in a general sense living things construct the planet as the niche in which life can be sustained. Would we call Gaia a living organism? Would we call a smaller self-sustaining ecological system an organism? If you take the view that for an entity to be alive it has to display heredity and reproduction, then Gaia and smaller self-sustaining ecological communities are not alive. If you prefer the metabolic definition of life, according to which an entity is alive if it is a system that sustains its organization over time through the dynamic control of the flow of energy and material through it, then Gaia and some ecological communities can be thought of as alive. As we see it, the arguments about Gaia's living/nonliving status are neither very fruitful nor very interesting. What we do believe is enormously important and fascinating is the fact that the planet has provided an incredible self-sustaining environment for living things over huge spans of time—over billions of years. Life on Earth has continued because Earth's systems have kept its temperature and atmospheric composition stable, even though the sun's luminosity has increased by 30 percent since life began 3 billion years ago. By pointing out this incredible dynamic stability, Lovelock has highlighted a great biological question—the question of how the global system self-regulates. We have yet to find adequate answers to that.

I.M.: Is work on epigenetic inheritance tied up at all with these ecological concerns? With the ideas and work on Gaia, or with smaller-scale ecological communities?

M.E.: Not much, as far as we know. There is a lot of work and even more talk about the conservation of species and of communities, and people study the interaction between geological and biotic factors, but not from an epigenetic point of view. The epigenetic aspects of ecology are hardly ever discussed, let alone studied. Nevertheless, we expect a new field, epigenetic ecology, will emerge in the near future. So far, when looking at the effects of ecological changes and trying to predict the outcomes of conservation policies, people have been focusing on genetic variability. No work has been done on heritable epigenetic variations in natural populations. Yet there are many fundamental questions to which we do not have answers. For example, although a typical non–sex chromosome spends on average 50 percent of its time in males and 50 percent in females, in a large population there will be a distribution of chromosomes with different transmission histories. Just by chance a few chromosomes will have been transmitted for many generations only through males, and others only through females, although most will have had very few generations of consecutive transmission through a single sex.

How do such differing transmission histories affect the marks the chromosomes carry? We really have no idea, yet we need an answer to this and many similar questions, because they have relevance for both ecology and medicine. As we said, medics are already looking at the epigenetic aspects of disease, and recognize that knowledge of DNA alone cannot deliver all the futuristic promises that were part of the hype around the Human Genome Project. The medical and agricultural implications of EISs are now so obvious that more and more people are working on them. The implications for ecology are less obvious, and for most people, less urgent. People are reluctant to change their lifestyles to deal with the environmental problems that we already know about, let alone worry about unfamiliar ones. But once the basic work is done, the relevance of epigenetic inheritance to conservation will be recognized. It is just a matter of time.

I.M.: There may be no time, I'm afraid. Forgive my cynicism, but I think you are incredibly naive. You pointed out in chapter 2 that if solving the health problems of the world were a real concern, then providing everyone with food, clean water, and clean air would solve most problems. And it would not be very expensive either. Richard Lewontin has been saying this for years. He has drawn attention to the fact that the way in which genetics (his own field) is being applied in both agriculture and forensic science is related to political and economic interests, and to various social prejudices. The interests of most funding bodies, certainly most government agencies and big corporations, do not include improving the welfare of the poor people of the world. Look at the tobacco industry! The same is true when it comes to ecological problems. There are many possible ways of sustaining ecological systems, and some have been tried successfully. Preserving the most significant areas of ecological diversity is economically feasible, but it is unlikely to happen. I find it very difficult to believe that a lot of money will be spent on figuring out how to prevent the effects of starvation, pollution, and ecological devastation from having epigenetic consequences for the environment of future generations!

M.E.: We disagree. We are all living in an increasingly polluted and impoverished world, and many of the problems that we experience, although felt most acutely by the poor, cut across ethnic and economic groups. Cancer, with its epigenetic component, is one obvious example. There is also much more ecological awareness now than there was even thirty years ago, and with this comes a degree of political power. So funding for research into EISs and the other nongenetic hereditary effects of various environmental agents and conditions will be found.

I.M.: As long as the big corporations continue to make big profits, all will be well, no doubt. You didn't mention the behavioral and cultural/symbolic systems. What is happening there?

M.E.: Behavioral traditions in animals are already getting much more attention than they used to, and cultural evolution has always been of interest. After many years of careful isolation from the social sciences, the biologically oriented community is now contributing ideas about human social and cultural evolution.

I.M.: I am not sure that the biologists' contributions are that wonderful: sociobiology on the one hand, and memetics on the other. It's not a very appetizing diet!

M.E.: The influence of one discipline on the other is inevitable. The boundaries between the social sciences and biology are being broken down. People are aware that neither social nor biological evolution can be studied in isolation.

I.M.: This may be merely a transient cultural trend, just as it was in the second half of the nineteenth century when Darwinism and other evolutionary ideas made people think about biological and social continuity. Herbert Spencer had his grand unified vision of an evolving universe, which in many ways was more sophisticated than some of the things I read today, even though it was tainted by some horrible racial, sexual, and class prejudice.

M.E.: Spencer has such a bad reputation as a nasty sexist, racist social Darwinian that few people now bother to read his work. He was probably no more racist or sexist than Darwin and most of his other contemporaries, which does not absolve him, but we should see him in his cultural and social context. And just for the record and for the sake of historical justice, he was all his life a staunch opponent of slavery and colonialism. He was a very interesting thinker, and we can learn a lot from his work. The cultural spiral is now at a point where there is a renewed interest in an evolutionary view of life, just as there was in the second half of the nineteenth century, when Spencer was the most eloquent advocate of this way of seeing things.

I.M.: This is somewhat related to my second major concern, the philosophical aspects of your views. There is a great deal of interest in evolution among what are called "educated laypersons"—people like me. And as you said in your first chapter, there is also lively controversy among evolutionists. So I would like to know who you see as your allies and adversaries, although I must say that to me, a nonbiologist, many of the disagreements among you evolutionists seem rather minor. After all, Dawkins and Gould have both said that natural selection is important, that genes are important, that the constraints of development and ecology are important, and that chance plays a role in the grand scheme of things. And you, too, agree with all of this. It seems to me that a lot of the controversy is about where the emphasis should be put. It's a bit like the Bolshevik/Menshevik disputes, and so many of the arguments between radical left-wing groups—hairsplitting, with small differences being enormously magnified. To the eyes and ears of an outsider, this is both amusing and annoying. It seems to me that there is more that unites you with the selfish-geners than separates you from them, and certainly you are not far from Gould's position on evolution.

M.E.: As far as our attitude to so-called creation science is concerned, we are certainly all singing from the same hymn sheet: we all think it is nonsense. And there is truth in what you say about our differences: part of the dynamics of a controversy, whether political, scientific, literary, or whatever, is the exaggeration of small differences in position. We can see the similarity and points of agreement between our own position and that of others very clearly. Our 4D view emphasizes heredity, and the focus is very much on hereditary variations—on their production and their evolutionary effects. In this we share Dawkins's and Maynard Smith's position, for example, but we differ from them in our focus on the developmental aspects of heredity—in our belief that epigenetic and behavioral variation also play an important role in evolution, both directly and indirectly.

Other people whose general approach to biology we share, and who, like us, refuse to think in terms of gene selection alone, have a somewhat different emphasis from ours. They emphasize development, with the life cycle as an integrated unit whose multiple contributory causes cannot be teased apart. This, for example, is the perspective of Susan Oyama and a whole group of biologists and philosophers of biology who hold a view known as developmental systems theory (DST for short). We agree with their criticisms of the gene-centered, nondevelopmental views of evolution, and share many of their conclusions. However, unlike them, we focus on heredity rather than on development, and think about evolution in terms of different types of hereditary variations. We believe that the origin of these variations needs to be explicitly recognized, so, unlike the DST people, we tease apart the different heredity systems, making more or less sharp distinctions between them. We think it is necessary to look at the evolutionary effects of each of them before trying to reintegrate them into the whole again.

We also share many of Lewontin's views, particularly his views about the complexity of the mapping between genes and characters, and the active role of the organism in constructing its environment. However we feel that he is wrong to neglect nongenetic inheritance, and disagree with his position on cultural change, for which he believes evolutionary reasoning is inappropriate. We also accept the importance that Gould attributed to developmental and historical constraints, believing it to be fundamental to any evolutionary explanation, but again we would argue that an analysis in terms of heredity systems and nongenetic variations is necessary for a proper understanding of both evolution and development.

So you see, there is a lot of agreement between our ideas and those of other evolutionary biologists. We have tried not to enter too much into the aggressive polemics that characterizes a lot of evolutionary theorizing, because we think that it is not helpful. There are genuine overlaps between our position and those of other evolutionists, but as you yourself noted, there are some profound differences too. The major differences stem from our focus on the origins of hereditary variations, some of which we maintain are semidirected, not entirely blind changes. It is this that leads to our claim that evolution has to be seen in terms of instructive, as well as selective, processes.

I.M.: I noticed that you avoided using the words "reductionism" or "reductionist" when discussing the selfish gene and meme views of the world, which you obviously oppose. The charge of "reductionism" is commonly heard when these views are under attack. Why did you avoid it?

M.E.: We really did not find it necessary or useful. We advocate an approach that is synthetic, but starts with analysis, and hence requires breaking down integrated wholes into more or less well-defined parts or subsystems. This analytical aspect is reductionist in the sense that we believe that breaking down heredity and analyzing separate systems of inheritance yields important information about it. Such analysis does not yield full information, but it is methodologically necessary. That is why we discussed the different systems of inheritance separately before putting them together again.

I.M.: But you don't like the replicator concept, which is a sharp analytical tool. Don't you think you might be able to use it?

M.E.: You keep returning to this point, so we'll try again to pinpoint our reasons for not using it. If, as you acknowledge, there are well-defined nongenetic systems that can transmit variations that arise during the physical or cognitive development of organisms, then development and heredity impinge on each other. This means that development has an active role in evolution. It is not just a constraint defining the impossible, it is sometimes also a specifier. You cannot separate heredity from development, so a replicator concept for which the distinction between heredity (a property of replicators) and development (a property of vehicles) is central is unworkable.

I.M.: Can't you opt for a wider replicator concept? You can use Dawkins's original definition of replicators—"anything of which copies are made"—without worrying about the nature of the copying process. If you do that, then DNA sequences, methylation patterns, the 3D structures of prions, self-sustaining loops, physiological and behavioral patterns, ideas and artifacts can all be replicators. Why not?

M.E.: This option was adopted by Kim Sterelny and his colleagues, who developed the idea in an article entitled "The Extended Replicator." We have several problems with this idea. First, it is far from clear what the replicator actually is: even if we think about the genetic system, is it not clear whether the replicator is the single gene or a canalized network. Dawkins did not allow the single nucleotide or even the triplet codon the status of replicator, because they are not independent functional units; but in a highly canalized system, the gene is not an independent functional unit either. Our second problem is that most of the things we have been talking about, such as methylation patterns, metabolic loops, membranes, or patterns of behavior, are phenotypic traits that are the products of development. Yet the replicator concept strongly implies that replication or "copying" is distinct from the rest of development. We therefore think that it is best to avoid this concept and concentrate on the way variation is constructed and transmitted through developmental processes.

I.M.: I understand that your position is process-oriented rather than unit-oriented. But for some problems you need to focus on the entity that is the total outcome of all the developmental processes. The processes are, as you keep saying, interdependent, and act together to generate a coherent, stable whole. Metaphorically, I would say that you need to see things from a less close-up perspective. You really need to focus on a unit of some kind, so that you can ask general questions about organisms and their evolution. Such questions are difficult to answer from a pure process-oriented point of view, where you are always seeing the developmental matrix rather than what is generated. The unit that evolutionary theory needs may not be the replicator or the vehicle, but something is needed to allow discussion of the evolution and development of entities. Do you have an alternative?

M.E.: There are alternatives to the replicator and the vehicle. As you know, our own preference is to focus on "heritably varying traits," rather than replicators. These are the units that develop and are selected during evolution—the units whose stability and changeability we try to understand. By thinking in terms of heritably varying traits, we avoid the problems that individual-based or gene-based views raise, and gain insights into the developmental aspects of the traits' generation and canalization. So the heritably varying trait is our alternative to the replicator. Another type of unit, an alternative to the individual, has been suggested by the American philosopher James Griesemer. He suggested the "reproducer" as the biological target of selection, and we think that this is a brilliant concept. Griesemer's reproducer is a unit of multiplication, development, and hereditary variation. His notion of reproduction involves material overlap between parents and offspring: parts of the parent entity are transferred to the offspring entity and confer developmental capacities on it, minimally the developmental capacity required for further multiplication. The reproducer therefore unites development and heredity. It avoids the dichotomies that the replicator/vehicle concept created—the dichotomy between heredity and development, and the dichotomy between development and evolution. A reproducer can be a replicating RNA molecule in the RNA world, a cell, a multicellular organism, a society. It very naturally allows one to think about variations at different levels of organization, and allows for any mix of selective and instructive processes in evolution.

I.M.: I can see how the reproducer fits your view. I can also see how you regard development, as an agent of evolutionary change. But I am not sure how your 4D view *illuminates* development. What do we learn about development that we did not know before?

M.E.: It is a different way of looking at things. Geneticists who take a developmental approach to evolutionary problems are traditionally interested in explaining two complementary aspects of development—canalization and plasticity. Canalization refers to the resistance of development to genetic and environmental variation: organisms can maintain a typical phenotype in spite of quite different genotypes and environments. Plasticity describes the way developing organisms can react to different conditions by producing a change in phenotype: the same genotype can produce several different phenotypes. Plasticity and canalization both show that genetic variation and phenotypic variation can be decoupled. This is all well-known, and biologists like Conrad Waddington in Britain and Ivan Schmalhausen in the Soviet Union were discussing it more than fifty years ago, but in the last decade or so there has been a resurgence of interest in the plasticity of development.

In her recent and important book *Developmental Plasticity and Evolution*, Mary Jane West-Eberhard has suggested that plasticity is one of the keys to understanding adaptive evolution. Although her starting point is development, whereas ours is heredity, there are many similarities between her way of seeing evolution and ours. Like us, she believes that neither development nor evolution can be reduced to genes or genomes; she also stresses that there is phenotypic continuity between generations, something that we express in terms of our four dimensions of heredity. West-Eberhard's analysis shows how the evolved plasticity of development allows the evolution of new adaptive phenotypes without major genetic change, because variations can be self-perpetuating if the environmental inputs into development remain the same. However, West-Eberhard focuses her discussion on the plastic responses that occur during the life cycle of the organism, and, with the exception of maternal effects, she does not extend the notion of plasticity temporally, as we do, to effects that last for many generations even when the environment no longer induces the phenotype. We think that focusing on the reproduction and reconstruction of phenotypes through the various inheritance systems emphasizes the importance of *all* sources of information in development, as well as evolution.

I.M.: It's not just the organism's phenotype that is being reproduced or reconstructed, is it? It's also the conditions in which it lives. Do biologists agree that niche construction has been a significant factor in evolution?

M.E.: Lewontin has been stressing the importance of niche construction for years, but recently more biologists have taken up his ideas and extended them. Some ecologists are now talking about organisms "engineering" the whole ecosystem, controlling the flow of energy and materials through it. It's not simply that organisms construct their own niches, it's also that their niches are often inherited, because the initial conditions and resources for reconstructing them are transmitted to their progeny. Theoretical biologists are now looking at the effects of niche construction, showing how powerful it can be in altering the direction and dynamics of genetic evolution. Of course, niche construction is only one of the construction processes that involve and affect organisms. There is also the construction and reconstruction of developmental pathways, and of the preferences and skills that are part of animals' social niches. These interdependent processes of reconstruction all have evolutionary implications. This has been one of the main messages of this book. Inevitably there is selection among the heritable developmental and behavioral options, and developmental stability or flexibility has itself been modified by natural selection. Genetic changes are not necessary for all evolutionary change: epigenetic and behavioral inheritance systems and self-sustaining networks of ecological interaction can do a lot, although usually the genetic system becomes involved too.

I.M.: Can you model this? One of the great attractions of the genetic view is that you can model hereditary transmission and evolution. You can take the basic rules of Mendelian genetics, add factors such as selection, mutation, migration, chance, and so on, and from this you can get an idea of how evolution proceeds. Population geneticists have been doing this successfully for a long time: you said in chapter 1 that it was a big part of the Modern Synthesis of evolution, which occurred in the 1930s. The trouble with what you suggest is that everything is interdependent, and hinges on local conditions. According to you, selection and the generation of variation commonly go together, and if you are dealing with behavioral variants, both are affected by migration. Worse than that, your different inheritance systems all behave rather differently. So what are we left with? No general models? If you cannot provide alternative general models, how do you expect people to accept your views?

M.E.: Maybe we should not expect a single, universal type of model. We have four types of heredity, rather than just a single one, and each heredity system requires its own models. If someone finds one unifying model for all the dimensions of heredity and evolution, it will be wonderful, but it's not necessary. Plurality of models may be what is required. Niche construction can be, and is, modeled; so are certain aspects of cultural evolution; there are even a few models of epigenetic inheritance and evolution. Don't forget that there are many types of models—descriptive models and computer simulations, as well as the type of mathematical models used in classical population genetics. We don't see modeling as a major stumbling block. Once you understand some central features of an inheritance system, you can model it, and people do. In our opinion, the empirical evaluation of the 4D view and its practical implications are far more important than formal modeling at this stage.

I.M.: I think that models are absolutely necessary for understanding complex systems. You need them to ask focused questions, so I hope that your optimism about the feasibility of modeling is justified. I want to get to my last concern, however—the moral implications of your views. Please don't tell me that any theory can be used in different, even contradictory, ways. I know that. I am also well aware of the "is" and "ought" distinction—the difference between what we know and what we ought to do with what we know in the moral sense. But we are not living and acting in some debating society; we live and act in a painfully concrete social and intellectual environment, and this is where you are presenting your ideas. In the context of today's social and ethical attitudes and concerns, what are the ethical implications of your view?

M.E.: We don't want to go deeply into questions of moral philosophy, but we do agree that what people know affects how they believe they should behave. It affects what people judge to be moral or amoral behavior, although how it does so depends on their socially constructed ideology and beliefs. If people who have been brought up according to the Judeo-Christian cultural tradition suddenly discovered that cows are in many ways more like "us" in terms of their emotions and their intellect, it would probably make a difference in the way they treat cows. It might not make a great deal of difference to Buddhists, who have a very different view of the world to start with, and treat cows very differently anyway. Nevertheless, because there is a relationship between what we think we know and our practical morality, scientists do have a public responsibility.

I.M.: I hope you are not trying to heap everything on the scientists' shoulders and blame them alone for the public's lack of involvement. Those of us brought up in the Judeo-Christian tradition already know enough about the feelings and the emotional lives of animals to stop tormenting them, but it hasn't stopped cruelty. Fox hunting in England comes to mind, and there are some horrors in the meat trade.

M.E.: People *are* using knowledge about the psychology of animals to try to stop cruel practices. They present both moral arguments and new information about animals to support their position. There are many economic and political interests that oppose this, of course. But it was the same with the abolition of slavery and the emancipation of women, which also took too long, given the state of knowledge. We do not claim that arguments and information are sufficient to make changes, but they are crucial.

I.M.: So what *are* the moral implications of your view?

M.E.: Biologists with different approaches to heredity and evolution commonly have similar general social goals and values. Most oppose racism; most want a better and more just world, and so on. The main problem is the public image of various biological ideas. Since many biologists emphasize the genetic aspect of human behavior, their views are frequently interpreted in a way that leads to the widespread belief that common behaviors (often rather objectionable ones) are "genetic," "natural," and, like simple monogenic diseases, inevitable. This is nonsense, but it is the way their ideas are perceived, and most of them do not do enough to try to counteract this perception. A broader view of heredity and evolution makes explicit the wealth of possibilities that are open to us, and the fact that our activities, as individuals and as groups, construct the world in which we live. In particular, recognizing that we have a history and can plan our future, that we are able to construct shared imaginary worlds and systematically explore them and strive for them, greatly expands our freedom. The plasticity of human behavior is enormous. On the basis of present biological knowledge, there is no way one can dismiss the power of historical social construction and explain the social and behavioral status quo in terms of genes or memes. We cannot transfer explanatory power and responsibility to these entities!

I.M.: This is a criticism of human sociobiology, isn't it?

M.E.: It is a criticism of the "public persona" of that discipline, which is to a large extent the sociobiologists' responsibility. We want to be fair and clear—most sociobiologists *do not* believe that we are the slaves of our genes. The problem is that some of them do tend to promote a vulgar public image of genetically determined evolved "tendencies." To that end, they ridicule their opponents, erecting straw men and then triumphantly destroying them, and interpret every pattern of behavior, from joking to raping, as the manifestation of an evolved adaptation that was selected in the distant past. Thornhill and Palmer's *A Natural History of Rape* is a prime example of the genre. They don't say you can't override the actual manifestation of the behavior, such as the evolved tendency to rape, but imply that it is not easy, because it is deeply embedded in an evolved module of the mind. Needless to say, there is no shred of evidence for these evolutionary claims. They are mere just so stories.

I.M.: It seems to me that "just so story" is a term you evolutionary biologists use for any hypothesis you don't like! Yet you all tell just so stories—it's part of your trade.

M.E.: You are right, of course, but in the particular case of rape the story is based on a rather questionable analysis of data, and it is not clear what the claims made about the evolved module are supposed to mean. Do they imply that it is impossible to change the behavioral tendency through education and social change? The advocates of these views would say certainly not, that on the contrary they help you know how to shape society and educate people to overcome the problems associated with the unpleasant side of our evolved behaviors. But we are not told how we are supposed to construct a society in which genetically evolved raping tendencies will not be manifest. Beyond providing juicy and best-selling descriptions of sexual behavior and some platitudes about preventing or controlling inappropriate urges—by giving courses about sexual behavior to male adolescents before they get their driving license, for example, or by advising young women to dress modestly (the male chastity belt has not been advocated)—no insight is gained. There is little actual content in this "scientific" soft pornography. This is an extreme example, of course; not all human sociobiological stories are so empty. The problem is that the opposition to the stories is not coming as strongly as it should be from the ranks of the many more restrained sociobiologists. As we noted earlier, some sociobiological hypotheses are very reasonable and stimulating, and might be right. But even so, with the arguable exception of language, they lack significant empirical support. Yet these hypotheses are presented as the only serious theories in town. And in their vulgarized versions they are extraordinarily popular.

I.M.: Why are they so popular? What kind of needs do they satisfy? Maybe this can provide a clue about the kind of worldview they reflect.

M.E.: There is probably no single answer. Maybe they satisfy a wish to think in terms of single causes, like in classical physics. Newton's laws explain the movements of the heavenly bodies, and Mendelian genes explain human behavior. Complexity is simply and scientifically explained. But there is another, contrasting side to the fascination with genes. Genes are seen as links to our ancient past, to our ancestors, which govern us in an irrational and mysterious manner. There is something very romantic in this notion—in the eternal dark and deep guiding force of the genes. And this peculiar combination of the romantic and the scientific is embodied in many of the human sociobiologists' evolutionary stories.

Maybe this is why people find these gene-based explanations of human behavior so attractive.

I.M.: They seem to feed into what you earlier called "genetic astrology"— seeing genes both as fate and as the magical yet scientific key to human nature. You know, I sometimes get very uneasy about these magical genes. They seem to dominate so much of biological thinking and research. One of my favorite books is Lewis Carroll's long poem, the story of a great and absurd quest, *The Hunting of the Snark*. You may remember how the hunters tried to catch the elusive snark:

They sought it with thimbles, they sought with care;
They pursued it with forks and hope;
They threatened its life with a railway-share;
They charmed it with smiles and soap.

Today it would have to be a biotech-share, I suppose. Carroll ends the poem on a hilarious but dark note. One of the hunters, the valiant and hopeful Baker, unknowingly found instead of one of the common snarks (which "do no manner of harm") the dreaded boojum.

In the midst of the word he was trying to say,
In the midst of his laughter and glee,
He had softly and suddenly vanished away—
For the Snark *was* a Boojum, you see.

Reading the newspapers, I sometimes feel that today's hopeful scientists may find a boojum too. The major motivation for many molecular projects seems to be economic: they are sponsored by large biotech companies and serve their interests. The political and ideological claims made about the importance of the projects are indirectly derived from the economic interests. That this might give rise to various boojums worries me.

M.E.: It worries most people. A lot of molecular biology does revolve around big business economic interests, so many of the boojums are likely to be social and political ones, although straight genetic risks cannot be ruled out. Biotechnology is expensive, so it is inevitable that much of it is financed by commercial companies, but it is worrying that so many molecular biologists now have a personal financial stake in the biotech industry. We have to hope that their approach and outlook will not be distorted by the needs of their companies to make profits. It is almost impolite to talk about ideology these days, and nonreligious social utopias are not fashionable. Fortunately, however, human curiosity is impossible to tame, and some kind of absurd and hopeful snark hunting always goes on, wherever science is done. In the long run, curiosity is stronger than self-interest or dogma.

I.M.: Maybe, but in the short run, that of my own life span, I shall still keep a wary eye open for boojums. And I still think there is too much emphasis on genes. Wouldn't you agree that "The Hunting of the Gene" has pauperized biological research?

M.E.: Not really. You should remember that it has been the search for the genetic philosopher's stone that has directly or indirectly led to many of the findings that have begun to undermine naive beliefs in genetic wizardry. Genes are no longer seen as the sole source of hereditary information. In our opinion, a profound change in biological outlook is now taking place. The 4D view of heredity that we have been advocating is not just something for the future. Although we hear less about it, people are already studying all of the different types of inheritance, developing different methodologies and putting them into practice. We expect that as they do so, nongenetic inheritance will be incorporated more fully into evolutionary studies. We hope that our book will encourage more people to take this path. For as Lu Hsun, the great Chinese writer, said, "When many men pass one way, a road is made."

Notes

1 The Transformations of Darwinism

Many of the primary sources for this chapter, including books and papers by Bateson, Darwin, de Vries, Galton, Weismann, Mendel, Morgan, Wright, and others, can be found through the Electronic Scholarly Publishing Project's web-based version of Sturtevant's *A History of Genetics* (1965) at http://www.esp.org/books/sturt/history.

Page 10. Among the best of the many accounts of the debates surrounding evolutionary theories are those by Bowler (1989a), who gives a good historical overview of the changing fortunes of ideas about evolution, particularly since the mid-nineteenth century, and by Depew and Weber (1995), who take a more philosophical approach and include recent ideas about the evolution of complex systems.

Page 11. Maynard Smith's generalization of evolution through natural selection can be found in his *The Problems of Biology* (1986), chap. 1. A somewhat different formulation was given by Lewontin (1970). Griesemer (2000a) discusses the differences between Maynard Smith's and Lewontin's approaches, and puts their views into a wider philosophical and biological context.

Page 13. Darwin discussed the heritable effects of use and disuse and environmentally induced changes in the 1st ed. of *The Origin* (1859), particularly in chap. 4. He developed this discussion further in the 5th and 6th eds., in which he responded to criticisms of his theory.

Page 13. Lamarck set out his evolutionary ideas in *Philosophie zoologique* (1809), but revised some of them in later publications. A summary of Lamarck's theory and the changes he introduced can be found in Burkhardt (1977), chap. 6. Lamarck's views were not well received during his lifetime, and *Philosophie zoologique* was not translated into English until 1914. Georges Cuvier, Lamarck's colleague and one of the most influential biologists of the time, ridiculed Lamarck's ideas, especially in the "eulogy" written after Lamarck's death in 1829. This "eulogy," read in 1832, was widely disseminated and was the source of major misrepresentations of Lamarck's ideas for decades to come. A translation of Cuvier's "Éloge de M. Lamarck" can be found in the 1984 reprint of *Zoological Philosophy* (the English translation of *Philosophie zoologique*). Disparaging comments about Lamarckian views are still the norm today, when they are usually patronizingly presented as reflecting a failure to understand Darwinism, developmental biology, and basic logic. For a representative example, see Cronin (1991) pp. 35–47.

Page 14. For his provisional hypothesis of pangenesis, see Darwin (1868) vol. 2, chap. 27. Robinson (1979) reviews other nineteenth-century pangenesis-like theories of heredity. According to Darwin's letter to his cousin Francis Galton, he started developing his pangenesis theory in the early 1840s. The often-repeated claim that Darwin developed Lamarckian ideas only as a result of criticism, and against his own better judgment, is a myth refuted by both his own letters and a reading of the 1st ed. of *The Origin*.

Page 16. Weismann's ideas, which evolved over the years, can be found in his many clearly written and well-translated books and essays. His heredity–development theory is described in *The Germ-Plasm* (1893a); his mature thoughts on heredity, development, and evolution are presented in *The Evolution Theory* (1904), which includes his ideas about the origin of variations through changes in the quality and quantity of determinants (vol. 2, chaps. 25 and 26) and his views on levels of selection (vol. 2, chap. 36).

Page 16. Darwin discusses Virchow's cell theory and alternative views about cell formation in chap. 27 of the 1st ed. (1868) of *The Variation* (vol. 2, p. 370), where he states, "As I have not especially attended to histology, it would be presumptuous in me to express an opinion on the two opposed doctrines." In the 2nd ed. (1883), his discussion is much the same, but this sentence is omitted.

Page 19. Burt (2000) evaluated Weismann's ideas about the significance of sexual reproduction in the production of variation. In the light of modern evolutionary theory, he concluded that they are basically correct.

Page 21. According to Bowler's (1983, 1988) very readable accounts of the debates raging between neo-Lamarckians and neo-Darwinians in the late nineteenth and early twentieth centuries, the term "neo-Lamarckism" was coined by the American biologist Alpheus Packard in 1885. By the end of the nineteenth century, Lamarckism had become very different from Lamarck's original theory because, like Darwinism, it had undergone some interesting transformations. One notable difference was a greater emphasis on environmentally induced variations that do not involve overt use and disuse, but are brought about by effects on the embryo. This trend reflects the influence of Étienne Geoffroy Saint Hilaire, a younger colleague of Lamarck, who in the 1820s developed an evolutionary theory according to which heritable variation is caused by changes induced during embryonic development. Samuel Butler, who is better known as the author of *Erewhon* and *The Way of All*

Flesh than for his scientific writings, supported the views of Lamarck and Geoffroy Saint Hilaire in his *Evolution, Old and New* (1879). Spencer's conviction that natural selection and the inheritance of acquired characters are both important is at the core of his evolutionary philosophy, and appears in most of his many books. The famous debate between Spencer and Weismann, the flavor of which can be savored in Spencer (1893a,b) and Weismann (1893b), has been analyzed by Churchill (1978).

Page 22. Galton's experiments are described in Galton (1871), and an account of his heredity theory, which stresses transmission through a substance akin to Weismann's germ plasm, is given in Galton (1875). The rabbit experiments and Darwin's argument with Galton about their interpretation are described in Gayon (1998), chap. 4 and Gillham (2001), chap. 13. Both books also explain Galton's heredity theory.

Page 22. A notorious example of early twentieth-century experimental results that supposedly demonstrated Lamarckian evolution is that involving the Viennese biologist Paul Kammerer. He forced midwife toads, which normally mate on dry land, to mate in moist conditions. The males developed dark swellings on their forelimbs, which resembled the nuptial pads which, in water-mating species of toads, help males grasp the females. According to Kammerer, the adaptive pads acquired by his experimental toads were inherited. Unfortunately, most of Kammerer's material was lost during the First World War, and when in 1926 other scientists examined the single specimen that remained, they found that the pad area had recently been injected with ink. Soon after this, Kammerer shot himself, thereby reinforcing the suspicion that he had forged the results of the toad experiments, and casting doubts on his other evidence for Lamarckian inheritance. The whole story was documented by Arthur Koestler (1971). Fascinating accounts of American neo-Lamarckism in the late nineteenth and early twentieth century are given by Pfeifer (1965), Greenfield (1986), and Cook (1999). Persell (1999) describes Lamarckism in France in the late nineteenth and early twentieth centuries.

Page 23. See W. Bateson (1894, 1909) and de Vries (1909–10) for the idea that new species originate through sudden large changes (which de Vries called "mutations") rather than by natural selection of small differences. The debate between the mutationists and the Darwinians is summarized in Provine (1971), chap. 3.

Page 24. A translation of Mendel's original paper, the papers of the rediscoverers, and other papers relating to the early days of genetics can be found in Stern and Sherwood (1966). The background to Mendel's discoveries is given in Olby (1985) and Orel (1996).

Page 27. The history of classical (Mendelian) genetics and developments in the study of heredity during the first third of the twentieth century are documented in many books, e.g., Jacob (1989), chap. 4; Dunn (1965); Sturtevant (1965); and Bowler (1989b), chaps. 5–7.

Page 28. The definitions of *genotype, phenotype,* and *gene* are given in Johannsen (1911), a seminal paper in genetics.

Page 29. For an account of the Modern Synthesis by one of its founders, see Mayr (1982). The book edited by Mayr and Provine (1980) contains papers describing the route to the Synthesis in different countries; a more recent account of the Synthesis has been provided by Smocovitis (1996).

Page 30. Our account of the beginnings of molecular genetics is necessarily sketchy. For reasons of space we have been unable to describe the experiments that pointed to nucleic acids, rather than proteins, being the main constituents of genes, or to discuss why the significance of these experiments took so long to be appreciated. Good accounts of the history of molecular biology can be found in Jacob (1989), chaps. 4 and 5; Judson (1996); Morange (1998); and Olby (1994). Watson's *The Double Helix* (1968) provides a very personal account of the discovery of the structure of DNA.

Page 30. Although many of the commonly studied bacteria (e.g., *E. coli* and *Bacillus subtilis*), have a single circular chromosome, others have more than one and some have linear chromosomes.

Page 31. For the original statement of the central dogma, see Crick (1958); the idea is clarified in Crick (1970). The way in which "central dogma" has become an article of faith and stamp of authority can be seen from Dawkins's (1982) use of it in statements such as "violation of the 'central dogma' of the non-inheritance of acquired characteristics" (p. 97), and "the inheritance of an instructively acquired adaptation would violate the 'central dogma' of embryology" (p. 173). When using the central dogma in its original sense (p. 168), he puts no quotes around the term.

Page 32. Twentieth-century challenges to the hegemony of the nuclear genes in heredity are described in Sapp (1987). Lewontin (1974) discusses the consequences for evolutionary theory of the large amounts of variation that had been revealed using new molecular techniques. A description of the neutral theory of molecular evolution can be found in Kimura (1983). Sewall Wright's ideas about the importance of sampling errors in small populations are described in detail in vol. 2 of his four-volume masterpiece *Evolution and the Genetics of Populations* (1968–1978), and are discussed in an accessible way in Provine (1971), chap. 5.

Page 32. The idea that a lot of DNA is "junk," "parasitic," or "selfish" was put forward by Doolittle and Sapienza (1980), and by Orgel and Crick (1980). Both papers can be found in the collection edited by Maynard Smith (1982). Some of the strong reaction to the suggestion that a lot of DNA is junk can be found in letters in *Nature* 285: 617–620, 1980, and 288: 645–648, 1980.

Page 33. For a discussion of the concept of a genetic program, see Keller (2000), chap. 3. The metaphor of the genotype as a recipe and phenotype as a cake can be found in Dawkins (1986), pp. 295–298. Maynard Smith used the metaphor of the genotype as the plan for an airplane and the phenotype as the "plane that is built" in a lecture given to the Linnaean Society, which was broadcast by the BBC in 1982.

Page 34. The debate about levels and units of selection was initiated by Wynne-Edwards's (1962) claim that group selection played an important role in the evolution of the mechanisms through which animal numbers are regulated. Maynard Smith was one of the first to try to clarify the question and show mathematically what group selection involves (1964; reviewed in 1978), but the sharpest critique came from Williams (1966). Wilson (1983) argued for the feasibility of certain types of group selection. Sober and Wilson (1998) have reviewed the whole group selection debate, and analyzed the misconceptions and misunderstandings that accompanied it. Hamilton's theory of kin selection can be found in Hamilton (1964a,b), which are reprinted with a helpful commentary in Hamilton (1996).

Page 35. The selfish gene perspective is evident in most of Dawkins's books and papers, but especially in *The Selfish Gene* (1976), *The Extended Phenotype* (1982), and *The Blind Watchmaker* (1986). The replicator concept is discussed fully in *The Extended Phenotype*, chap. 5.

Page 36. Hull (1980) proposed an alternative to Dawkins's "vehicle"—the "interactor," which is an entity that interacts with the environment and responds to it as a coherent whole. The term avoids the passivity implied by the vehicle concept, and recognizes that sometimes a replicator and an interactor may be one and the same entity. However, since for both Dawkins and Hull a replicator cannot be a unit of development, and a vehicle/interactor cannot be a unit of hereditary variation, from our point of view the dichotomy suggested by both authors is similar and equally unacceptable. In the rest of this book we will talk about the vehicle rather than the interactor, since this term is more familiar and is widely used.

Page 38. Gould's massive *The Structure of Evolutionary Theory* (2002) details his position on the importance of historical and developmental constraints (especially in chaps. 10 and 11), and summarizes his arguments against the selfish-gene approach (pp. 613–644). A very readable and informative popular account of the battles between Gould and Dawkins is that of Sterelny (2001). Brown (1999) has also provided an introduction to the disputes about selfish genes, memes, etc. that is aimed at nonprofessionals. Sociological aspects of the debate are described in Segerstråle (2000). We must stress that Dawkins certainly recognizes the importance of the constraints and contingencies mentioned by Gould. The difference is that Dawkins asks, How is this kind of adaptation at all possible?, whereas Gould asks, Why do we see this particular adaptation? The different perspectives make Dawkins focus on natural selection (for it alone can explain the evolution of complex adaptation), and Gould on the sum of all the factors that affect a trait's evolution.

Page 40. For the "definition of life" issue and the scientific approaches stemming therefrom, see Fry (2000). The notion of limited heredity and the conditions for an evolutionary increase in organizational complexity are discussed in Maynard Smith and Szathmáry (1995).

Page 41. See Jablonka (2004) for a criticism of the replicator concept and a discussion of the value of "the trait" as a unit of evolution.

Page 43. Lindegren (1966) examined the biases in Anglo-Saxon genetics during the first half of the twentieth century. Krementsov (1997) and Soyfer (1994) have described the effects of Lysenko's ideas on Soviet biology. The horrors of twentieth-century eugenics in Germany have been documented by Müller-Hill (1988) in his tellingly titled *Murderous Science: Elimination by Scientific Selection of Jews, Gypsies, and Others, Germany 1933–1945*. His evidence shows that scientists were happy to use their knowledge in the Nazi cause, and that by doing so many benefited from the jobs "vacated" by their Jewish colleagues and from the "research material" available in the death camps.

Page 43. See Lindegren (1949), chap. 20, pp. 6–7 for the statement that two-thirds of the variants in *Neurospora* did not show Mendelian segregation.

Page 45. Sapp (1987) discusses how the Mendelist-Morganist view came to dominate the science of heredity.

2 From Genes to Characters

The term "character" is not easy to define, as can be seen from Schwartz's (2002) historical analysis of what geneticists have meant by characters and the problems they encountered. We use "character" to mean a feature or attribute that has arisen through development and can vary from individual to individual. An excellent description and discussion of the history of the concept of the gene and what it means today can be found in Keller (2000); the practical problems of defining genes in the genomic era are discussed by Snyder and Gerstein (2003). From his interesting analysis of the origins and use of the gene concept, Lenny Moss (2003) concluded that "the gene" is used in two different ways in modern biology—as a predictor of phenotypes and as a developmental resource.

Information about the genetic diseases we mention in this and later chapters can be found in McKusick's *Mendelian Inheritance in Man* (1998), which is available at http://www.ncbi.nlm.nih.gov/entrez/query.fcgi?db=OMIM.

Page 47. Prior to the Human Genome Project, scientists had guessed that humans have about 100,000 genes. When the first drafts of the sequence of human DNA were published, they were accompanied by commentaries in which the estimated number of genes was said to be around 31,000 according to one group of sequencers, and 39,000 according to the other (*Nature* 409: 819, 2001). More recently, it has been suggested that the number of genes is nearer 25,000 (*Nature* 423: 576, 2003). These numbers are based on sequences that are thought to code for protein products. If a gene is defined more broadly—as a DNA sequence whose transcription provides functional information—the number of genes is likely to be very much higher, because it is becoming clear that many of the large number of non–protein-coding RNAs have roles in regulatory control systems (see Eddy, 2001).

Page 48. The original paper describing the structure of DNA was published in *Nature* in *April* of 1953 (Watson and Crick, 1953a). In May it was followed by a second paper (1953b) in which Watson and Crick dealt with the genetic implications of the DNA structure they had proposed. A detailed account of today's ideas about DNA replication and the flow of information from DNA to proteins is given in Alberts et al. (2002) *Molecular Biology of the Cell*, chaps. 5 and 6, and many similar textbooks.

Page 49. The interactions of molecular biologists with biochemists, communication engineers, and physicists, and their influence in shaping the concept of molecular information and organization are described in Keller (1995) and Kay (2000).

Page 52. Two key historical papers on genetic regulation are those by Jacob and Monod (1961a,b). For a clear and beautiful description of the role of genes in development and the complexity of gene regulation, see Coen (1999).

Page 52. Interest in semantic information in biology was rekindled by an article by Maynard Smith (2000) in *Philosophy of Science*, which was followed by responses from several philosophers of biology. Oyama (2000) has given a detailed and influential critique of the concept of information in biology from the developmental systems theory (DST) perspective. Although we agree with much of Oyama's critique, we find it impossible to avoid expressions such as "the transfer of information," "expression of information," etc. without resorting to tortuous language. However, like Oyama, we regard the implication that information is something that is distinct from the process of interpretation as a fallacy. Our concept of information is discussed more extensively in Jablonka (2002).

Page 55. The analogy between a photocopying machine and DNA replication is restricted to the indifference both copying mechanisms have to content; a photocopying machine is not, of course, an *active* replicator in Dawkins's sense.

Page 56. Accounts of the basic molecular biology of sickle cell anemia can be found in almost every textbook of genetics. More details can be found in McKusick (1998).

Page 58. Scriver and Waters (1999) and Badano and Katsanis (2002) show how the distinction between monogenic diseases and more complex ones is an oversimplification, because even so-called monogenic traits involve interactions between several genes. In spite of this, some popular books, e.g., Hamer and Copeland's *Living with Our Genes* (1998), continue to present the effects of genes in simple deterministic ways (although they pay lip service to the complex relations between genotype and phenotype). In contrast, Morange's *The Misunderstood Gene* (2001), which is also written for a general audience, describes very well how complicated the relationship between genes and diseases really is.

Page 59. See Lewontin (1997) for a discussion of the reaction of politicians and others in the United States to cloning. This article, first published in *The New York Review of Books*, is reprinted with additional commentaries in Lewontin (2000a), chap. 8. Keller and Ahouse (1997) show how the way in which the results from the first successful cloning experiment were presented in both the popular and scientific media reinforced a nuclear-DNA-centered view of development.

Page 59. A popular account of "the gene for novelty seeking" is given in Hamer and Copeland (1998), chap. 1. Some of the original data can be found in Benjamin et al. (1996).

Page 61. The role of *APOE* in coronary artery disease is discussed in Templeton (1998). Sing et al. (1995) give some of the technical details of the analysis that Templeton considered.

Page 62. We use the term "plasticity" in the same way as West-Eberhard (2003, p. 33), i.e., as the ability of an organism to react to internal or external environmental inputs with a change in state, form, movement, or rate of activity. This notion of plasticity includes adaptive and nonadaptive, reversible and irreversible, active and passive, continuous and discontinuous responses.

Page 63. Over the years, Waddington modified the form of his epigenetic landscapes, but they are explained in detail in his *The Strategy of the Genes* (1957). For a discussion of the use and present meaning of canalization, plasticity, and related terms, see Debat and David (2001).

Page 63. Knockout mutations are reviewed and discussed in Morange (2001), chap. 5. Wagner (2000) discusses the way in which gene networks can explain the lack of effects of knockouts in yeast, and Gu et al. (2003) analyze the role of duplicate genes. Siegel and Bergman (2002) have shown through mathematical models that where you have complex genetic networks, selection for developmental stability leads to the evolution of canalization.

Page 66. For a review of alternative splicing, see Maniatis and Tasic (2002). Black (1998) gives a brief review of work on the *cSlo* gene.

Page 68. Developmental changes in DNA are described in Watson et al. (1988), chaps. 22 and 23, and in Jablonka and Lamb (1995), chap. 3.

Page 68. Weismann's comments on the behavior of *Ascaris* chromosomes are in the final footnote (pp. 415–416) of vol. 1 of *The Evolution Theory* (1904).

Page 74. Chakravarti and Little (2003) discuss the problems of the bar code approach to human diseases. Weatherall's (1998) appraisal of the contribution of molecular genetics to medical problems was published in the widely read *Times Literary Supplement*; the figure for the incidence of thalassemia comes from this article.

Page 75. Petronis (2001) discusses the role of heritable non-DNA factors in human disease.

Page 76. Estimates of the numbers of genes in flies and worms come from *Nature* 409: 819, table 1 (2001).

Page 77. Strohman (1997) has suggested that biology is about to undergo a paradigm shift, because the present reductionist molecular-genetic paradigm is unable to accommodate the complexity of the networks of interactions that are now being revealed.

3 Genetic Variation: Blind, Directed, Interpretive?

Page 79. The paper that sparked the debate about directed mutation was Cairns et al. (1988). It was accompanied by a summary and opinion by Stahl (1988), and was followed shortly after by letters commenting on the observations and their interpretation (*Nature* 336: 218, 525, 1988; *Nature* 337: 123, 1989). Some of the reactions to the original work and additional experiments carried out during the next five years are reviewed in Jablonka and Lamb (1995), chap. 3. More recent research on directed mutations has been reviewed by Foster (1999, 2000) and Rosenberg (2001). Work described by Hendrickson et al. (2002) and Slechta et al. (2002) suggests that in the case described by Cairns et al. there probably was no preferential induction of adaptive mutations.

Page 82. The origin and advantages of sexual reproduction have been discussed in many books and articles. An indication of the range of ideas that have been considered can be seen from the papers in Michod and Levin (1988). For a later review of the origin of sex, see Maynard Smith and Szathmáry (1995), chap. 9.

Page 83. The variety of modes of sexual reproduction and their possible significance are described and discussed in Maynard Smith (1978), Bell (1982), Burt (2000), and in several papers in Michod and Levin (1988). Birdsell and Wills (2003) review and evaluate most of the theories. One observation for which theories about the adaptive value of sex have not provided a good explanation is the long-term, exclusively asexual reproduction of bdelloid rotifers.

Page 83. For a short, popular account of the evolutionary significance of the life cycle of aphids, see Blackman (2000).

Page 84. See Maynard Smith (1978) and Michod and Levin (1988) for discussions of the evolution of recombination rates, and Bernstein and Bernstein (1991) for the relationship between sex, recombination, and DNA repair.

Page 86. An account of the origin of mutations through accidental mistakes in DNA replication and maintenance can be found in Alberts et al. (2002), chap. 5. The values for error rates during DNA replication are taken from Radman and Wagner (1988).

Page 87. Drake et al. (1998) discuss the evolutionary factors that have shaped the mutation rates of different organisms.

Page 88. Details of Barbara McClintock's ideas about the mechanisms through which the genome can be restructured can be found in Fedoroff and Botstein's *The Dynamic Genome* (1992), in which McClintock's key papers are reprinted and her work discussed by her former students and colleagues. Jim Shapiro was among the few scientists who stressed the regulated nature of genomic changes long before it became the fashionable topic that it now is (e.g., see Shapiro 1983, 1992). For recent views on adaptive genetic changes, see *Molecular Strategies in Biological Evolution*, a collection of papers edited by Caporale (1999).

Page 89. For a general discussion of stress-induced mutation, see Velkov (2002); for evidence of stress-induced mutation in natural populations of bacteria, see Bjedov et al. (2003).

Page 94. Moxon et al. (1994), and Moxon and Wills (1999) discuss the evolution of mutational "hot spots" in contingency genes.

Page 96. Caporale (2000) gives brief details of highly mutable loci in snails and snakes.

Page 97. See B. E. Wright et al. (1999) for details of her experimental work with *E. coli*, and B. E. Wright (2000) for the evolutionary interpretation of the phenomena she and her colleagues have studied. The work of Datta and Jinks-Robertson (1995) with yeast shows that the association of increased mutation rates with high levels of transcription is not confined to bacteria.

Page 98. Schneeberger and Cullis (1991) found an increase in regional mutation following stress in flax, and Waters and Schaal (1996) obtained similar results with *Brassica*.

Page 101. For short reviews on the evolution of adaptive mutation rates, see Metzgar and Wills (2000) and Caporale (2000).

Page 101. "Chance favors the prepared genome" is the title of Caporale's introduction to *Molecular Strategies in Biological Evolution* (1999).

Page 102. For the distinction between instructive and selective processes in evolution, see Jablonka and Lamb (1998a), and for proximate and ultimate causes in biology, see Mayr (1982).

Page 103. The effects of environmental conditions on recombination rates are reviewed in Hoffmann and Parsons (1997); Grell (1971, 1978) has described the effects of heat stress on recombination in *Drosophila*.

Page 104. For his assertion that Lamarckian inheritance is incompatible with the central dogma, see Maynard Smith (1966), p. 66. The similar assertion by Ernst Mayr was made in an interview with Adam Wilkins; see Wilkins (2002a), p. 965.

4 The Epigenetic Inheritance Systems

Jablonka and Lamb (1995, 1998b) give more detailed information about most of the topics discussed in this chapter. Waddington coined the term "epigenetics" in the 1940s, suggesting that it was "a suitable name for the branch of biology which studies the causal interactions between genes and their products which bring the phenotype into being" (see Waddington 1968, p. 9). The subsequent transformations of the concept have been described by Holliday (1994), Wu and Morris (2001), and Jablonka and Lamb (2002). We need to stress here that "epigenetics" and "epigenetic inheritance" are not the same, and in this chapter we focus on epigenetic inheritance, not on epigenetics, which is a much broader topic. We look at some of the more general aspects of epigenetics in chapter 7.

Page 113. See Holliday and Pugh (1975) and Riggs (1975) for the first suggestions that DNA methylation is a cell memory system. A good overview of the history and current state of epigenetics research can be found in Urnov and Wolffe (2001).

Page 114. There are good evolutionary reasons why most Jaynus creatures reproduce through a single-cell stage. Discussion of the advantages of a single-cell stage in the life cycle can be found in Jablonka and Lamb (1995), chap. 8, and Dawkins (1982), chap. 14.

Page 119. One of the first substantial discussions of the role of cell heredity in development was that of Holliday (1990). Jablonka et al. (1992) divided epigenetic inheritance systems into three categories, but the discovery of RNA interference has forced us to add a fourth EIS, and in view of the daily developments in molecular biology, further updates may be necessary.

Page 119. For an outline of early ideas about the inheritance of self-sustaining feedback loops, see Jablonka and Lamb (1995), chap. 4. Jablonka et al. (1992) have modeled the inheritance of simple self-sustaining loops in cell lineages, and Thieffry and Sánchez (2002) have used models and computer simulations to describe the behavior of regulatory loops during the development of *Drosophila*. Complex self-sustaining regulatory networks are described in Kauffman (1993).

Page 121. For some of the early work on structural inheritance in *Paramecium*, see Beisson and Sonneborn (1965). More recently a seemingly similar instance of structural inheritance has been found in yeast by Chen et al. (2000). For an excellent review of structural inheritance in general, see Grimes and Aufderheide (1991). Hyver and Le Guyader (1995) have developed a model that suggests how cortical inheritance in ciliates may take place.

Page 122. See Cavalier-Smith (2000, 2004) for introductions to that author's extensive work on membrane heredity. The quotation comes from Cavalier-Smith (2004).

Page 123. Rhodes (1997) gives a popular account of kuru and other prion diseases. Durham (1991, pp. 393–414) discusses the anthropological aspects of cannibalism and kuru among the Fore people. The genetic interpretation of kuru is given in Bennett et al. (1959).

Page 123. An outline of Gajdusek's work on prion diseases can be found in Gajdusek (1977), the lecture he gave when he received the Nobel Prize.

Page 124. For a discussion of the molecular nature of prions and the mechanisms of their propagation, see Prusiner (1995) and his Nobel Prize lecture (1998). Collinge (2001) gives an excellent review of all aspects of prion diseases in humans and other animals. The protein-only hypothesis of prion propagation was not accepted by all workers in the field, and some still maintain that nucleic acids must be involved somewhere. However, the work of King and Diaz-Avalos (2004) and Tanaka et al. (2004) leaves little doubt that, at least in yeast, prion propagation does not involve nucleic acids.

Page 125. For a brief account of prions in yeast and *Podospora*, see Couzin (2002b). The prion-like properties of a protein that is important in memory in *Aplysia* are described in Si et al. (2003a,b).

Page 126. A clear description of how DNA is packaged into chromatin can be found in Alberts et al. (2002), chap. 4. For a more detailed description of chromatin by one of the leading workers in the field, see B. M. Turner (2001).

Page 128. See Holliday (1996) for an account of the first twenty years of research on DNA methylation, and Bird (2002) for a review of current ideas about its role in cell memory. Several papers in which various aspects of methylation are reviewed can be found in *Science* vol. 293 (2001), of which the August 10 issue is devoted to epigenetics. For evidence of highly specific localized methylation patterns that are associated with gene activity, see Yokomori et al. (1995) and Futscher et al. (2002).

Page 130. Holliday suggested as long ago as 1979 that changes in DNA methylation are involved in the development of cancer. Work described by Jones and Baylin (2002), Feinberg and Tycko (2004), and papers in *Epigenetics in Cancer Prevention* (Verma et al., 2003) show that he was right. The possible role of heritable methylation changes in aging is discussed in Lamb (1994), and Issa (2000) provides evidence of progressive changes in methylation associated with cancer and with age. Evidence that mice embryos lacking any of the three known methyltransferase genes develop abnormally can be found in Li et al. (1992) and Okano et al. (1999).

Page 131. Protein marks and their propagation are described in Lyko and Paro (1999), Cavalli (2002), and Henikoff et al. (2004).

Page 131. Cell-heritable histone modifications are described in B.M. Turner (2001), and Jenuwein and Allis (2001) discuss ideas about the histone code. McNairn and Gilbert (2003) describe the way epigenetic marks are replicated.

Page 132. Urnov and Wolffe's (2001) review is a comprehensive account of the development of ideas about epigenetic marks. For a review of the inheritance of chromatin marks in mammals, see Rakyan et al. (2001), and for an account of the way in which chromatin marks are changed by environmental conditions, see Jaenisch and Bird (2003).

Page 132. RNA interference is a new and fast-moving field, and it is difficult to follow the changing ideas and terminology, but a short overview and history of the subject is available at http://www.ambion.com. In December 2002, the journal *Science* (298: 2296–2297) named small RNAs (such as siRNA) the "breakthrough of the year." Good short reviews of RNAi can be found in Matzke et al. (2001), Hannon (2002), Voinnet (2002), and Novina and Sharp (2004).

Page 134. Information about the movement of siRNAs in plants is given in Jorgensen (2002).

Page 135. For RNAi as a defense system, see Plasterk (2002).

Page 135. The developmental aspects of microRNAs are discussed in Pasquinelli and Ruvkun (2002), and Banerjee and Slack (2002).

Page 136. Dykxhoorn et al. (2003) review the application of RNAi in medicine and other areas of biology.

Page 137. The mechanistic overlap of EISs is exemplified by the case described by Roberts and Wickner (2003) in which a yeast prion is propagated through a self-sustaining feedback loop. The mature, functional form of the protein is a protease that is necessary for the conversion of the immature form to the mature one. Thus, as with conventional prions, one form of a protein (the mature form) converts another (the immature form) into its own image.

Page 138. Work by Hirschbein and her colleagues on epigenetic inheritance in bacteria is described in Grandjean et al. (1998); Klar (1998) discusses epigenetic inheritance in yeast; Casadesús and D'Ari (2002) review other work on epigenetic transmission in microorganisms.

Page 139. The term "imprint" was first used in Crouse's (1960) description of her work on the chromosomes of the fly *Sciara*. For a review of early work on imprinting, see Jablonka and Lamb (1995), chap. 5, and for more recent observations and ideas, see Ferguson-Smith and Surani (2001) and Haig (2002). Up-to-date information on im-printed genes in mammals is available at http://www.mgu.har.mrc.ac.uk/research/imprinting.

Page 140. The history of the peloric variant of *Linaria* is given in Gustafsson (1979), which is the source of the quotation from Linnaeus. Experiments showing the epigenetic nature of peloric variants are described in Cubas et al. (1999). Other heritable epigenetic variations are known in plants; e.g., see Jacobsen and Meyerowitz (1997).

Page 142. The term "epimutation" was introduced by Holliday (1987), who used it for an inherited change based on DNA modifications such as methylation. It is now used more widely and is applied to any heritable epigenetic modification.

Page 142. The role of epimutations in the inheritance of the yellow phenotype in mice is described in Morgan et al. (1999), Whitelaw and Martin (2001), and Rakyan et al. (2001). For the effect of methyl supplements on the heritable coat color phenotypes, see Wolff et al. (1998). Rakyan et al. (2003) found that another classic mouse mutant, *Fused*, is also an epimutation. Sutherland et al. (2000) give a general account of the reactivation of heritably silenced transgenes in mice.

Page 144. Transgenerational effects of injected double-stranded RNA are described in Fire et al. (1998) and Grishok et al. (2000).

Page 145. The role of EISs in speciation is discussed in Jablonka and Lamb (1995), chap. 9. See also Pikaard (2001) for arguments about the importance of heritable gene silencing in speciation through polyploidy.

Page 145. For the work on Mongolian gerbils, see Clark et al. (1993) and Clark and Galef (1995).

Page 150. McLaren (2000) has given an interesting historical review of the development of cloning techniques. Solter (2000) and Kang et al. (2003) discuss the practical difficulties involved in reprogramming the mammalian genome during cloning, and Rhind et al. (2003) review the biological problems associated with cloning humans.

Page 151. Pál and Hurst (2004) have argued (incorrectly, we believe) that since the transmissibilities of epigenetic variants can be greater or less than 50 percent, the role of epigenetic inheritance in adaptive evolution is limited. It is certainly possible for transmissibility to be more than 50 percent: it can occur through paramutation, in which one allele in a heterozygote converts the other allele into its own epigenetic image. This phenomenon, which was reviewed by Brink (1973), is now being studied at the molecular level, and seems to involve DNA methylation (see Hollick et al., 1997; Chandler et al., 2000).

Page 152. Steele's hypothesis is described in his 1981 book. A more recent and popular account of it can be found in Steele et al., *Lamarck's Signature* (1998). Zhivotovsky (2002) has developed models that show the evolutionary advantages of the type of mechanisms that Steele proposed.

Page 153. The quotation comes from Crick (1970), p. 563.

5 The Behavioral Inheritance Systems

A large part of this chapter is based on Avital and Jablonka's *Animal Traditions* (2000).

Page 155. Books that stress the genetic basis of human sexual behavior include those by Baker (1996), Buss (1994, 1999), Miller (2000), and Thornhill and Palmer (2000).

Page 156. The tarbutnik thought experiment is a modified version of that given in Avital and Jablonka (2000), chap. 1.

Page 159. An adaptive change in an auditory signal that is probably socially learned has been reported by Slabbekoorn and Peet (2003). They found that in urban areas, where the normal songs of great tits are masked by the noise, the tits sing at a higher pitch.

Page 160. For a discussion of the definition and use of the term "culture" in ethology, see Mundinger (1980), Avital and Jablonka (2000), pp. 21–24, and Rendell and Whitehead (2001). For the definitions used in anthropology, see Kuper (2002).

Page 161. See Avital and Jablonka (2000), chap. 3, for a discussion of learning processes and the mechanisms of social learning. Boakes's *From Darwin to Behaviourism* (1984) gives an excellent overview of the history of experimental psychology and ideas about learning, and the books edited by Heyes and Galef (1996) and Box and Gibson (1999) provide good summaries of recent studies of social learning in animals.

Page 162. Experimental work on the transmission of food preferences in rabbits is described in Bilkó et al. (1994).

Page 163. The carrot juice experiment in humans is described in Mennella et al. (2001).

Page 165. In addition to food preferences, there is evidence that other types of behavior are strongly influenced by information transmitted prenatally and through maternal care. For example, through cross-fostering, Francis et al. (1999, 2003) found that the pre- and postnatal environments provided by mothers influence strain differences in maternal behavior and the stress responses of rats, and in the learning ability, anxiety, and perceptual acuity of mice.

Page 167. The word "imprinting" causes biologists problems, because it is used for two totally different phenomena—the genomic imprinting we described in chapter 4, and the behavioral imprinting we discuss here. Usually it is clear from the context which type is being discussed; it is only with computer searches that life gets difficult! For a definition and discussion of behavioral imprinting, see Immelmann (1975). The original paper in which Spalding (1873) described filial imprinting is reproduced in Haldane (1954). For Lorenz's work on imprinting, see Lorenz (1970).

Page 168. In their excellent review, ten Cate and Vos (1999) have described the complexities of sexual imprinting and its wide distribution in birds. Laland (e.g., 1994) has developed theoretical models showing some of the evolutionary consequences of the transmission of information through sexual imprinting.

Page 169. Heyes (1993) has discussed the mechanisms of social learning and the lack of evidence for widespread imitation in animals.

Page 169. The spread of the milk bottle–opening habit in birds is described in Fisher and Hinde (1949). Sherry and Galef (1984) discuss the type of learning involved in this behavior.

Page 170. The work on the new tradition found in Israeli black rats is described in Aisner and Terkel (1992) and Terkel (1996).

Page 172. For discussions of imitation, see Whiten and Ham (1992), Heyes (1993), Tomasello (1999), Byrne (2002), and Sterelny (2003). Catchpole and Slater (1995) discussed imitative learning in songbirds; evidence for imitation in whales and dolphins is reviewed in Rendell and Whitehead (2001).

Page 176. The idea that is now called "niche construction" was discussed by Darwin (1881) in his book on worms, and later in books by Waddington (1959a) and Hardy (1965), and in an influential article by Lewontin (1978). For more recent work on the theory of niche construction, see Odling-Smee (1988) and Odling-Smee et al. (1996, 2003).

Page 177. Work on chimpanzee cultures is summarized in Whiten et al. (2001). Van Schaik et al. (2003) describe similar cultural diversity in orangutans.

Page 178. The observations and ideas of the Japanese primatologists who have studied the Koshima macaques are described in Hirata et al. (2001). De Waal (2001) has examined the research tradition of the Japanese school of behavioral ecology, and provided an enlightening analysis of their work and the debates surrounding it.

Page 179. The way in which group organization and social construction may lead to cultural evolution in animal societies is discussed in more detail in Avital and Jablonka (2000), chap. 8.

Page 181. The adoption of a new nesting habit by Mauritius kestrels is described in Collias and Collias (1984).

Page 183. Sept and Brooks (1994) have documented early observations of culture in chimpanzees.

Page 184. For a discussion of the role of behavioral imprinting in speciation, see Immelmann (1975), Irwin and Price (1999), and ten Cate (2000). The experiments showing the establishment of reproductive isolation through host imprinting in parasitic birds are described in Payne et al. (2000). Many possible cases of environmentally induced speciation events in plants and animals are described in West-Eberhard (2003) chap. 27. Gottlieb (2002) gives a detailed interpretation of observations (Bush, 1974) that suggest that sympatric speciation occurred in *Rhagoletis* fruit flies. He argues that speciation in this case was initiated by migration to a new niche and the adoption of changed habits, which persisted.

Page 186. Experiments on sexual imprinting in quails were described and discussed by P. P. G. Bateson (1982).

Page 186. The role of adopters and helpers in the transmission of behavioral information is discussed in Avital and Jablonka (1994, 1996, 2000). Avital et al. (1998) have developed a model of the cultural spread of adoption.

Page 188. Huffman (1996) has discussed what may be a "bad habit" in animals—the seemingly nonadaptive and possibly damaging "stone-handling" tradition of some Japanese macaques. It would be interesting to know whether the frequently reported "drunkenness" of elephants and monkeys that eat fermented fruit is also the result of a tradition that has damaging effects.

Page 190. The work on fire ants is described in Keller and Ross (1993). Other evidence for behavioral inheritance in insects can be found in Avital and Jablonka (2000), pp. 353–356.

6 The Symbolic Inheritance System

Articles in *Technological Innovation as an Evolutionary Process*, edited by Ziman (2000), and *The Evolution of Cultural Entities*, edited by Wheeler et al. (2002), give an idea of the ways in which evolutionary theory is being applied in studies of human culture. Different approaches to cultural evolution can be found in Dunbar et al., *The Evolution of Culture* (1999).

Page 193. "The third chimpanzee" is how Jared Diamond (1991) described humans in a book discussing both our continuity with other animal species and our uniqueness.

Page 193. Cassirer's philosophy, which is based on the analysis of symbolic systems, is described in his three-volume masterpiece, *The Philosophy of Symbolic Forms* (published in German 1923–1929, and in English translation 1953–1957). In addition, he wrote a beautiful and accessible book, *An Essay on Man* (1944), especially for an English-speaking audience.

Page 195. The thought experiment on which ours is based can be found in Spalding (1873), which is reproduced in Haldane (1954); we return to it in chapter 8. We need to stress that our parrot story does not reflect the true learning abilities and communication of parrots. Pepperberg's (1999) account of her work with gray parrots shows that this species has amazing cognitive abilities and can be taught some symbolic communication, but we know little about parrots' communication in the wild.

Page 198. Lakoff and Johnson (1999) discuss the fundamental importance of metaphors in the development of thought and language.

Page 200. In this chapter our focus is on symbols and symbolic systems, which are uniquely human, so we do not discuss the complexities of the different types of signs, such as iconic and indexical signs. Interested readers can turn to Sebeok (1994) for a discussion of this topic. For a discussion of language as a symbolic system, see Deacon (1997).

Page 201. In the early days of molecular genetics, the organization of information in DNA was constantly likened to the organization of information in language. It is interesting to see that the analogy now works the other way around. In their recent paper on the evolution of language, Hauser et al. (2002) wrote that a Martian might note that "the human faculty of language appears to be organized like the genetic code—hierarchical, generative, recursive, and virtually limitless with respect to its scope of expression" (p. 1569).

Page 202. The genetic basis of courtship dances and songs in *Drosophila* is described in Hall (1994). Song learning by male songbirds is reviewed in Catchpole and Slater (1995).

Page 205. Cavalli-Sforza and Feldman (1981) and Boyd and Richerson (1985, 1988) have constructed different types of models of cultural evolution. The former assume that there are discrete units of culture that are transmitted from generation to generation through copying-like processes; the latter do not think in terms of discrete cultural units, focusing instead on phenotypic change over time. For a recent general review of these and other models, see Laland and Brown (2002), chap. 7. Sperber's criticism of models of cultural evolution that are based on the assumption of copying can be found in Sperber (1996) chap. 5.

Page 206. For a helpful exposition of the memetic view of culture, see Dennett (1995, 2001). The "selfish" meme view can be found in Blackmore (1999, 2000). Aunger (2002) argues that memes are dynamic neural circuits that are maintained and replicated in the brain.

Page 208. Different views and criticisms of the meme concept are presented in Aunger (2000). Additional critical discussions of the concept's usefulness in studies of cultural evolution can be found in Rose and Rose (2000).

Page 210. The transmission of mother's psychopathological behavior was described by Peter Molnar of the Semmelweis Medical School, Budapest, in a lecture given at an interdisciplinary symposium held at Bielefeld University in Germany in 1991.

Page 211. Some delightful examples of nonrandom, context-sensitive changes in children's rhymes can be found in Opie and Opie (1959), where it is shown how rhymes have been modified to fit the geographic area and political climate. For example, English rhymes that 350 years ago featured the King of France, in the early twentieth century featured Kaiser Bill.

Page 212. Representative examples of the rapidly growing literature (which already includes several textbooks) describing human behavioral evolution from the sociobiology–evolutionary psychology perspective are Barkow et al. (1992), Buss (1994,

1999), Plotkin (1997), and Miller (2000). Cosmides and Tooby (1997) have provided online "a primer" for evolutionary psychology.

Page 213. This thought experiment was first used by Jablonka and Rechav (1996), and later was expanded in Avital and Jablonka (2000), chap. 10.

Page 215. Pinker's *The Language Instinct* (1994) is an accessible account of arguments for the existence of a language module.

Page 216. The case for a module for the detection of social rule-breaking is presented by Cosmides and Tooby (1997). Arguments for the existence of a mate-choice module can be found in Buss (1994, 1999). The sexual-selection-for-male-creativity explanation of sex differences is presented in Miller (2000).

Page 218. The experiments with ducklings are described in Gottlieb (1997). In *Thought in a Hostile World* (2003), chaps. 10 and 11, Sterelny gives a detailed critique of the "massive modularity view" supported by most evolutionary psychologists. By focusing on "the natural history module"—a module that is supposed to explain the cross-cultural ability of humans to identify different animals in a way that corresponds to the species category—he illustrates some of the problems associated with the extreme modularity view.

Page 218. See Lenneberg (1967) and Chomsky (1968) for their early views on innate language capacities.

Page 220. For Mary Midgley's criticism of the usual selectionist approach to cultural change, see Midgley (2002). Plotkin (1997, 2000) shares many of her concerns, and in spite of his sympathy for a (reformed) meme-oriented view of culture, stresses its developmental aspects.

Page 221. Teubner (2002) discusses how legal and cultural institutions constrain each other.

Page 224. Donald's scenario of human evolution is described in his book *Origins of the Modern Mind* (1991).

Page 225. I.M. is referring to the transmission of developmental legacies in gerbils, which was discussed in chapter 4, p. 145.

Page 227. Here I.M. adopts the view of Fracchia and Lewontin (1999), who argued that historical change is not evolutionary in any of the accepted senses of the term.

Page 227. Dobzhansky's definition of evolution as "a change in the genetic composition of populations" can be found in his *Genetics and the Origin of Species* (1937), p. 11.

Page 229. For David Hull's position on Lamarckism, see Hull (2000). Mayr's definition of soft inheritance can be found in Mayr (1982), p. 959.

Between the Acts: an Interim Summary

Page 237. Jared Diamond (1997) has discussed the importance of ecological and geographic factors in the origin and spread of major cultural innovations. For example, he argues that the origin of domestication depended primarily on the availability of suitable species in certain (limited) regions. Domestication gave the populations that practiced it enormous benefits over their neighbors, and these benefits, as well as the geographic opportunities for expansion, determined the way agriculture spread out over the continents.

7 Interacting Dimensions—Genes and Epigenetic Systems

Reviews and discussions about the interactions between genetic and epigenetic systems can be found in Jablonka and Lamb (1995), chaps. 7–9, and Jablonka and Lamb (1998b). Many of the papers in which Waddington describes and discusses his work on genetic assimilation can be found in the collection *The Evolution of an Evolutionist* (Waddington, 1975a).

Page 245. The well-known remark of Solly Zuckerman is reported in (among other places) Lewontin (1993), p. 9.

Page 247. For a general discussion of the way in which chromatin structure biases DNA sequence changes, see Jablonka and Lamb (1995), chap. 7.

Page 248. The possible role of epimutations in cancer was first suggested by Holliday (1979). Cancer epigenetics is now a very active field of research, as can be seen from the papers and reviews in Verma et al. (2003). The interplay between the epigenetic and genetic systems in tumorogenesis has been described by Baylin and Herman (2000).

Page 249. For McClintock's idea about the role of stress in promoting transposition and molding the genome, see McClintock (1984). The molecular mechanisms underlying some types of transposition are described in Raina et al. (1998).

Page 249. The role of transposable elements in adaptive evolution and the evolution of genome structure is the subject of much speculation and an increasing amount of experimental work. Capy et al. (2000), Kidwell and Lisch (2000), and Jordan et al. (2003) review some of the recent data and ideas about the actual and potential beneficial effects of transposable elements on host evolution and on patterns of gene regulation. A strong case for the adaptive significance of repetitive DNA elements, including transposons, has been made by Sternberg (2002). Evidence suggesting how transposable elements are involved, through the mediation of epigenetic processes, in the formation of new plant species through polyploidization has been reviewed by Pikaard (2001).

Page 251. For simplicity, we have ignored the way that multicellular organisms maintain structures through interactions between cells and the extracellular matrix.

The maintenance of animal morphology during growth seems to involve complex 3D templating processes that are as yet poorly understood (see Ettinger and Doljanski, 1992).

Page 251. A well-known example of cells switching from one type to another is the transdetermination of larval cells seen in *Drosophila* (see Hadorn, 1978).

Page 254. Early work on imprinting is summarized in Jablonka and Lamb (1995) chap. 5, where it is suggested that imprints originated as a by-product of the different ways chromatin is packaged in male and female gametes. De la Casa-Esperón and Sapienza (2003) have developed this idea further, and suggested that the persistence of imprints in the germ line might be associated with the need for homologous maternal and paternal chromosomes to recognize each other, pair, and recombine. Holliday (1984) also argued that epigenetic differences (specifically methylation differences) between homologous chromosomes are signals for recombination. Many of the evolutionary theories of genomic imprinting have been described and evaluated by Hurst (1997) and Wilkins and Haig (2003). Other good sources of information on various aspects of imprinting are Ohlsson (1999), Ferguson-Smith and Surani (2001), and Haig (2002).

Page 256. The inactivation or elimination of chromosomes in the cells of male scale insects is described in White (1973), chap. 14. Ohno (1967) provides an excellent summary of the early work on sex chromosomes. Recently it has been found that the Y chromosome has more genes than previously assumed (seventy-eight protein-coding genes altogether, although they code for only twenty-seven distinct proteins), but their number is still small compared to the large number of genes on the X chromosome (Skaletsky et al., 2003).

Page 256. Lyon presented her famous hypothesis in a short paper in *Nature* in 1961. Recent articles focusing on the mechanism of X chromosome inactivation include Lyon (1998) and Plath et al. (2002). Lyon's 1999 review discusses imprinting and nonrandom X chromosome inactivation in extraembryonic tissues and in marsupials. Park and Kuroda (2001) emphasize the epigenetic aspects of X chromosome inactivation.

Page 257. Haig's hypothesis and its ramifications and implications are described in his collected papers (Haig, 2002).

Page 258. The idea that the asymmetrical transmission of X chromosomes could lead to the paternal and maternal X chromosomes being differentially marked was suggested by Jablonka and Lamb (1990a), p. 265, and later extended and modeled by Iwasa and Pomiankowski (1999).

Page 259. The Russian group's work on silver foxes is described in Belyaev (1979) and Belyaev et al. (1981b). A later account of the history, results, and state of the study was given by Trut (1999). Belyaev's interpretation of patterns of inheritance in terms of dormant genes can be found in Belyaev et al. (1981a,b). His views on domestication as a model for the role of stress in adaptive evolution are summarized in his 1979 paper.

Page 261. The ways in which Waddington's term "epigenetics" and the term "development" are used overlap, and this is particularly evident when they are used adjectivally. An "epigenetic change" is a "developmental change" that leads to an altered phenotype. The adjective "epigenetic" does not mean that the change is heritable, i.e., "epigenetic change" is not identical with "heritable epigenetic change." Waddington did not address the question of the mechanism of cell heredity in any detail; the concept of "epigenetic inheritance" came later.

Page 262. For his ideas on canalization, see Waddington (1942, 1957, 1975a).

Page 264. See Waddington (1961) for an outline of some of his experimental results and arguments for thinking about the evolution of adaptations in terms of genetic assimilation.

Page 264. A simple and idealized example of the way genetic assimilation might occur is as follows. Imagine that only three genes (A, B, and C) each with two alleles (A^1 and A^2, B^1 and B^2, C^1 and C^2) contribute to the development of a trait such as crossvein formation in fruit flies. The frequency of each allele A^1, B^1, and C^1 (call them type 1 alleles) in the population is 1/10, and of A^2, B^2, and C^2 (type 2 alleles) is 9/10. In a normal environment, without a heat shock, abnormal crossveins develop only if a fly is $A^1A^1B^1B^1C^1C^1$. This occurs with a frequency of $(1/10)^6$, i.e., 1 in 1 million, so the chances of finding a fly showing a crossveinless phenotype are very small. However, if pupae with genotypes with two or more type 1 alleles (e.g., $A^1A^2B^2B^2C^2C^1$, $A^2A^2B^2B^2C^1C^1$, $A^2A^2B^1B^1C^2C^1$, etc.—there will be 17.3 percent of these in the population) develop the crossveinless phenotype when exposed to a heat shock, and if flies with such a phenotype are systematically selected as parents of the next generation, the frequency of the type 1 alleles will increase. As the population becomes enriched with type 1 alleles, the probability of encountering the rare, heat shock–independent genotype ($A^1A^1B^1B^1C^1C^1$) will increase too.

Page 265. One of the persons who made creative use of Waddington's ideas was Matsuda (1987). He argued that changes in life histories, such as whether or not an animal has a larval stage, were initiated by environmentally induced changes in development that were subsequently genetically assimilated. West-Eberhard (2003) discusses Matsuda's work and gives many examples of such life history changes.

Page 266. For reviews of the biological role of Hsp90 chaperones, see Buchner (1999) and Mayer and Bukau (1999). The experiments on the effects of Hsp90 in *Drosophila* are described in Rutherford and Lindquist (1998). McLaren (1999) discusses these experiments in the more general context of other observations showing the transition from developmentally induced to uninduced characters, including Waddington's genetic assimilation experiments. She suggested that the famous pads of the midwife toad, which were the focus of so much controversy (see note for page 22), could have arisen through a process of genetic assimilation.

Page 268. The effects of a shortage of Hsp90 in *Arabidopsis thaliana* are described in Queitsch et al. (2002).

Page 269. For the experiments on assimilation through the selection of epigenetic variations in isogenic lines of *Drosophila*, see Sollars et al. (2003). Ho et al. (1983) described an earlier experiment in which an induced trait was incorporated into the hereditary makeup of an inbred line of *Drosophila*. Rutherford and Henikoff (2003) have provided an interesting discussion about the contribution of epigenetic variations to quantitative heritable traits.

Page 271. The behavior of the yeast [PSI$^+$] prion is reviewed in Serio and Lindquist (2000); its heritable effects are described in True and Lindquist (2000). Chernoff (2001) discusses the general evolutionary significance of inherited variations in protein structure.

Page 274. The experiments on salt resistance and the assimilation of anal papilla changes are described in Waddington (1959b). Scharloo's (1991) review of ideas about canalization is focused on Waddington's work and includes a reappraisal of the salt resistance experiments.

Page 279. The suggestion that the paternal X chromosome is inactivated by maternal factors was made by Moore et al. (1995). For evidence that the paternal X chromosome in the zygote is at first partially active, and is later reactivated in the embryo but further inactivated in the extraembryonic tissues, see Huynh and Lee (2003). Evolutionary biologists have suggested many possible scenarios leading to small Y chromosomes with few genes; often degeneration of the Y chromosome is posited to be associated with a lack of recombination between it and the X chromosome. The scenario suggested here, in which the X and Y chromosomes would have different chromatin conformations, is consistent with many of these hypotheses, because differences in chromatin structure would lead to a reduction in recombination between the chromosomes (Jablonka and Lamb, 1990a).

Page 281. West-Eberhard (2003) has emphasized that newly induced responses are often adaptive adjustments, and genetic changes follow rather than precede the environmentally induced changes. This view is shared by Schlichting and Pigliucci (1998), who have provided a detailed analysis of canalization and plasticity. Theoretical models showing how selection leads to canalization are discussed by Meiklejohn and Hartl (2002) and Siegel and Bergman (2002).

8 Genes and Behavior, Genes and Language

The first part of this chapter is based on Avital and Jablonka's discussion of the evolution of learning in their *Animal Traditions* (2000), chap. 9. The section on the interplay between genes and culture in the evolution of the linguistic capacity is based on Dor and Jablonka (2000).

Page 285. Ideas and models of niche construction are discussed in Odling-Smee (1988) and Odling-Smee et al. (1996, 2003). J. S. Turner (2000) has taken the idea of niche construction in a different direction, arguing that in many cases biological interactions within the niche lead to the construction of an "extended-organism," which is made up of several closely interacting and developing species, and grows and develops as a coherent whole.

Page 286. More information on the nonlearned responses of hyenas and other animals can be found in Avital and Jablonka (2000), chap. 9.

Page 288. Cronin (1991) describes the history of the idea of evolution through sexual selection.

Page 289. It was Simpson (1953) who popularized the term "Baldwin effect" for the mechanism of evolution described independently by Osborn, Lloyd Morgan, and Baldwin. All three scholars developed the idea at more or less the same time and communicated about it; in Groos's book *The Play of Animals* (1898), there is an appendix by Baldwin which includes a statement about what we now call the Baldwin effect which was "prepared in consultation with Principal Morgan and Professor Osborn." The quotation about "congenital variations" is from this appendix. Details of the original sources of the ideas embodied in the Baldwin effect can be found in Simpson (1953), who was generally dismissive of it, and in Hardy (1965). Hardy (chap. 6) gives a more sympathetic account of the idea and its subsequent treatment, and also compares it with genetic assimilation. Waddington discussed the supposed differences between genetic assimilation and the Baldwin effect in *The Strategy of the Genes* (1957), chap. 5. He stressed the importance of the selection of hereditary factors that determine the capacity to respond to the environment, rather than the selection of factors that simulated a particular induced response, and believed that this made genetic assimilation different from the Baldwin effect. However, Lloyd Morgan in particular certainly recognized that changes in the plasticity of a response would be one of the effects of selection. Historical and philosophical analyses of the Baldwin effect can be found in the book edited by Weber and Depew (2003). Ancel (1999) has developed a useful model of the Baldwin effect.

Page 290. For a discussion of the genetic assimilation of behavioral traits, the assimilate-stretch principle, and the evolution of categorization through partial assimilation, see Avital and Jablonka (2000), chap. 9.

Page 293. Baldwin's interest in social heredity is evident in many of his writings, including his 1896 paper.

Page 293. "Coevolution" is now used for two somewhat different processes. It was originally used for the interdependent genetic evolution of two different species, such as a parasite and its host, but is now also used to describe the interdependent selection of genes and culture in human evolution. Durham discusses the coevolution of genes and culture in his *Coevolution: Genes, Culture, and Human Diversity*

(1991), which gives a detailed account of the genetic and cultural aspects of lactose absorption.

Page 293. "Drinka pinta milka day" was a very successful advertising slogan devised by the National Dairy Council in Britain in the late 1960s. Thanks to cultural evolution, "pinta" became a recognized synonym for a pint bottle of milk.

Page 296. For a more recent analysis of the role of cultural processes in determining the frequency of lactose absorbers, see Holden and Mace (1997). Swallow and Hollox (2001) provide a comprehensive discussion of all aspects of lactose absorption.

Page 297. For a useful and readable review of the evolutionary aspects of human genetic diseases, including sickle cell anemia and Tay-Sachs disease, see Diamond and Rotter (2002).

Page 297. Arranged marriage is practiced in ultrareligious Jewish communities, and premarriage genetic counseling is quite common.

Page 298. In her *Developmental Plasticity and Evolution* (2003), West-Eberhard provides excellent analyses of many evolutionary adaptations that were initially environmentally induced or learned.

Page 298. A good account of language and its development and origins can be found in Aitchison (1996). Since the 1990s, there has been a greatly increased interest in the evolution of language, and this is reflected in the many conferences and the publications resulting from them, which include the books edited by Hurford et al. (1998), Knight et al. (2000), Wray (2002), and Christiansen and Kirby (2003). The book edited by Trabant and Ward (2001) includes more historical and semiotic approaches to language evolution, and that edited by Briscoe (2002) focuses on computational models of language.

Page 298. Chomsky's views on the language organ can be found in Chomsky (1975, 2000). Pinker's popular book *The Language Instinct* (1994) provides an accessible summary of the Chomskian position.

Page 302. Many evolutionary biologists have been critical of Chomsky's saltational position regarding the evolution of language. See, e.g., Pinker and Bloom (1990), Dor and Jablonka (2000), and papers in Christiansen and Kirby (2003). Chomsky's recent view of language evolution is presented in Hauser, Chomsky, and Fitch (2002).

Page 303. The functionalist position has many versions: an influential functionalist analysis by a linguist is that of Givón (1995); Elman et al. (1996) present a functionalist approach from a computational and psychological point of view; Deacon (1997) argues for a functionalist analysis of the evolution of language, stressing brain evolution, and Lieberman (2000) emphasizes the evolution of the motor system.

Page 305. The study of the development of the Nicaraguan sign language is described in detail by Kegl et al. (1999). Helmuth (2001) gives a brief account of this sign language and the controversies about what its development means.

Page 305. Dor's views (e.g., see Dor, 2000) stem from the semantically oriented approach to syntax that is described in Frawley (1992), Levin (1993), and Levin and Rappaport Hovav (1995).

Page 307. The idea that genetic assimilation may have been important in the evolution of language was discussed briefly by Waddington (1975b). Dor and Jablonka (2000) extended the idea that partial genetic assimilation plays a crucial role in the evolution of language. Briscoe (2003) has also stressed the importance of genetic assimilation in language evolution.

Page 307. Most innovations, whether genetic, epigenetic, behavioral, or symbolic, are ephemeral, even when they are potentially very beneficial. The conditions for the establishment and regular transmission of an innovation are quite demanding. For example, for a cultural innovation in humans to be accepted and perpetuated, the innovator must be able to convince others that the innovation is worthwhile, must have the right connections, and so on. Most good ideas are never implemented, and many great inventions are doomed to oblivion.

Page 308. The cultural aspects of language evolution have been discussed and modeled by Kirby (1999, 2002).

Page 308. A gene whose normal function is relevant to the development of linguistic proficiency (among other behaviors) has been isolated by Lai et al. (2001), but the role of this gene in the evolution of language remains unclear.

Page 311. For a neural network–based model of the genetic assimilation of learned behavior, see Hinton and Nowlan (1987).

Page 312. West-Eberhard (2003) explains the idea of genetic accommodation in chap. 6 of her book.

Page 313. The idea that linguistic rules evolved through selecting arbitrary conventions was suggested by Pinker and Bloom (1990).

9 Lamarkism Evolving: The Evolution of the Educated Guess

In the previous chapters we have not specifically discussed the evolutionary origins of the systems we described, but the literature we cited often did. We therefore refer here only to work that explicitly includes ideas relevant to origins.

Page 320. The idea that new structures and functions arise when by-products of existing phenotypes are co-opted and selected in changed conditions has its origins in Darwin's writings and appears in almost all books on evolution. For example, it is used creatively by Maynard Smith and Szathmáry in *The Major Transitions in Evolution* (1995), and by Wilkins in *The Evolution of Developmental Pathways* (2002b).

Page 320. Fry (2000) has provided an excellent summary and discussion of different ideas about the origin of life, which includes a historical introduction and a review of current theories. An interesting and original approach to the evolution of the first biological entities, which is compatible with our view of the parallel and interdependent evolution of the genetic and the epigenetic systems, was developed by the Hungarian chemist and theoretical biologist Tibor Gánti in the 1960s (see Gánti, 2003). He constructed a very elegant abstract model of the simplest biological entity, which he called the "chemoton." The chemoton is made up of three interconnected subsystems, which together are self-maintaining. The subsystems correspond to the cytoplasm (self-perpetuating metabolic cycles), the cell membrane (which is based on self-assembly of three-dimensional structures), and the genetic material (a replicating linear polymer).

Page 322. The SOS system is described in Alberts et al. (2002), chap. 5. Radman's ideas can be found in Radman (1999), Taddei et al. (1997), and Radman et al. (1999).

Page 323. The evolution of mutation rates in different regions of the genome and global genomic stress responses are discussed by Moxon and Wills (1999).

Page 324. Nanney (1960) explored the advantages of epigenetic inheritance in erratically changing environments. The effect of transgenerational memory in such conditions was modeled by Jablonka et al. (1995).

Page 324. Lachmann and Jablonka (1996) discussed and modeled the evolution of transgenerational epigenetic inheritance in organisms that go through several generations in each phase of an environmental cycle.

Page 327. See Kauffman (1993) for his views of regulatory networks.

Page 328. See Liu and Lindquist (1999) for details of their experimental manipulations of prion-forming proteins.

Page 328. See Cavalier-Smith (2004) for his ideas about membrane evolution.

Page 329. For the idea that DNA methylation evolved as a genomic defense system, see Bestor (1990) and Yoder et al. (1997).

Page 331. For evidence that DNA methylation follows rather than initiates silencing, see Mutskov and Felsenfeld (2004).

Page 331. On the basis of the distribution of methylation in different taxa, Regev et al. (1998) suggested that the regulation of gene activity is an ancient function of DNA methylation, which preceded or evolved in parallel with the defense function. The variety of the functions of DNA methylation, which include inhibiting transcription initiation, arresting transcription elongation, acting as an imprinting signal, and the suppression of homologous recombination, led Colot and Rossignol

(1999) to conclude that DNA methylation is highly conserved because it has so many different functions. Bird (2002) compared patterns of DNA methylation in different taxa, and interpreted the differences between invertebrate and vertebrate patterns in terms of different strategies of genome and gene regulation.

Page 332. The term "stubborn mark" was used by Jablonka and Lamb (1995) to describe chromatin marks that are not readily erased during embryogenesis or gametogenesis; it was argued that commonly they were carried by DNA sequences such as those rich in CG nucleotides or repeated motifs. DNA sequence changes that are generated and spread by the processes that Dover calls "molecular drive" (see Dover, 2000a) can produce enormous variation in the size and composition of repeated sequences, even without selection, and these may be important for the evolution of epigenetic memory. Rakyan et al. (2002) have described alleles capable of carrying heritable marks that are liable to change (e.g., the mouse yellow allele we describe on p. 142) as "metastable epialleles," and suggested that typically they are associated with the presence of transposon sequences.

Page 332. The origin of RNAi as a defense system is discussed by Plasterk (2002). In a discussion of the many functions associated with RNAi, Cerutti (2003) speculated that the regulatory role of microRNAs may be very ancient. There is now so much evidence that non–protein-coding RNAs are common and play such key roles in cell regulation and development (see Mattick, 2003) that their evolutionary origins and roles will probably need to be totally reassessed. A concise account of the RNA world that may have existed before DNA became the major information carrier is given in Maynard Smith and Szathmáry (1995), chap. 5.

Page 333. The origin and evolution of social learning in different contexts is discussed in Avital and Jablonka (2000), chap. 9.

Page 337. Sue Savage-Rumbaugh's studies of language acquisition by bonobos are described in Savage-Rumbaugh and Lewin (1994) and Savage-Rumbaugh et al. (1998).

Page 337. The life and the biology of bonobos is lovingly described and illustrated in De Waal and Lanting's *Bonobo: The Forgotten Ape* (1997).

Page 339. The origin of linguistic communication in a society of highly social hominids with a bonobo-like intelligence is discussed by most researchers interested in the evolution of language, but they focus on different aspects. For example, Lieberman (2000) stresses the evolution of motor control; Dunbar (1996), the role of group size; Bickerton (2002) focuses on ecological factors; and Tomasello (1999) highlights the central importance of intentional imitation.

Page 341. Transitions to new levels of biological organization and the evolution of new types of information are discussed by Jablonka (1994), Jablonka and Szathmáry (1995), Maynard Smith and Szathmáry (1995), and Jablonka et al. (1998).

Page 344. Data relating methylation levels in different invertebrate taxa to the importance of cell memory in their development and tissue maintenance are given in Regev et al. (1998).

Page 345. The definition of evolvability is based on that of Gerhart and Kirschner (1997), p. 597, who defined it as "an increased ability to generate nonlethal, relevant, phenotypic variation on which selection can act."

Page 346. The idea that sexual reproduction in mammals is (partially) maintained by imprinting was suggested by Solter (1987), but was criticized by Haig (e.g., see Haig, 2002, paper 10).

Page 346. The work on mutability in natural populations of *E. coli* is described by Bjedov et al. (2003). Experiments suggesting the evolution-enhancing potential of mutator systems involving the SOS response are described in McKenzie et al. (2000) and Yeiser et al. (2002). Good brief discussions of the view that organisms can speed up their own evolution by boosting genetic variations have been given by Chicurel (2001) and Rosenberg and Hastings (2003). Poole et al. (2003) provide an interesting review and discussion of evolvability, particularly stress-induced responses, in both prokaryotes and eukaryotes.

Page 347. Eaglestone et al. (1999) have shown that in stress conditions yeast cells with the prion variant [PSI⁺] have an enhanced growth rate compared with those that lack it.

Page 347. Differing views about the role of the decanalizing factors Hsp90 and [PSI⁺] in evolvability can be found in Dickinson and Seger (1999), Dover (2000b), Lindquist (2000), and Partridge and Barton (2000). Masel and Bergman (2003) have argued that the ability to form a prion is probably the result of selection for evolvability. Their mathematical models show that as long as environmental changes make partial readthrough of stop codons adaptive once every million years, yeast strains having the ability to form prions have a selective advantage.

Page 347. See Belyaev (1979) for his view about the value of domestication as a model of evolution. Discussions of the role of stress in speciation can be found in Hoffmann and Parsons (1997) and Jablonka and Lamb (1995), chap. 9.

Page 350. The example of Kanzi's language comprehension is taken from Savage-Rumbaugh and Rumbaugh (1993).

Page 351. See Maynard Smith and Szathmáry (1995), p. 114 for their views on the origin of chromosomes.

Page 352. The collection of papers edited by Nitecki (1988) provides useful discussions of the philosophical, historical, and biological aspects of evolutionary progress.

10   A Last Dialogue

Page 360. See Weismann (1893a), chap. 13, and Hull (2000) for their positions on Lamarckism. Although labeling a biologist as a "Lamarckist" is still usually intended to be derogatory, there are signs of a slight change in attitude. For example, a cautious acknowledgment that some evolution is Lamarckian can be seen in Balter's (2000) "focus article" in *Science*, a journal that generally reflects acceptable, "respectable" opinions.

Page 362. Feldman and Laland (1996) have shown how estimates of IQ and other behavioral and personal traits are affected when cultural inheritance is factored in. Kisdi and Jablonka (in preparation) have developed a way of estimating the component of heritable variation that can be attributed to epigenetic inheritance.

Page 363. A simple and approachable discussion of heritability and its limitations, especially with respect to IQ, can be found in Rose et al. (1984), chap. 5.

Page 364. The significance of inherited epigenetic defects in human disease was pointed out by Holliday (1987). Verma et al. (2003) discuss the epigenetic aspects of cancer, and Murphy and Jirtle (2000) discuss the role of defects in imprinting in human diseases. Dennis (2003) provides a short account of the more general role of epigenetic variations in human disease, including the way they may cause differences between identical twins. Jablonka (2004) discusses the importance of epigenetic inheritance for epidemiology.

Page 365. The long-term effects of prenatal experience in humans, including those found in the Dutch famine study, are discussed in Barker's *Mothers, Babies and Health in Later Life* (1998), and in a brief article by Couzin (2002a). Evidence that there are paternally transmitted transgenerational effects in humans was given by Kaati et al. (2002), and the general implications of this finding were discussed by Pembrey (2002). Holliday (1998) considered the possibility that thalidomide may affect later generations. Early studies showing transgenerational effects in animals were reviewed and discussed by Campbell and Perkins (1988); for references to later work, see the notes to pages 137–145.

Page 366. Check (2003) discusses the potential uses of RNAi-based technology in medicine and the growing commercial interest in RNAi. Information about the project to map the methylation profile of the human genome can be found at http://www.epigenome.org.

Page 367. Errors in epigenetic reprogramming that lead to impaired development of cloned embryos are described in Solter (2000) and Rhind et al. (2003).

Page 368. The best-known example of what happens to language development when a child is not exposed to language is the tragic case of Genie, a girl who was deprived of human contact, including language, between babyhood and puberty (see Curtiss, 1977).

Page 369. See Margulis (1998) for her general views on the role of symbiosis in evolution.

Page 369. The Easter Island extinction is described by Daily (1999).

Page 370. The importance of social learning for the successful reintroduction of endangered bird and mammal species is discussed in Avital and Jablonka (2000) chap. 10.

Page 370. For a comprehensive account of the Gaia hypothesis by its originator, see Lovelock (1995). Among those who have taken up the idea are Lynn Margulis, who has integrated it with her symbiotic view of evolution (e.g., see Margulis, 1998), and Markoš (2002), who describes the crucial role of prokaryotes in Gaia.

Page 371. For examples of Lewontin's views on the way in which political and social factors interact with biological research, see Lewontin (1993, 2000a).

Page 372. Nineteenth-century evolution-based views of the world are described by Bowler (1989a). Spencer's views can be found in his *First Principles* (1862).

Page 373. Dawkins's emphasis on heredity through genes is seen clearly in *The Extended Phenotype* (1982) and *The Blind Watchmaker* (1986). Maynard Smith's similar viewpoint is evident in his *Evolutionary Genetics* (1989). Oyama presented the DST perspective in *The Ontogeny of Information* (1985/2000), and it was developed by Griffiths and Gray (1994, 2000). Sterelny and Griffiths (1999) give a summary and balanced evaluation of DST. Depew and Weber (1995) also argue for the need to adopt a developmental perspective in evolutionary theory, and emphasize the importance of self-organization. The book edited by Oyama et al. (2001) contains papers that discuss and criticize various aspects of DST. Lewontin's views can be seen in his *The Doctrine of DNA* (1993) and *The Triple Helix* (2000b); for his views on cultural evolution, see Fracchia and Lewontin (1999). Gould's position is described very fully in his *The Structure of Evolutionary Theory* (2002).

Page 375. For the extended-replicator concept, see Sterelny et al. (1996).

Page 376. Jablonka (2004) gives a critique of the replicator concept, and suggests "heritably varying traits" as units of evolutionary variation. Griesemer (2000a,b) argues that the target of selection and the unit of evolutionary transitions should be the reproducer. His assertion that material overlap is necessary for reproduction is based on empirical observations, and is not an axiom. Material overlap ensures the same developmental context for "parent" and "offspring" entities, so transmission is more reliable; if the resources available for development are reliable anyway, there is no need for material overlap. That is why retroviruses, which have the reliable environment of the host cell as a resource, can do without it (Griesemer, personal communication, 2003).

Page 377. The revival of interest in the significance of plasticity in evolution stems from the work of people like Matsuda (1987), Schlichting and Pigliucci (1998), and

Stearns (1992), but West-Eberhard (2003) has given the most comprehensive summary and arguments for its importance. A somewhat different and very important approach, which also centers on phenotypic plasticity, is that of Newman and Müller (2000).

Page 377. Odling-Smee et al. (2003) describe and evaluate niche construction, which they call "the neglected process in evolution." The idea that organisms are ecosystem engineers has been taken much further by J. S. Turner (2000).

Page 380. Pinker's bestseller, *The Blank Slate* (2002), warns against the pitfalls of the vulgarized version of sociobiology, yet itself vulgarizes and ridicules the positions of those who take opposing views. Pinker takes an almost exclusively gene-centered view of human heredity and evolution, one that minimizes the role of cultural construction and cultural evolution, and ignores their effects on the genetic evolution of human behavior.

Page 380. Critical responses of anthropologists, evolutionary biologists, philosophers, sociologists, and psychologists to Thornhill and Palmer's ideas can be found in Travis (2003).

Page 382. The quotations are from the beginning of the Sixth Fit and end of the Eighth Fit of Lewis Carroll's *The Hunting of the Snark* (1876).

Page 383. Lu Hsun's words come from a beautiful paragraph at the end of his story "My Old Home" (1972): "As I dozed, a stretch of jade-green seashore spread itself before my eyes, and above a round golden moon hung in a deep blue sky. I thought: hope cannot be said to exist, nor can it be said not to exist. It is just like roads across the earth. For actually the earth had no roads to begin with, but when many men pass one way, a road is made."

Bibliography

Aisner, R., and J. Terkel. 1992. Ontogeny of pine cone opening behaviour in the black rat, *Rattus rattus*. *Animal Behaviour* 44: 327–336.

Aitchison, J. 1996. *The Seeds of Speech: Language Origin and Evolution*. Cambridge, UK: Cambridge University Press.

Alberts, B., A. Johnson, J. Lewis, M. Raff, K. Roberts, and P. Walter. 2002. *Molecular Biology of the Cell*, 4th ed. New York: Garland Science.

Ancel, L. W. 1999. A quantitative model of the Simpson–Baldwin effect. *Journal of Theoretical Biology* 196: 197–209.

Aunger, R., ed. 2000. *Dawinizing Culture: The Status of Memetics as a Science*. Oxford: Oxford University Press.

Aunger, R., 2002. *The Electric Meme: A New Theory of How We Think*. New York: Free Press.

Avital, E., and E. Jablonka. 1994. Social learning and the evolution of behaviour. *Animal Behaviour* 48: 1195–1199.

Avital, E., and E. Jablonka. 1996. Adoption, memes and the Oedipus complex: A reply to Hansen. *Animal Behaviour* 51: 476–477.

Avital, E., and E. Jablonka. 2000. *Animal Traditions: Behavioural Inheritance in Evolution*. Cambridge, UK: Cambridge University Press.

Avital, E., E. Jablonka, and M. Lachmann. 1998. Adopting adoption. *Animal Behaviour* 55: 1451–1459.

Badano, J. L., and N. Katsanis. 2002. Beyond Mendel: An evolving view of human genetic disease transmission. *Nature Reviews Genetics* 3: 779–789.

Baker, R. 1996. *Sperm Wars: Infidelity, Sexual Conflict and Other Bedroom Battles*. London: Fourth Estate.

Baldwin, J. M. 1896. A new factor in evolution. *American Naturalist* 30: 441–451, 536–553.

Balter, M. 2000. Was Lamarck just a little bit right? *Science* 288: 38.

Banerjee, D., and F. Slack. 2002. Control of developmental timing by small temporal RNAs: A paradigm for RNA-mediated regulation of gene expression. *BioEssays* 24: 119–129.

Barker, D. J. P. 1998. *Mothers, Babies and Health in Later Life*, 2nd ed. Edinburgh: Churchill Livingstone.

Barkow, J. H., L. Cosmides, and J. Tooby. 1992. *The Adapted Mind: Evolutionary Psychology and the Generation of Culture*. New York: Oxford University Press.

Bateson, P. P. G. 1982. Preferences for cousins in Japanese quail. *Nature* 295: 236–237.

Bateson, W. 1894. *Materials for the Study of Variation, Treated with Especial Regard to Discontinuity in the Origin of Species*. London: Macmillan.

Bateson, W. 1909. *Mendel's Principles of Heredity*. Cambridge, UK: Cambridge University Press.

Baylin, S. B., and J. G. Herman. 2000. DNA hypermethylation in tumorigenesis: Epigenetics joins genetics. *Trends in Genetics* 16: 168–174.

Beisson, J., and T. M. Sonneborn. 1965. Cytoplasmic inheritance of the organization of the cell cortex in *Paramecium aurelia*. *Proceedings of the National Academy of Sciences of the United States of America* 53: 275–282.

Bell, G. 1982. *The Masterpiece of Nature: The Evolution and Genetics of Sexuality*. London: Croom Helm.

Belyaev, D. K. 1979. Destabilizing selection as a factor in domestication. *Journal of Heredity* 70: 301–308.

Belyaev, D. K., A. O. Ruvinsky, and P. M. Borodin. 1981a. Inheritance of alternative states of the fused gene in mice. *Journal of Heredity* 72: 107–112.

Belyaev, D. K., A. O. Ruvinsky, and L. N. Trut. 1981b. Inherited activation-inactivation of the star gene in foxes. *Journal of Heredity* 72: 267–274.

Benjamin, J., L. Li, C. Patterson, B. D. Greenberg, D. L. Murphy, and D. H. Hamer. 1996. Population and familial association between the D4 dopamine receptor gene and measures of novelty seeking. *Nature Genetics* 12: 81–84.

Bennett, J. H., F. A. Rhodes, and H. N. Robson. 1959. A possible genetic basis for kuru. *American Journal of Human Genetics* 11: 169–187.

Bernstein, C., and H. Bernstein. 1991. *Aging, Sex, and DNA Repair*. San Diego: Academic Press.

Bestor, T. H. 1990. DNA methylation: Evolution of a bacterial immune function into a regulator of gene expression and genome structure in higher eukaryotes. *Philosophical Transactions of the Royal Society of London Series B: Biological Sciences* 326: 179–187.

Bickerton, D. 2002. Foraging versus social intelligence in the evolution of protolanguage. In A. Wray, ed., *The Transition to Language*. Oxford: Oxford University Press, pp. 207–225.

Bilkó, Á., V. Altbäcker, and R. Hudson. 1994. Transmission of food preference in the rabbit: The means of information transfer. *Physiology and Behavior* 56: 907–912.

Bird, A. 2002. DNA methylation patterns and epigenetic memory. *Genes and Development* 16: 6–21.

Birdsell, J. A., and C. Wills. 2003. The evolutionary origin and maintenance of sexual recombination: A review of contemporary models. *Evolutionary Biology* 33: 27–138.

Bjedov, I., O. Tenaillon, B. Gérard, V. Souza, E. Denamur, M. Radman, F. Taddei, and I. Matic. 2003. Stress-induced mutagenesis in bacteria. *Science* 300: 1404–1409.

Black, D. L. 1998. Splicing in the inner ear: A familiar tune, but what are the instruments? *Neuron* 20: 165–168.

Blackman, R. L. 2000. The cloning experts. *Antenna* 24: 206–214.

Blackmore, S. 1999. *The Meme Machine*. Oxford: Oxford University Press.

Blackmore, S. 2000. The power of memes. *Scientific American* 283(4): 52–61.

Boakes, R. 1984. *From Darwin to Behaviourism: Psychology and the Minds of Animals*. Cambridge, UK: Cambridge University Press.

Bowler, P. J. 1983. *The Eclipse of Darwinism: Anti-Darwinian Evolution Theories in the Decades around 1900*. Baltimore: Johns Hopkins University Press.

Bowler, P. J. 1988. *The Non-Darwinian Revolution: Reinterpreting a Historical Myth*. Baltimore: Johns Hopkins University Press.

Bowler, P. J. 1989a. *Evolution: The History of an Idea*, revised ed. Berkeley: University of California Press.

Bowler, P. J. 1989b. *The Mendelian Revolution: The Emergence of Hereditarian Concepts in Modern Science and Society*. London: Athlone Press.

Box, O. H., and K. R. Gibson, eds. 1999. *Mammalian Social Learning: Comparative and Ecological Perspectives*. Cambridge, UK: Cambridge University Press.

Boyd, R., and P. J. Richerson. 1985. *Culture and the Evolutionary Process*. Chicago: University of Chicago Press.

Boyd, R., and P. J. Richerson. 1988. An evolutionary model of social learning: The effects of spatial and temporal variation. In T. R. Zentall and B. G. Galef, Jr., eds., *Social Learning: Psychological and Biological Perspectives*. Hillsdale, NJ: Lawrence Erlbaum, pp. 29–48.

Brink, R. A. 1973. Paramutation. *Annual Review of Genetics* 7: 129–152.

Briscoe, T., ed. 2002. *Linguistic Evolution through Language Acquisition*. Cambridge, UK: Cambridge University Press.

Briscoe, T. 2003. Grammatical assimilation. In M. H. Christiansen and S. Kirby, eds., *Language Evolution*. Oxford: Oxford University Press, pp. 295–316.

Brown, A. 1999. *The Darwin Wars: The Scientific Battle for the Soul of Man*. London: Simon & Schuster.

Buchner, J. 1999. Hsp90 & Co.—a holding for folding. *Trends in Biochemistry* 24: 136–141.

Burkhardt, R. W., Jr. 1977. *The Spirit of the System: Lamarck and Evolutionary Biology*. Cambridge, MA: Harvard University Press.

Burt, A. 2000. Sex, recombination, and the efficacy of selection—was Weismann right? *Evolution* 54: 337–351.

Bush, G. L. 1974. The mechanism of sympatric host race formation in the true fruit flies (Tephritidae). In M. J. D. White, ed., *Genetic Mechanisms of Speciation in Insects*. Sydney: Australia and New Zealand Book Company, pp. 3–23.

Buss, D. M. 1994. *The Evolution of Desire: Strategies of Human Mating*. New York: Basic Books.

Buss, D. M. 1999. *Evolutionary Psychology: The New Science of the Mind*. Boston: Allyn & Bacon.

Butler, S. 1879. *Evolution, Old and New; or The Theories of Buffon, Dr. Erasmus Darwin, and Lamarck, as Compared with That of Mr. Charles Darwin*. London: Hardwicke & Bogue.

Byrne, R. W. 2002. Imitation of novel complex actions: What does the evidence from animals mean? *Advances in the Study of Behavior* 31: 77–105.

Cairns J., J. Overbaugh, and S. Miller. 1988. The origin of mutants. *Nature* 335: 142–145.

Campbell, J. H., and P. Perkins. 1988. Transgenerational effects of drug and hormonal treatments in mammals: A review of observations and ideas. *Progress in Brain Research* 73: 535–553.

Caporale, L. H., ed. 1999. *Molecular Strategies in Biological Evolution*. Proceedings of the New York Academy of Sciences, vol. 870.

Caporale, L. H. 2000. Mutation is modulated: Implications for evolution. *BioEssays* 22: 388–395.

Capy, P., G. Gasperi, C. Biémont, and C. Bazin. 2000. Stress and transposable elements: Co-evolution or useful parasites? *Heredity* 85: 101–106.

Carroll, L. 1876. *The Hunting of the Snark*. London: Macmillan. Reprinted 1995 by Penguin Press. Harmondsworth, Middlesex, UK: Penguin Books.

Casadesús, J., and R. D'Ari. 2002. Memory in bacteria and phage. *BioEssays* 24: 512–518.

Cassirer, E. 1944. *An Essay on Man: An Introduction to a Philosophy of Human Culture*. New Haven, CT: Yale University Press.

Cassirer, E. 1953–1957. *The Philosophy of Symbolic Forms*, 3 vols. New Haven, CT: Yale University Press. Translation by R. Manheim of *Philosophie der Symbolischen Formen*, 3 vols. 1923–1929. Berlin: Bruno Cassirer.

Catchpole, C. K., and P. J. B. Slater. 1995. *Bird Song: Biological Themes and Variations.* Cambridge, UK: Cambridge University Press.

Cavalier-Smith, T. 2000. Membrane heredity and early chloroplast evolution. *Trends in Plant Science* 5: 174–182.

Cavalier-Smith, T. 2004. The membranome and membrane heredity in development and evolution. In R. P. Hirt and D. S. Horner, eds., *Organelles, Genomes and Eukaryote Phylogeny: An Evolutionary Synthesis in the Age of Genomics.* Boca Raton: CRC Press, pp. 335–351.

Cavalli, G. 2002. Chromatin as a eukaryotic template of genetic information. *Current Opinion in Cell Biology* 14: 269–278.

Cavalli-Sforza, L. L., and M. W. Feldman. 1981. *Cultural Transmission and Evolution.* Princeton, NJ: Princeton University Press.

Cerutti, H. 2003. RNA interference: Traveling in the cell and gaining functions? *Trends in Genetics* 19: 39–46.

Chakravarti, A., and P. Little. 2003. Nature, nurture and human disease. *Nature* 421: 412–414.

Chandler, V. L., W. B. Eggleston, and J. E. Dorweiler. 2000. Paramutation in maize. *Plant Molecular Biology* 43: 121–145.

Check, E. 2003. RNA to the rescue? *Nature* 425: 10–12.

Chen, T., T. Hiroko, A. Chaudhuri, F. Inose, M. Lord, S. Tanaka, J. Chant, and A. Fujita. 2000. Multigenerational cortical inheritance of the Rax2 protein in orienting polarity and division in yeast. *Science* 290: 1975–1978.

Chernoff, Y. O. 2001. Mutation processes at the protein level: Is Lamarck back? *Mutation Research* 488: 39–64.

Chicurel, M. 2001. Can organisms speed their own evolution? *Science* 292: 1824–1827.

Chomsky, N. 1968. *Language and Mind.* New York: Harcourt, Brace & World.

Chomsky, N. 1975. *Reflections on Language.* New York: Pantheon.

Chomsky, N. 2000. *New Horizons in the Study of Language and Mind.* Cambridge, UK: Cambridge University Press.

Christiansen, M. H., and S. Kirby, eds. 2003. *Language Evolution.* Oxford: Oxford University Press.

Churchill, F. B. 1978. The Weismann-Spencer controversy over the inheritance of acquired characters. In E. G. Forbes, ed., *Proceedings of the XVth International*

Congress of the History of Science. Edinburgh: Edinburgh University Press, pp. 451–468.

Clark, M. M., and B. G. Galef, Jr. 1995. Prenatal influences on reproductive life history strategies. *Trends in Ecology and Evolution* 10: 151–153.

Clark, M. M., P. Karpiuk, and B. G. Galef, Jr. 1993. Hormonally mediated inheritance of acquired characteristics in Mongolian gerbils. *Nature* 364: 712.

Coen, E. 1999. *The Art of the Genes: How Organisms Make Themselves.* Oxford: Oxford University Press.

Collias, N. E., and E. C. Collias. 1984. *Nest Building and Bird Behavior.* Princeton, NJ: Princeton University Press.

Collinge, J. 2001. Prion diseases of humans and animals: Their causes and molecular basis. *Annual Review of Neuroscience* 24: 519–550.

Colot, V., and J.-L. Rossignol. 1999. Eukaryotic DNA methylation as an evolutionary device. *BioEssays* 21: 402–411.

Cook, G. M. 1999. Neo-Lamarckian experimentalism in America: Origins and consequences. *Quarterly Review of Biology* 74: 417–437.

Cosmides, L., and J. Tooby. 1997. Evolutionary psychology: A primer. Available at http://www.psych.ucsb.edu/research/cep/primer.html.

Couzin, J. 2002a. Quirks of fetal environment felt decades later. *Science* 296: 2167–2169.

Couzin, J. 2002b. In yeast, prions' killer image doesn't apply. *Science* 297: 758–761.

Crick, F. H. C. 1958. On protein synthesis. *Symposia of the Society for Experimental Biology* 12: 138–163.

Crick, F. H. C. 1970. Central dogma of molecular biology. *Nature* 227: 561–563.

Cronin, H. 1991. *The Ant and the Peacock: Altruism and Sexual Selection from Darwin to Today.* Cambridge, UK: Cambridge University Press.

Crouse, H. V. 1960. The controlling element in sex chromosome behavior in *Sciara*. *Genetics* 45: 1429–1443.

Cubas, P., C. Vincent, and E. Coen. 1999. An epigenetic mutation responsible for natural variation in floral symmetry. *Nature* 401: 157–161.

Curtiss, S. 1977. *Genie: A Psycholinguistic Study of a Modern-day "Wild Child."* New York: Academic Press.

Daily, G. C. 1999. Developing a scientific basis for managing Earth's life support systems. *Conservation Ecology* 3(2): 14. Available at http://www.consecol.org/vol3/iss2/art14.

Darwin, C. 1859. *On the Origin of Species by Means of Natural Selection, or the Preservation of Favoured Races in the Struggle for Life.* London: John Murray. Fascimile reprint with an introduction by E. Mayr, 1964. Cambridge, MA: Harvard University Press.

Darwin, C. 1868. *The Variation of Animals and Plants under Domestication*, 2 vols. London: John Murray. 2nd ed. (1883) reprinted with a new foreword by H. Ritvo, 1998. Baltimore: John Hopkins University Press.

Darwin, C. 1871. *The Descent of Man, and Selection in Relation to Sex*. London: John Murray. Fascimile reprint with an introduction by J. T. Bonner and R. M. May, 1981. Princeton, NJ: Princeton University Press.

Darwin, C. 1881. *The Formation of Vegetable Mould, through the Action of Worms, with Observations on Their Habits*. London: John Murray. Fascimile reprint with a foreword by S. J. Gould, 1985. Chicago: University of Chicago Press.

Datta, A., and S. Jinks-Robertson. 1995. Association of increased spontaneous mutation rates with high levels of transcription in yeast. *Science* 268: 1616–1619.

Dawkins, R. 1976. *The Selfish Gene*. Oxford: Oxford University Press.

Dawkins, R. 1982. *The Extended Phenotype: The Gene as the Unit of Selection*. Oxford: Freeman.

Dawkins, R. 1986. *The Blind Watchmaker*. Harlow, UK: Longman.

Deacon, T. W. 1997. *The Symbolic Species: The Co-evolution of Language and the Brain*. New York: Norton.

Debat, V., and P. David. 2001. Mapping phenotypes: Canalization, plasticity and developmental stability. *Trends in Ecology and Evolution* 16: 555–561.

De la Casa-Esperón, E., and C. Sapienza. 2003. Natural selection and the evolution of genome imprinting. *Annual Review of Genetics* 37: 349–370.

Dennett, D. C. 1995. *Darwin's Dangerous Idea: Evolution and the Meaning of Life*. New York: Simon & Schuster.

Dennett, D. C. 2001. The evolution of culture. *The Monist* 84: 305–324.

Dennis, C. 2003. Altered states. *Nature* 421: 686–688.

Depew, D. J., and B. H. Weber. 1995. *Darwinism Evolving: Systems Dynamics and the Genealogy of Natural Selection*. Cambridge, MA: MIT Press.

De Vries, H. 1909–10. *The Mutation Theory: Experiments and Observations on the Origin of Species in the Vegetable Kingdom*, 2 vols. Chicago: Open Court. Translation by J. B. Farmer and A. D. Darbishire of *Die Mutationstheorie: Versuche und Beobachtungen über die Entstehung der Arten im Pflanzenreich*. 1901–1903. Leipzig: Veit.

De Waal, F. 2001. *The Ape and the Sushi Master: Cultural Reflections by a Primatologist*. New York: Basic Books.

De Waal, F., and F. Lanting. 1997. *Bonobo: The Forgotten Ape*. Berkeley: University of California Press.

Diamond, J. 1991. *The Rise and Fall of the Third Chimpanzee*. London: Radius.

Diamond, J. 1997. *Guns, Germs and Steel: The Fates of Human Societies*. London: Jonathan Cape.

Diamond, J., and J. I. Rotter. 2002. Evolution of human genetic diseases. In R. A. King, J. I. Rotter, and A. G. Motulsky, eds., *The Genetic Basis of Common Diseases*, 2nd ed. New York: Oxford University Press, pp. 50–64.

Dickinson, W. J., and J. Seger. 1999. Cause and effect in evolution. *Nature* 399: 30.

Dobzhansky, T. 1937. *Genetics and the Origin of Species*. New York: Columbia University Press. Reprinted 1982 with an introduction by S. J. Gould.

Donald, M. 1991. *Origins of the Modern Mind: The Stages in the Evolution of Culture and Cognition*. Cambridge, MA: Harvard University Press.

Doolittle, W. F., and C. Sapienza. 1980. Selfish genes, the phenotype paradigm and genome evolution. *Nature* 284: 601–603.

Dor, D. 2000. From the autonomy of syntax to the autonomy of linguistic semantics: Notes on the correspondence between the transparency problem and the relationship problem. *Pragmatics and Cognition* 8: 325–356.

Dor, D., and E. Jablonka. 2000. From cultural selection to genetic selection: A framework for the evolution of language. *Selection* 1: 33–55.

Dover, G. 2000a. *Dear Mr Darwin: Letters on the Evolution of Life and Human Nature*. Berkeley: University of California Press.

Dover, G. 2000b. Results may not fit well with current theories. . . . *Nature* 408: 17.

Drake, J. W., B. Charlesworth, D. Charlesworth, and J. F. Crow. 1998. Rates of spontaneous mutation. *Genetics* 148: 1667–1686.

Dunbar, R. I. M. 1996. *Grooming, Gossip and the Evolution of Language*. London: Faber & Faber.

Dunbar, R., C. Knight, and C. Power, eds. 1999. *The Evolution of Culture: An Interdisciplinary View*. Edinburgh: Edinburgh University Press.

Dunn, L. C. 1965. *A Short History of Genetics*. New York: McGraw-Hill.

Durham, W. H. 1991. *Coevolution: Genes, Culture, and Human Diversity*. Stanford, CA: Stanford University Press.

Dykxhoorn, D. M., C. D. Novina, and P. A. Sharp. 2003. Killing the messenger: Short RNAs that silence gene expression. *Nature Reviews Molecular Cell Biology* 4: 457–467.

Eaglestone, S. S., B. S. Cox, and M. F. Tuite. 1999. Translation termination efficiency can be regulated in *Saccharomyces cerevisiae* by environmental stress through a prion-mediated mechanism. *EMBO Journal* 18: 1974–1981.

Eddy, S. R. 2001. Non-coding RNA genes and the modern RNA world. *Nature Reviews Genetics* 2: 919–929.

Elman, J. L., E. A. Bates, M. H. Johnson, A. Karmiloff-Smith, D. Parisi, and K. Plunkett. 1996. *Rethinking Innateness: A Connectionist Perspective on Development.* Cambridge, MA: MIT Press.

Ettinger, L., and F. Doljanksi. 1992. On the generation of form by the continuous interactions between cells and their extracellular matrix. *Biological Reviews* 67: 459–489.

Fedoroff, N., and D. Botstein, eds. 1992. *The Dynamic Genome: Barbara McClintock's Ideas in the Century of Genetics.* Plainview, NY: Cold Spring Harbor Laboratory Press.

Feinberg, A. P., and B. Tycko. 2004. The history of cancer epigenetics. *Nature Reviews Cancer* 4: 143–153.

Feldman, M. W., and K. N. Laland. 1996. Gene-culture coevolutionary theory. *Trends in Ecology and Evolution* 11: 453–457.

Ferguson-Smith, A. C., and M. A. Surani. 2001. Imprinting and the epigenetic asymmetry between parental genomes. *Science* 293: 1086–1089.

Fire, A., S. Xu, M. K. Montgomery, S. A. Kostas, S. E. Driver, and C. C. Mello. 1998. Potent and specific genetic interference by double-stranded RNA in *Caenorhabditis elegans. Nature* 391: 806–811.

Fisher, J., and R. A. Hinde. 1949. The opening of milk bottles by birds. *British Birds* 42: 347–357.

Foster, P. L. 1999. Mechanisms of stationary phase mutation: A decade of adaptive mutation. *Annual Review of Genetics* 33: 57–88.

Foster, P. L. 2000. Adaptive mutation: Implications for evolution. *BioEssays* 22: 1067–1074.

Fracchia, J., and R. C. Lewontin. 1999. Does culture evolve? *History and Theory, Theme Issue* 38. *The Return of Science: Evolutionary Ideas and History,* pp. 52–78.

Francis, D., J. Diorio, D. Liu, and M. J. Meaney. 1999. Nongenomic transmission across generations of maternal behavior and stress responses in the rat. *Science* 286: 1155–1158.

Francis, D. D., K. Szegda, G. Campbell, W. D. Martin, and T. R. Insel. 2003. Epigenetic sources of behavioral differences in mice. *Nature Neuroscience* 6: 445–446.

Frawley, W. 1992. *Linguistic Semantics.* Hillsdale, NJ: Lawrence Erlbaum.

Fry, I. 2000. *The Emergence of Life on Earth: A Historical and Scientific Overview.* New Brunswick, NJ: Rutgers University Press.

Futscher, B. W., M. M. Oshiro, R. J. Wozniak, N. Holtan, C. L. Hanigan, H. Duan, and F. E. Domann. 2002. Role for DNA methylation in the control of cell type–specific maspin expression. *Nature Genetics* 31: 175–179.

Gajdusek, D. C. 1977. Unconventional viruses and the origin and disappearance of kuru. *Science* 197: 943–960.

Galton, F. 1871. Experiments in pangenesis, by breeding from rabbits of a pure variety, into whose circulation blood taken from other varieties had previously been largely transfused. *Proceedings of the Royal Society of London* 19: 393–410.

Galton, F. 1875. A theory of heredity. *Contemporary Review* 27: 80–95.

Gánti, T. 2003. *The Principles of Life*. Oxford: Oxford University Press.

Gayon, J. 1998. *Darwinism's Struggle for Survival: Heredity and the Hypothesis of Natural Selection*. Cambridge, UK: Cambridge University Press.

Gerhart, J., and M. Kirschner. 1997. *Cells, Embryos, and Evolution: Towards a Cellular and Developmental Understanding of Phenotypic Variation and Evolutionary Adaptability*. Malden, MA: Blackwell.

Gillham, N. W. 2001. *A Life of Sir Francis Galton: From African Exploration to the Birth of Eugenics*. New York: Oxford University Press.

Givón, T. 1995. *Functionalism and Grammar*. Philadelphia: John Benjamins.

Gottlieb, G. 1997. *Synthesizing Nature-Nurture: Prenatal Roots of Instinctive Behavior*. Mahwah, NJ: Lawrence Erlbaum.

Gottlieb, G. 2002. Developmental-behavioral initiation of evolutionary change. *Psychological Review* 109: 211–218.

Gould, S. J. 2002. *The Structure of Evolutionary Theory*. Cambridge MA: Harvard University Press.

Grandjean, V., Y. Hauck, C. Beloin, F. Le Hégarat, and L. Hirschbein. 1998. Chromosomal inactivation of *Bacillus subtilis* exfusants: A prokaryotic model of epigenetic regulation. *Biological Chemistry* 379: 553–557.

Greenfield, T. J. 1986. *Variation, Heredity, and Scientific Explanation in the Evolutionary Theories of Four American Neo-Lamarckians, 1867–1897*. Ph.D. dissertation, University of Wisconsin, Madison.

Grell, R. F. 1971. Heat-induced exchange in the fourth chromosome of diploid females of *Drosophila melanogaster*. *Genetics* 69: 523–527.

Grell, R. F. 1978. A comparison of heat and interchromosomal effects on recombination and interference in *Drosophila melanogaster*. *Genetics* 89: 65–77.

Griesemer, J. 2000a. The units of evolutionary transition. *Selection* 1: 67–80.

Griesemer, J. 2000b. Reproduction and the reduction of genetics. In P. J. Beurton, R. Falk, and H-J. Rheinberger, eds., *The Concept of the Gene in Development and Evolution: Historical and Epistemological Perspectives*. Cambridge: Cambridge University Press, pp. 240–285.

Griffiths, P. E., and R. D. Gray. 1994. Developmental systems and evolutionary explanation. *Journal of Philosophy* 91: 277–304.

Griffiths, P. E., and R. D. Gray. 2001. Darwinism and developmental systems. In S. Oyama, P. E. Griffiths, and R. D. Gray, eds., *Cycles of Contingency*. Cambridge, MA: MIT Press, pp. 195–218.

Grimes, G. W., and K. J. Aufderheide. 1991. *Cellular Aspects of Pattern Formation: The Problem of Assembly. Monographs in Developmental Biology*, vol. 22. Basel: Karger.

Grishok, A., H. Tabara, and C. C. Mello. 2000. Genetic requirements for inheritance of RNAi in *C. elegans. Science* 287: 2494–2497.

Groos, K. 1898. *The Play of Animals. A Study of Animal Life and Instinct*. London: Chapman & Hall.

Gu, Z., L. M. Steinmetz, X. Gu, C. Scharfe, R. W. Davis, and W.-H. Li. 2003. Role of duplicate genes in genetic robustness against null mutations. *Nature* 421: 63–66.

Gustafsson, Å. 1979. Linnaeus' peloria: The history of a monster. *Theoretical and Applied Genetics* 54: 241–248.

Hadorn, E. 1978. Transdeterminaton. In M. Ashburner and T. R. F. Wright, eds., *The Genetics and Biology of Drosophila*, vol. 2c. London: Academic Press, pp. 555–617.

Haig, D. 2002. *Genomic Imprinting and Kinship*. New Brunswick, NJ: Rutgers University Press.

Haldane, J. B. S. 1954. Introducing Douglas Spalding. *British Journal of Animal Behaviour* 2: 1–11.

Hall, J. C. 1994. The mating of a fly. *Science* 264: 1702–1714.

Hamer, D., and P. Copeland. 1998. *Living with Our Genes: Why They Matter More Than You Think*. New York: Doubleday.

Hamilton, W. D. 1964a. The genetical evolution of social behaviour. I. *Journal of Theoretical Biology* 7: 1–16.

Hamilton, W. D. 1964b. The genetical evolution of social behaviour. II. *Journal of Theoretical Biology* 7: 17–52.

Hamilton, W. D. 1996. *Narrow Roads of Gene Land. The Collected Papers of W. D. Hamilton*, vol. 1. *Evolution of Social Behaviour*. Oxford, UK: Freeman.

Hannon, G. J. 2002. RNA interference. *Nature* 418: 244–251.

Hardy, A. 1965. *The Living Stream: A Restatement of Evolution Theory and Its Relation to the Spirit of Man*. London: Collins.

Hauser, M. D., N. Chomsky, and W. T. Fitch. 2002. The faculty of language: What is it, who has it, and how did it evolve? *Science* 298: 1569–1579.

Helmuth, L. 2001. From the mouths (and hands) of babes. *Science* 293: 1758–1759.

Hendrickson, H., E. S. Slechta, U. Bergthorsson, D. I. Andersson, and J. R. Roth. 2002. Amplification-mutagenesis: Evidence that "directed" adaptive mutation and general hypermutability result from growth with a selected gene amplification. *Proceeding of the National Academy of Sciences of the United States of America* 99: 2164–2169.

Henikoff, S., T. Furuyama, and K. Ahmad. 2004. Histone variants, nucleosome assembly and epigenetic inheritance. *Trends in Genetics* 20: 320–326.

Heyes, C. M. 1993. Imitation, culture and cognition. *Animal Behaviour* 46: 999–1010.

Heyes, C. M., and B. G. Galef, Jr., eds. 1996. *Social Learning in Animals: The Roots of Culture*. San Diego: Academic Press.

Hinton, G. E., and S. J. Nowlan. 1987. How learning can guide evolution. *Complex Systems* 1: 495–502.

Hirata, S., K. Watanabe, and M. Kawai. 2001. "Sweet-potato washing" revisited. In T. Matsuzawa, ed., *Primate Origins of Human Cognition and Behavior*. Tokyo: Springer-Verlag, pp. 487–508.

Ho, M. W., C. Tucker, D. Keeley, and P. T. Saunders. 1983. Effects of successive generations of ether treatment on penetrance and expression of the *bithorax* phenocopy in *Drosophila melanogaster*. *Journal of Experimental Zoology* 225: 357–368.

Hoffmann, A. A., and P. A. Parsons. 1997. *Extreme Environmental Change and Evolution*. Cambridge, UK: Cambridge University Press.

Holden, C., and R. Mace. 1997. Phylogenetic analysis of the evolution of lactose digestion in adults. *Human Biology* 69: 605–628.

Hollick, J. B., J. E. Dorweiler, and V. L. Chandler. 1997. Paramutation and related allelic interactions. *Trends in Genetics* 13: 302–308.

Holliday, R. 1979. A new theory of carcinogenesis. *British Journal of Cancer* 40: 513–522.

Holliday, R. 1984. The biological significance of meiosis. In C. W. Evans and H. G. Dickinson, eds., *Controlling Events in Meiosis, Symposia of the Society for Experimental Biology*, vol. 38. Cambridge, UK: Company of Biologists, pp. 381–394.

Holliday, R. 1987. The inheritance of epigenetic defects. *Science* 238: 163–170.

Holliday, R. 1990. Mechanisms for the control of gene activity during development. *Biological Reviews* 65: 431–471.

Holliday, R. 1994. Epigenetics: An overview. *Developmental Genetics* 15: 453–457.

Holliday, R. 1996. DNA methylation in eukaryotes: 20 years on. In V. E. A. Russo, R. A. Martienssen, and A. D. Riggs, eds., *Epigenetic Mechanisms of Gene Regulation*. Plainview, NY: Cold Spring Harbor Laboratory Press, pp. 5–27.

Holliday, R. 1998. The possibility of epigenetic transmission of defects induced by teratogens. *Mutation Research* 422: 203–205.

Holliday, R., and J. E. Pugh. 1975. DNA modification mechanisms and gene activity during development. *Science* 187: 226–232.

Huffman, M. A. 1996. Acquisition of innovative cultural behaviors in nonhuman primates: A case study of stone handling, a socially transmitted behavior in Japanese macaques. In C. M. Heyes and B. G. Galef, eds., *Social Learning in Animals: The Roots of Culture.* San Diego: Academic Press, pp. 267–289.

Hull, D. L. 1980. Individuality and selection. *Annual Review of Ecology and Systematics* 11: 311–332.

Hull, D. L. 2000. Taking memetics seriously: Memetics will be what we make of it. In R. Aunger, ed., *Darwinizing Culture: the Status of Memetics as a Science.* Oxford: Oxford University Press, pp. 43–67.

Hurford, J. R., M. Studdert-Kennedy, and C. Knight, eds. 1998. *Approaches to the Evolution of Language: Social and Cognitive Bases.* Cambridge, UK: Cambridge University Press.

Hurst, L. D. 1997. Evolutionary theories of genomic imprinting. In W. Reik and A. Surani, eds., *Genomic Imprinting.* Oxford: IRL Press, pp. 211–237.

Huynh, K. D., and J. T. Lee. 2003. Inheritance of a pre-inactivated paternal X chromosome in early mouse embryos. *Nature* 426: 857–862.

Hyver, C., and H. Le Guyader. 1995. Cortical memory in *Paramecium*: A theoretical approach to the structural heredity. *Comptes Rendus de l'Académie des Sciences, Série III: Sciences de la Vie* 318: 375–380.

Immelmann, K. 1975. The evolutionary significance of early experience. In G. Baerends, C. Beer, and A. Manning, eds., *Function and Evolution in Behaviour: Essays in Honour of Professor Niko Tinbergen, FRS.* Oxford: Clarendon Press, pp. 243–253.

Irwin, D. E., and T. Price. 1999. Sexual imprinting, learning and speciation. *Heredity* 82: 347–354.

Issa, J.-P. 2000. CpG island methylation in aging and cancer. *Current Topics in Microbiology and Immunology* 249: 101–118.

Iwasa, Y., and A. Pomiankowski. 1999. Sex specific X chromosome expression caused by genomic imprinting. *Journal of Theoretical Biology* 197: 487–495.

Jablonka, E. 1994. Inheritance systems and the evolution of new levels of individuality. *Journal of Theoretical Biology* 170: 301–309.

Jablonka, E. 2002. Information: Its interpretation, its inheritance, and its sharing. *Philosophy of Science* 69: 578–605.

Jablonka, E. 2004. From replicators to heritably varying traits: The extended phenotype revisited. *Biology and Philosophy* 19: 353–375.

Jablonka, E. 2004. Epigenetic epidemiology. *International Journal of Epidemiology* 33: 929–935.

Jablonka, E., and M. J. Lamb. 1989. The inheritance of acquired epigenetic variations. *Journal of Theoretical Biology* 139: 69–83.

Jablonka, E., and M. J. Lamb. 1990a. The evolution of heteromorphic sex chromosomes. *Biological Reviews* 65: 249–276.

Jablonka, E., and M. J. Lamb. 1995. *Epigenetic Inheritance and Evolution: The Lamarckian Dimension.* Oxford: Oxford University Press.

Jablonka, E., and M. J. Lamb. 1998a. Bridges between development and evolution. *Biology and Philosophy* 13: 119–124.

Jablonka, E., and M. J. Lamb. 1998b. Epigenetic inheritance in evolution. *Journal of Evolutionary Biology* 11: 159–183.

Jablonka, E., and M. J. Lamb. 2002. The changing concept of epigenetics. *Annals of the New York Academy of Sciences* 981: 82–96.

Jablonka, E., and G. Rechav. 1996. The evolution of language in light of the evolution of literacy. In J. Trabant, ed., *Origins of Language.* Budapest: Collegium Budapest, pp. 70–88.

Jablonka, E., and E. Szathmáry. 1995. The evolution of information storage and heredity. *Trends in Ecology and Evolution* 10: 206–211.

Jablonka, E., M. Lachmann, and M. J. Lamb. 1992. Evidence, mechanisms and models for the inheritance of acquired characters. *Journal of Theoretical Biology* 158: 245–268.

Jablonka, E., B. Oborny, E. Molnár, E. Kisdi, J. Hofbauer, and T. Czárán. 1995. The adaptive advantage of phenotypic memory in changing environments. *Philosophical Transactions of the Royal Society of London. Series B: Biological Sciences* 350: 133–141.

Jablonka E., M. J. Lamb, and E. Avital. 1998. "Lamarckian" mechanisms in darwinian evolution. *Trends in Ecology and Evolution* 13: 206–210.

Jacob, F. 1989. *The Logic of Life: A History of Heredity.* Harmondsworth, Middlesex, UK: Penguin Books. Translation by B. E. Spillman of *La Logique du vivant: Une histoire de l'hérédité,* 1970. Paris: Editions Gallimard.

Jacob, F., and J. Monod. 1961a. On the regulation of gene activity. *Cold Spring Harbor Symposia on Quantitative Biology* 26: 193–211.

Jacob, F., and J. Monod. 1961b. Genetic regulatory mechanism in the synthesis of proteins. *Journal of Molecular Biology* 3: 318–356.

Jacobsen, S. E., and E. M. Meyerowitz. 1997. Hypermethylated *SUPERMAN* epigenetic alleles in *Arabidopsis. Science* 277: 1100–1103.

Jaenisch, R., and A. Bird. 2003. Epigenetic regulation of gene expression: How the genome integrates intrinsic and environmental signals. *Nature Genetics* 33(Suppl.): 245–254.

Jenuwein, T., and C. D. Allis. 2001. Translating the histone code. *Science* 293: 1074–1080.

Johannsen, W. 1911. The genotype conception of heredity. *American Naturalist* 45: 129–159.

Jones, P. A., and S. B. Baylin. 2002. The fundamental role of epigenetic events in cancer. *Nature Reviews Genetics* 3: 415–428.

Jordan. I. K., I. B. Rogozin, G. V. Glazko, and E. V. Koonin. 2003. Origin of a substantial fraction of human regulatory sequences from transposable elements. *Trends in Genetics* 19: 68–72.

Jorgensen, R. A. 2002. RNA traffics information systematically in plants. *Proceedings of the National Academy of Sciences of the United States of America* 99: 11561–11563.

Judson, H. F. 1996. *The Eighth Day of Creation: Makers of the Revolution in Biology*, 2nd ed. New York: Cold Spring Harbor Laboratory Press.

Kaati, G., L. O. Bygren, and S. Edvinsson. 2002. Cardiovascular and diabetes mortality determined by nutrition during parents' and grandparents' slow growth period. *European Journal of Human Genetics* 10: 682–688.

Kang, Y.-K., K.-K. Lee, and Y.-M. Han. 2003. Reprogramming DNA methylation in the preimplantation stage: Peeping with Dolly's eyes. *Current Opinion in Cell Biology* 15: 290–295.

Kauffman, S. A. 1993. *The Origin of Order: Self-Organization and Selection in Evolution*. New York: Oxford University Press.

Kay, L. E. 2000. *Who Wrote the Book of Life? A History of the Genetic Code*. Stanford, CA: Stanford University Press.

Kegl, J., A. Senghas, and M. Coppola. 1999. Creation through contact: Sign language emergence and sign language change in Nicaragua. In M. DeGraff, ed., *Language Creation and Language Change*. Cambridge, MA: MIT Press, pp. 179–237.

Keller, E. F. 1995. *Refiguring Life: Metaphors of Twentieth-Century Biology*. New York: Columbia University Press.

Keller, E. F. 2000. *The Century of the Gene*. Cambridge, MA: Harvard University Press.

Keller, E. F, and J. C. Ahouse. 1997. Writing and reading about Dolly. *BioEssays* 19: 741–742.

Keller, L., and K. G. Ross. 1993. Phenotypic plasticity and "cultural transmission" of alternative social organizations in the fire ant *Solenopsis invicta*. *Behavioral Ecology and Sociobiology* 33: 121–129.

Kidwell, M. G., and D. R. Lisch. 2000. Transposable elements and host genome evolution. *Trends in Ecology and Evolution* 15: 95–99.

Kimura, M. 1983. *The Neutral Theory of Molecular Evolution*. Cambridge, UK: Cambridge University Press.

King, C.-Y., and R. Diaz-Avalos. 2004. Protein-only transmission of three yeast prion strains. *Nature* 428: 319–323.

Kirby, S. 1999. *Function, Selection, and Innateness: The Emergence of Language Universals.* New York: Oxford University Press.

Kirby, S. 2002. Learning, bottlenecks and the evolution of recursive syntax. In T. Briscoe, ed., *Linguistic Evolution through Language Acquisition.* Cambridge, UK: Cambridge University Press, pp. 173–204.

Klar, A. J. S. 1998. Propagating epigenetic states through meiosis: Where Mendel's gene is more than a DNA moiety. *Trends in Genetics* 14: 299–301.

Knight, C., M. Studdert-Kennedy, and J. R. Hurford, eds. 2000. *The Evolutionary Emergence of Language: Social Function and the Origins of Linguistic Form.* Cambridge, UK: Cambridge University Press.

Koestler, A. 1971. *The Case of the Midwife Toad.* London: Hutchinson.

Krementsov, N. L. 1997. *Stalinist Science.* Princeton, NJ: Princeton University Press.

Kuper, A. 2002. Culture. *Proceedings of the British Academy* 112: 87–102.

Lachmann, M., and E. Jablonka. 1996. The inheritance of phenotypes: An adaptation to fluctuating environments. *Journal of Theoretical Biology* 181: 1–9.

Lai, C. S. L., S. E. Fisher, J. A. Hurst, F. Vargha-Khadem, and A. P. Monaco. 2001. A forkhead-domain gene is mutated in a severe speech and language disorder. *Nature* 413: 519–523.

Lakoff, G., and M. Johnson. 1999. *Philosophy in the Flesh: The Embodied Mind and Its Challenge to Western Thought.* New York: Basic Books.

Laland, K. N. 1994. On the evolutionary consequences of sexual imprinting. *Evolution* 48: 477–489.

Laland, K. N., and G. R. Brown. 2002. *Sense and Nonsense: Evolutionary Perspectives on Human Behaviour.* Oxford: Oxford University Press.

Lamarck, J. B. 1809. *Philosophie zoologique, ou Exposition des considérations relatives à l'histoire naturelle des animaux.* Paris: Dentu. English translation by H. Elliot, 1914, *Zoological Philosophy: An Exposition with Regard to the Natural History of Animals.* London: Macmillan. Reprinted by University of Chicago Press, 1984.

Lamb, M. J. 1994. Epigenetic inheritance and aging. *Reviews in Clinical Gerontology* 4: 97–105.

Lenneberg, E. H. 1967. *Biological Foundations of Language.* New York: Wiley.

Levin, B. 1993. *English Verb Classes and Alternations: A Preliminary Investigation.* Chicago: University of Chicago Press.

Levin, B., and M. Rappaport Hovav. 1995. *Unaccusativity: At the Syntax-Lexical Semantics Interface.* Cambridge, MA: MIT Press.

Lewontin, R. C. 1970. The units of selection. *Annual Review of Ecology and Systematics* 1: 1–18.

Lewontin, R. C. 1974. *The Genetic Basis of Evolutionary Change*. New York: Columbia University Press.

Lewontin, R. C. 1978. Adaptation. *Scientific American* 239(3): 156–169.

Lewontin, R. C. 1993. *The Doctrine of DNA: Biology as Ideology*. Harmondsworth, Middlesex, UK: Penguin Books.

Lewontin, R. C. 1997. The confusion over cloning, *The New York Review of Books*, October 23, pp. 18–23. Reprinted in Lewontin 2000a, pp. 273–291.

Lewontin, R. C. 2000a. *It Ain't Necessarily So: The Dream of the Human Genome and Other Illusions*. London: Granta Books.

Lewontin, R. C. 2000b. *The Triple Helix*. Cambridge, MA: Harvard University Press.

Li, E., T. H. Bestor, and R. Jaenisch. 1992. Targeted mutation of the DNA methyltransferase gene results in embryonic lethality. *Cell* 69: 915–926.

Lieberman, P. 2000. *Human Language and Our Reptilian Brain: The Subcortical Bases of Speech, Syntax, and Thought*. Cambridge, MA: Harvard University Press.

Lindegren, C. C. 1949. *The Yeast Cell, its Genetics and Cytology*. St. Louis: Educational Publishers.

Lindegren, C. C. 1966. *The Cold War in Biology*. Ann Arbor, MI: Planarian Press.

Lindquist, S. 2000. . . . but yeast prion offers clues about evolution. *Nature* 408: 17–18.

Liu, J.-J., and S. Lindquist. 1999. Oligopeptide-repeat expansions modulate "protein-only" inheritance in yeast. *Nature* 400: 573–576.

Lorenz, K. 1970. *Studies in Animal and Human Behaviour*, vol. 1. London: Methuen.

Lovelock, J. 1995. *The Ages of Gaia: A Biography of Our Living Earth*, 2nd ed. Oxford: Oxford University Press.

Lu Hsun. 1972. *Selected Stories of Lu Hsun*, 3rd ed. Translated by Y. Hsien-yi, and G. Yang. Peking: Foreign Language Press.

Lyko, F., and R. Paro. 1999. Chromosomal elements conferring epigenetic inheritance. *BioEssays* 21: 824–832.

Lyon, M. F. 1961. Gene action in the X-chromosome of the mouse (*Mus musculus* L.). *Nature* 190: 372–373.

Lyon, M. F. 1998. X-chromosome inactivation: A repeat hypothesis. *Cytogenetics and Cell Genetics* 80: 133–137.

Lyon, M. F. 1999. Imprinting and X-chromosome inactivation. In R. Ohlsson, ed., *Genomic Imprinting: An Interdisciplinary Approach*. Berlin: Springer-Verlag, pp. 73–90.

Maniatis, T., and B. Tasic. 2002. Alternative pre-mRNA splicing and proteome expansion in metazoans. *Nature*, 418: 236–243.

Margulis, L. 1998. *The Symbiotic Planet: A New Look at Evolution.* London: Weidenfeld & Nicolson.

Markoš, A. 2002. *Readers of the Book of Life: Contextualizing Developmental Evolutionary Biology.* New York: Oxford University Press.

Masel, J., and A. Bergman. 2003. The evolution of the evolvability properties of the yeast prion *[PSI⁺].* *Evolution* 57: 1498–1512.

Matsuda, R. 1987. *Animal Evolution in Changing Environments with Special Reference to Abnormal Metamorphosis.* New York: Wiley.

Mattick, J. S. 2003. Challenging the dogma: The hidden layer of non-protein-coding RNAs in complex organisms. *BioEssays* 25: 930–939.

Matzke, M., A. J. M. Matzke, and J. M. Kooter. 2001. RNA: Guiding gene silencing. *Science* 293: 1080–1083.

Mayer, M. P., and B. Bukau. 1999. Molecular chaperones: the busy life of Hsp90. *Current Biology* 9: R322–R325.

Maynard Smith, J. 1964. Group selection and kin selection. *Nature* 201: 1145–1147.

Maynard Smith, J. 1966. *The Theory of Evolution,* 2nd ed. Harmondsworth, Middlesex, UK: Penguin Books.

Maynard Smith, J. 1978. *The Evolution of Sex.* Cambridge, UK: Cambridge University Press.

Maynard Smith, J., ed. 1982. *Evolution Now: A Century after Darwin.* London: Macmillan.

Maynard Smith, J. 1986. *The Problems of Biology.* Oxford: Oxford University Press.

Maynard Smith, J. 1989. *Evolutionary Genetics.* Oxford: Oxford University Press.

Maynard Smith, J. 2000. The concept of information in biology. *Philosophy of Science* 67: 177–194.

Maynard Smith, J., and E. Szathmáry. 1995. *The Major Transitions in Evolution.* Oxford: Freeman.

Mayr, E. 1982. *The Growth of Biological Thought: Diversity, Evolution, and Inheritance.* Cambridge, MA: Harvard University Press.

Mayr, E., and W. B. Provine, eds. 1980. *The Evolutionary Synthesis: Perspectives on the Unification of Biology.* Cambridge, MA: Harvard University Press.

McClintock, B. 1984. The significance of responses of the genome to challenge. *Science* 226: 792–801.

McKenzie, G. J., R. S. Harris, P. L. Lee, and S. M. Rosenberg. 2000. The SOS response regulates adaptive mutation. *Proceedings of the National Academy of Sciences of the United States of America* 97: 6646–6651.

McKusick, V. A. 1998. *Mendelian Inheritance in Man: A Catalog of Human Genes and Genetic Disorders*, 12th ed. Baltimore: John Hopkins University Press. Available at http://www.ncbi.nlm.nih.gov/entrez/query.fcgi?db=OMIM.

McLaren, A. 1999. Too late for the midwife toad: Stress, variability and Hsp90. *Trends in Genetics* 15: 169–171.

McLaren, A. 2000. Cloning: Pathways to a pluripotent future. *Science* 288: 1775–1780.

McNairn, A. J., and D. M. Gilbert. 2003. Epigenomic replication: Linking epigenetics to DNA replication. *BioEssays* 25: 647–656.

Meiklejohn, C. D., and D. L. Hartl. 2002. A single mode of canalization. *Trends in Ecology and Evolution* 17: 468–473.

Mennella, J. A., C. P. Jagnow, and G. K. Beauchamp. 2001. Prenatal and postnatal flavor learning by human infants. *Pediatrics* 107: e88. Available at http://www.pediatrics.org/cgi/content/full/107/6/e88.

Metzgar, D., and C. Wills. 2000. Evidence for the adaptive evolution of mutation rates. *Cell* 101: 581–584.

Michod, R. E., and B. R. Levin, eds. 1988. *The Evolution of Sex: An Examination of Current Ideas*. Sunderland, MA: Sinauer.

Midgley, M. 2002. Choosing the selectors. *Proceedings of the British Academy* 112: 119–133.

Miller, G. 2000. *The Mating Mind: How Sexual Choice Shaped the Evolution of Human Nature*. London: Heinemann.

Moore, T., L. D. Hurst, and W. Reik. 1995. Genetic conflict and evolution of mammalian X-chromosome inactivation. *Developmental Genetics* 17: 206–211.

Morange, M. 1998. *A History of Molecular Biology*. Cambridge, MA: Harvard University Press.

Morange, M. 2001. *The Misunderstood Gene*. Cambridge, MA: Harvard University Press.

Morgan, H. D., H. G. E. Sutherland, D. I. K. Martin, and E. Whitelaw. 1999. Epigenetic inheritance at the agouti locus in the mouse. *Nature Genetics* 23: 314–318.

Moss, L. 2003. *What Genes Can't Do*. Cambridge, MA: MIT Press.

Moxon, E. R., and C. Wills. 1999. DNA microsatellites: agents of evolution? *Scientific American* 280(1): 72–77.

Moxon, E. R., P. B. Rainey, M. A. Nowak, and R. E. Lenski. 1994. Adaptive evolution of highly mutable loci in pathogenic bacteria. *Current Biology* 4: 24–33.

Müller-Hill, B. 1988. *Murderous Science: Elimination by Scientific Selection of Jews, Gypsies, and Others, Germany, 1933–1945*. Oxford: Oxford University Press.

Mundinger, P. C. 1980. Animal cultures and a general theory of cultural evolution. *Ethology and Sociobiology* 1: 183–223.

Murphy, S. K., and R. L. Jirtle. 2000. Imprinted genes as potential genetic and epigenetic toxicological targets. *Environmental Health Perspectives* 108(Suppl. 1): 5–11.

Mutskov, V., and G. Felsenfeld. 2004. Silencing of transgene transcription precedes methylation of promoter DNA and histone H3 lysine 9. *EMBO Journal* 23: 138–149.

Nanney, D. L. 1960. Microbiology, developmental genetics and evolution. *American Naturalist* 94: 167–179.

Newman, S. A., and G. B. Müller. 2000. Epigenetic mechanisms of character origination. *Journal of Experimental Zoology* 288: 304–317.

Nitecki, M. H., ed. 1988. *Evolutionary Progress.* Chicago: University of Chicago Press.

Novina, C. D., and P. A. Sharp. 2004. The RNAi revolution. *Nature* 430: 161–164.

Odling-Smee, F. J. 1988. Niche-constructing phenotypes. In H. C. Plotkin, ed., *The Role of Behavior in Evolution.* Cambridge, MA: MIT Press, pp. 73–132.

Odling-Smee, F. J., K. N. Laland, and M. W. Feldman. 1996. Niche construction. *American Naturalist* 147: 641–648.

Odling-Smee, F. J., K. N. Laland, and M. W. Feldman. 2003. *Niche Construction: The Neglected Process in Evolution.* Princeton, NJ: Princeton University Press.

Ohlsson, R., ed. 1999. *Genomic Imprinting: An Interdisciplinary Approach.* Berlin: Springer-Verlag.

Ohno, S. 1967. *Sex Chromosomes and Sex-linked Genes.* Berlin: Springer-Verlag.

Okano, M., D. W. Bell, D. A. Haber, and E. Li. 1999. DNA methyltransferases Dnmt3a and Dnmt3b are essential for de novo methylation and mammalian development. *Cell* 99: 247–257.

Olby, R. 1985. *Origins of Mendelism,* 2nd ed. Chicago: University of Chicago Press.

Olby, R. 1994. *The Path to the Double Helix: The Discovery of DNA,* 2nd ed. New York: Dover.

Opie, I., and P. Opie. 1959. *The Lore and Language of Schoolchildren.* Oxford: Oxford University Press.

Orel, V. 1996. *Gregor Mendel: The First Geneticist.* Oxford: Oxford University Press.

Orgel, L. E., and F. H. C. Crick. 1980. Selfish DNA: The ultimate parasite. *Nature* 284: 604–607.

Oyama, S. 2000. *The Ontogeny of Information: Developmental Systems and Evolution,* 2nd ed. Durham, NC: Duke University Press.

Oyama, S., P. E. Griffiths, and R. D. Gray, eds. 2001. *Cycles of Contingency.* Cambridge, MA: MIT Press.

Pál, C., and L. D. Hurst. 2004. Epigenetic inheritance and evolutionary adaptation. In R. P. Hirt and D. S. Horner, eds., *Organelles, Genomes and Eukaryote Phylogeny: An Evolutionary Synthesis in the Age of Genomics*. Boca Raton: CRC Press, pp. 353–370.

Park, Y., and M. I. Kuroda. 2001. Epigenetic aspects of X-chromosome dosage compensation. *Science* 293: 1083–1085.

Partridge, L., and N. H. Barton. 2000. Evolving evolvability. *Nature* 407: 457–458.

Pasquinelli, A. E., and G. Ruvkun. 2002. Control of developmental timing by micro-RNAs and their targets. *Annual Review of Cell and Developmental Biology* 18: 495–513.

Payne, R. B., L. L. Payne, J. L. Woods, and M. D. Sorenson. 2000. Imprinting and the origin of parasite-host species associations in brood-parasitic indigobirds, *Vidua chalybeata*. *Animal Behaviour* 59: 69–81.

Pembrey, M. E. 2002. Time to take epigenetic inheritance seriously. *European Journal of Human Genetics* 10: 669–671.

Pepperberg, I. M. 1999. *The Alex Studies: Cognitive and Communicative Abilities of Grey Parrots*. Cambridge, MA: Harvard University Press.

Persell, S. M. 1999. *Neo-Lamarckism and the Evolution Controversy in France, 1870–1920*. Lewiston, NY: The Edwin Mellen Press.

Petronis, A. 2001. Human morbid genetics revisited: Relevance of epigenetics. *Trends in Genetics* 17: 142–146.

Pfeifer, E. J. 1965. The genesis of American neo-Lamarckism. *Isis* 56: 156–167.

Pikaard, C. S. 2001. Genomic change and gene silencing in polyploids. *Trends in Genetics* 17: 675–677.

Pinker, S. 1994. *The Language Instinct*. London: Allen Lane.

Pinker, S. 2002. *The Blank Slate: The Modern Denial of Human Nature*. London: Allen Lane.

Pinker, S., and P. Bloom. 1990. Natural language and natural selection. *Behavioral and Brain Sciences* 13: 707–784.

Plasterk, R. H. A. 2002. RNA silencing: The genome's immune system. *Science* 296: 1263–1265.

Plath, K., S. Mlynarczyk-Evans, D. A. Nusinow, and B. Panning. 2002. Xist RNA and the mechanism of X chromosome inactivation. *Annual Review of Genetics* 36: 233–278.

Plotkin, H. 1997. *Evolution in Mind: An Introduction to Evolutionary Psychology*. London: Allen Lane.

Plotkin, H. 2000. Culture and psychological mechanisms. In R. Aunger, ed., *Darwinizing Culture: The Status of Memetics as a Science*. Oxford: Oxford University Press, pp. 69–82.

Poole, A. M., M. J. Phillips, and D. Penny. 2003. Prokaryote and eukaryote evolvability. *BioSystems* 69: 163–185.

Provine, W. B. 1971. *The Origins of Theoretical Population Genetics*. Chicago: University of Chicago Press.

Prusiner, S. B. 1995. The prion diseases. *Scientific American* 272(1): 48–57.

Prusiner, S. B. 1998. Prions. *Proceedings of the National Academy of Sciences of the United States of America* 95: 13363–13383.

Queitsch, C., T. A. Sangster, and S. Lindquist. 2002. Hsp90 as a capacitor of phenotypic variation. *Nature* 417: 618–624.

Radman, M. 1999. Enzymes of evolutionary change. *Nature* 401: 866–869.

Radman, M., and R. Wagner. 1988. The high fidelity of DNA duplication. *Scientific American* 259(2): 24–30.

Radman, M., I. Matic, and F. Taddei. 1999. Evolution of evolvability. *Annals of the New York Academy of Sciences* 870: 146–155.

Raina, R., M. Schläppi, and N. Fedoroff. 1998. Epigenetic mechanisms in the regulation of the maize *Suppressor-mutator* transposon. In *Novartis Foundation Symposium 214: Epigenetics*. Chichester, UK: Wiley, pp. 133–143.

Rakyan, V. K., J. Preis, H. D. Morgan, and E. Whitelaw. 2001. The marks, mechanisms and memory of epigenetic states in mammals. *Biochemical Journal* 356: 1–10.

Rakyan, V. K., M. E. Blewitt, R. Druker, J. I. Preis, and E. Whitelaw. 2002. Metastable epialleles in mammals. *Trends in Genetics* 18: 348–351.

Rakyan, V. K., S. Chong, M. E. Champ, P. C. Cuthbert, H. D. Morgan, K. V. K. Luu, and E. Whitelaw. 2003. Transgenerational inheritance of epigenetic states at the murine *Axin*^{Fu} allele occurs after maternal and paternal transmission. *Proceedings of the National Academy of Sciences of the United States of America* 100: 2538–2543.

Regev, A., M. J. Lamb, and E. Jablonka. 1998. The role of DNA methylation in invertebrates: Developmental regulation or genome defense? *Molecular Biology and Evolution* 15: 880–891.

Rendell, L., and H. Whitehead. 2001. Culture in whales and dolphins. *Behavioral and Brain Sciences* 24: 309–382.

Rhind, S. M., J. E. Taylor, P. A. De Sousa, T. J. King, M. McGarry, and I. Wilmut. 2003. Human cloning: Can it be made safe? *Nature Reviews Genetics* 4: 855–864.

Rhodes, R. 1997. *Deadly Feasts: Tracking the Secrets of a Terrifying New Plague*. New York: Simon & Schuster.

Riggs, A. D. 1975. X inactivation, differentiation and DNA methylation. *Cytogenetics and Cell Genetics* 14: 9–25.

Roberts, B. T., and R. B. Wickner. 2003. Heritable activity: A prion that propagates by covalent autoactivation. *Genes and Development* 17: 2083–2087.

Robinson, G. 1979. *A Prelude to Genetics.* Lawrence, KS: Coronado Press.

Rose, H., and S. Rose, eds. 2000. *Alas, Poor Darwin.* London: Jonathan Cape.

Rose, S., L. J. Kamin, and R. C. Lewontin. 1984. *Not in Our Genes: Biology, Ideology and Human Nature.* Harmondsworth, Middlesex, UK: Penguin Books.

Rosenberg, S. M. 2001. Evolving responsively: Adaptive mutation. *Nature Reviews Genetics* 2: 504–515.

Rosenberg, S. M., and P. J. Hastings. 2003. Modulating mutation rates in the wild. *Science* 300: 1382–1383.

Rutherford, S. L., and S. Henikoff. 2003. Quantitative epigenetics. *Nature Genetics* 33: 6–8.

Rutherford, S. L., and S. Lindquist. 1998. Hsp90 as a capacitor for morphological evolution. *Nature* 396: 336–342.

Sapp, J. 1987. *Beyond the Gene: Cytoplasmic Inheritance and the Struggle for Authority in Genetics.* New York: Oxford University Press.

Savage-Rumbaugh, S., and R. Lewin. 1994. *Kanzi: The Ape at the Brink of the Human Mind.* New York: Doubleday.

Savage-Rumbaugh, S., and D. M. Rumbaugh. 1993. The emergence of language. In K. R. Gibson and T. Ingold, eds., *Tools, Language and Cognition in Human Evolution.* Cambridge, UK: Cambridge University Press, pp. 86–108.

Savage-Rumbaugh, S., S. G. Shanker, and T. J. Taylor. 1998. *Apes, Language, and the Human Mind.* New York: Oxford University Press.

Scharloo, W. 1991. Canalization: genetic and developmental aspects. *Annual Review of Ecology and Systematics* 22: 65–93.

Schlichting, C. D., and M. Pigliucci. 1998. *Phenotypic Evolution: A Reaction Norm Perspective.* Sunderland, MA: Sinauer.

Schmalhausen, I. I. 1949. *Factors of Evolution: The Theory of Stabilizing Selection.* Translated from the Russian by I. Dordick. Philadelphia: Blakiston.

Schneeberger, R. G., and C. A. Cullis. 1991. Specific DNA alterations associated with the environmental induction of heritable changes in flax. *Genetics* 128: 619–630.

Schwartz, S. 2002. Characters as units and the case of the presence and absence hypothesis. *Biology and Philosophy* 17: 369–388.

Scriver, C. R., and P. J. Waters. 1999. Monogenic traits are not simple: Lessons for phenylketonuria. *Trends in Genetics* 15: 267–272.

Sebeok, T. A. 1994. *An Introduction to Semiotics.* London: Pinter.

Segerstråle, U. 2000. *Defenders of the Truth: The Battle for Science in the Sociology Debate and Beyond.* Oxford: Oxford University Press.

Sept, J. M., and G. E. Brooks. 1994. Reports of chimpanzee natural history, including tool use, in 16th- and 17th-century Sierra Leone. *International Journal of Primatology* 15: 867–878.

Serio, T. R., and S. L. Lindquist. 2000. Protein-only inheritance in yeast: Something to get [PSI⁺]-ched about. *Trends in Cell Biology* 10: 98–105.

Shapiro, J. A., ed. 1983. *Mobile Genetic Elements*. New York: Academic Press.

Shapiro, J. A. 1992. Natural genetic engineering in evolution. *Genetica* 86: 99–111.

Shapiro, J. A. 1999. Genome system architecture and natural genetic engineering in evolution. *Annals of the New York Academy of Sciences* 870: 23–35.

Sherry, D. F., and B. G. Galef, Jr. 1984. Cultural transmission without imitation: Milk bottle opening by birds. *Animal Behaviour* 32: 937–938.

Si, K., M. Giustetto, A. Etkin, R. Hsu, A. M. Janisiewicz, M. C. Miniaci, J.-H. Kim, H. Zhu, and E. R. Kandel. 2003a. A neuronal isoform of CPEB regulates local protein synthesis and stabilizes synapse-specific long-term facilitation in *Aplysia*. *Cell* 115: 893–904.

Si, K., S. Lindquist, and E. R. Kandel. 2003b. A neuronal isoform of the *Aplysia* CPEB has prion-like properties. *Cell* 115: 879–891.

Siegel, M. L., and A. Bergman. 2002. Waddington's canalization revisited: Developmental stability and evolution. *Proceedings of the National Academy of Sciences of the United States of America* 99: 10528–10532.

Simpson, G. G. 1953. The Baldwin effect. *Evolution* 7: 110–117.

Sing, C. F., M. B. Haviland, A. R. Templeton, and S. L. Reilly. 1995. Alternative genetic strategies for predicting risk of atherosclerosis. In F. P. Woodford, J. Davignon, and A. Sniderman, eds., *Atherosclerosis X*. Amsterdam: Elsevier, pp. 638–644.

Skaletsky, H., T. Kuroda-Kawaguchi, P. J. Minx, and 37 more authors. 2003. The male-specific region of the human Y chromosome is a mosaic of discrete sequence classes. *Nature* 423: 825–837.

Slabbekoorn, H., and M. Peet. 2003. Birds sing at a higher pitch in urban noise. *Nature* 424: 267.

Slechta, E. S., J. Liu, D. I. Andersson, and J. R. Roth. 2002. Evidence that selected amplification of a bacterial *lac* frameshift allele stimulates Lac⁺ reversion (adaptive mutation) with or without general hypermutability. *Genetics* 161: 945–956.

Smocovitis, V. B. 1996. *Unifying Biology: The Evolutionary Synthesis and Evolutionary Biology*. Princeton, NJ: Princeton University Press.

Snyder, M., and M. Gerstein. 2003. Defining genes in the genomics era. *Science* 300: 258–260.

Sober, E., and D. S. Wilson. 1998. *Unto Others: The Evolution and Psychology of Unselfish Behavior*. Cambridge, MA: Harvard University Press.

Sollars, V., X. Lu, L. Xiao, X. Wang, M. D. Garfinkel, and D. M. Ruden. 2003. Evidence for an epigenetic mechanism by which Hsp90 acts as a capacitor for morphological evolution. *Nature Genetics* 33: 70–74.

Solter, D. 1987. Inertia of the embryonic genome in mammals. *Trends in Genetics* 3: 23–27.

Solter, D. 2000. Mammalian cloning: Advances and limitations. *Nature Reviews Genetics* 1: 199–207.

Soyfer, V. N. 1994. *Lysenko and the Tragedy of Soviet Science*. New Brunswick, NJ: Rutgers University Press.

Spalding, D. 1873. Instinct with original observations on young animals. *Macmillan's Magazine* 27: 282–293. Reprinted with an introduction by J. B. S. Haldane in 1954 in the *British Journal of Animal Behaviour* 2: 1–11.

Spencer, H. 1862. *First Principles*. London: Williams & Norgate.

Spencer, H. 1893a. The inadequacy of natural selection. *Contemporary Review* 63: 153–166, 439–456.

Spencer, H. 1893b. A rejoinder to Professor Weismann. *Contemporary Review* 64: 893–912.

Sperber, D. 1996. *Explaining Culture: A Naturalistic Approach*. Oxford: Blackwell.

Stahl, F. W. 1988. A unicorn in the garden. *Nature* 335: 112–113.

Stearns, S. C. 1992. *The Evolution of Life Histories*. Oxford: Oxford University Press.

Steele, E. J. 1981. *Somatic Selection and Adaptive Evolution: On the Inheritance of Acquired Characters*, 2nd ed. Chicago: University of Chicago Press.

Steele, E. J., R. A. Lindley, and R. V. Blanden. 1998. *Lamarck's Signature: How Retrogenes Are Changing Darwin's Natural Selection Paradigm*. St. Leonards, NSW, Australia: Allen & Unwin.

Sterelny, K. 2001. *Dawkins vs. Gould: Survival of the Fittest*. Cambridge, UK: Icon Books.

Sterelny, K. 2003. *Thought in a Hostile World: The Evolution of Human Cognition*. Oxford: Blackwell.

Sterelny, K., and P. E. Griffiths. 1999. *Sex and Death: An Introduction to Philosophy of Biology*. Chicago: University of Chicago Press.

Sterelny, K., K. C. Smith, and M. Dickison. 1996. The extended replicator. *Biology and Philosophy* 11: 377–403.

Stern, C., and E. R. Sherwood, eds. 1966. *The Origin of Genetics: A Mendel Source Book*. San Francisco: Freeman.

Sternberg, R. v. 2002. On the roles of repetitive DNA elements in the context of a unified genomic-epigenetic system. *Annals of the New York Academy of Sciences* 981: 154–188.

Strohman, R. C. 1997. The coming Kuhnian revolution in biology. *Nature Biotechnology* 15: 194–200.

Sturtevant, A. H. 1965. *A History of Genetics*. New York: Harper & Row. Reprinted 2001, New York: Cold Spring Harbor Laboratory Press. Available at http://www.esp.org/books/sturt/history.

Sutherland, H. G. E., M. Kearns, H. D. Morgan, A. P. Headley, C. Morris, D. I. K. Martin, and E. Whitelaw. 2000. Reactivation of heritably silenced gene expression in mice. *Mammalian Genome* 11: 347–355.

Swallow, D. M., and E. J. Hollox. 2001. Genetic polymorphism of intestinal lactase activity in adult humans. In C. R. Scriver, A. L. Beaudet, W. S. Sly, and D. Valle, eds., *The Metabolic and Molecular Bases of Inherited Disease*, 8th ed., vol. 1. New York: McGraw-Hill, pp. 1651–1663.

Taddei, F., M. Vulić, M. Radman, and I. Matić. 1997. Genetic variability and adaptation to stress. In R. Bijlsma and V. Loeschcke, eds., *Environmental Stress, Adaptation and Evolution*. Basel: Birkhäuser-Verlag, pp. 271–290.

Tanaka, M., P. Chien, N. Naber, R. Cooke, and J. S. Weissman. 2004. Conformational variations in an infectious protein determine prion strain differences. *Nature* 428: 323–328.

Templeton, A. R. 1998. The complexity of the genotype-phenotype relationship and the limitations of using genetic "markers" at the individual level. *Science in Context* 11: 373–389.

ten Cate, C. 2000. How learning mechanisms might affect evolutionary processes. *Trends in Ecology and Evolution* 15: 179–181.

ten Cate, C., and D. R. Vos. 1999. Sexual imprinting and evolutionary processes in birds: A reassessment. *Advances in the Study of Behavior* 28: 1–31.

Terkel, J. 1996. Cultural transmission of feeding behavior in the black rat (*Rattus rattus*). In C. M. Heyes and B. G. Galef, Jr., eds., *Social Learning in Animals: The Roots of Culture*. San Diego: Academic Press, pp. 17–47.

Teubner, G. 2002. Idiosyncratic production regimes: Co-evolution of economic and legal institutions in the varieties of capitalism. *Proceedings of the British Academy* 112: 161–181.

Thieffry, D., and L. Sánchez. 2002. Alternative epigenetic states understood in terms of specific regulatory structures. *Annals of the New York Academy of Sciences* 981: 135–153.

Thornhill, R., and C. T. Palmer. 2000. *A Natural History of Rape: Biological Bases of Sexual Coercion*. Cambridge, MA: MIT Press.

Tomasello, M. 1999. *The Cultural Origins of Human Cognition*. Cambridge, MA: Harvard University Press.

Trabant, J., and S. Ward, eds. 2001. *New Essays on the Origin of Language*. Berlin: Mouton de Gruyter.

Travis, C. B., ed. 2003. *Evolution, Gender, and Rape*. Cambridge, MA: MIT Press.

True, H. L., and S. L. Lindquist. 2000. A yeast prion provides a mechanism for genetic variation and phenotypic diversity. *Nature* 407: 477–483.

Trut, L. N. 1999. Early canid domestication: The farm-fox experiment. *American Scientist* 87: 160–169.

Turner, B. M. 2001. *Chromatin and Gene Regulation: Mechanisms in Epigenetics*. Oxford: Blackwell.

Turner, J. S. 2000. *The Extended Organism: The Physiology of Animal-Built Structures*. Cambridge, MA: Harvard University Press.

Urnov, F. D., and A. P. Wolffe. 2001. Above and within the genome: Epigenetics past and present. *Journal of Mammary Gland Biology and Neoplasia* 6: 153–167.

Van Schaik, C. P., M. Ancrenaz, G. Borgen, B. Galdikas, C. D. Knott, I. Singleton, A. Suzuki, S. S. Utami, and M. Merrill. 2003. Orangutan cultures and the evolution of material culture. *Science* 299: 102–105.

Velkov, V. V. 2002. New insights into the molecular mechanisms of evolution: Stress increases genetic diversity. *Molecular Biology* 36: 209–215.

Verma, M., B. K. Dunn, and A. Umar, eds. 2003. *Epigenetics in Cancer Prevention: Early Detection and Risk Assessment. Annals of the New York Academy of Sciences*, vol. 983.

Voinnet, O. 2002. RNA silencing: Small RNAs as ubiquitous regulators of gene expression. *Current Opinion in Plant Biology* 5: 444–451.

Waddington, C. H. 1942. Canalization of development and the inheritance of acquired characters. *Nature* 150: 563–565.

Waddington, C. H. 1957. *The Strategy of the Genes*. London: Allen & Unwin.

Waddington, C. H. 1959a. Evolutionary adaptation. In S. Tax, ed., *Evolution after Darwin*, vol. 1. Chicago: University of Chicago Press, pp. 381–402. Reprinted in Waddington 1975a, pp. 36–59.

Waddington, C. H. 1959b. Canalization of development and genetic assimilation of acquired characters. *Nature* 183: 1654–1655.

Waddington, C. H. 1961. Genetic assimilation. *Advances in Genetics* 10: 257–293.

Waddington, C. H. 1968. The basic ideas of biology. In C. H. Waddington, ed., *Towards a Theoretical Biology*, vol. 1, *Prolegomena*. Edinburgh: Edinburgh University Press, pp. 1–32.

Waddington, C. H. 1975a. *The Evolution of an Evolutionist*. Edinburgh: Edinburgh University Press.

Waddington, C. H. 1975b. The evolution of altruism and language. In C. H. Waddington, ed., *The Evolution of an Evolutionist*. Edinburgh: Edinburgh University Press, pp. 299–307.

Wagner, A. 2000. Robustness against mutations in genetic networks of yeast. *Nature Genetics* 24: 355–361.

Waters, E. R., and B. A. Schaal. 1996. Heat shock induces a loss of rRNA-encoding DNA repeats in *Brassica nigra*. *Proceedings of the National Academy of Sciences of the United States of America* 93: 1449–1452.

Watson, J. D. 1968. *The Double Helix: A Personal Account of the Discovery of the Structure of DNA*. London: Weidenfeld & Nicolson.

Watson, J. D., and F. H. C. Crick. 1953a. A structure for deoxyribose nucleic acid. *Nature* 171: 737–738.

Watson, J. D., and F. H. C. Crick. 1953b. Genetical implications of the structure of deoxyribonucleic acid. *Nature* 171: 964–967.

Watson, J. D., N. H. Hopkins, J. W. Roberts, J. A. Steitz, and A. M. Weiner. 1988. *Molecular Biology of the Gene*, 4th ed. Menlo Park, CA: Benjamin/ Cummins.

Weatherall, D. J. 1998. How much has genetics helped? *Times Literary Supplement*, January 30, pp. 4–5.

Weber, B. H., and D. J. Depew, eds. 2003. *Evolution and Learning: The Baldwin Effect Reconsidered*. Cambridge, MA: MIT Press.

Weismann, A. 1891. *Essays upon Heredity and Kindred Biological Problems*, 2nd ed., vol. 1. Translated and edited by E. B. Poulton, S. Schönland, and A. E. Shipley. Oxford: Clarendon Press.

Weismann, A. 1893a. *The Germ-Plasm: A Theory of Heredity*. Translated from the 1892 German edition by W. Newton Parker and H. Rönnfeldt. London: Walter Scott.

Weismann, A. 1893b. The all-sufficiency of natural selection. *Contemporary Review* 64: 309–338, 596–610.

Weismann, A. 1904. *The Evolution Theory*, 2 vols. Translated from the 1904 2nd German edition by J. A. Thomson and M. R. Thomson. London: Edward Arnold.

West-Eberhard, M. J. 2003. *Developmental Plasticity and Evolution*. New York: Oxford University Press.

Wheeler, M., J. Ziman, and M. A. Boden, eds. 2002. *The Evolution of Cultural Entities*. Oxford: Oxford University Press.

White, M. J. D. 1973. *Animal Cytology and Evolution*, 3rd ed. Cambridge, UK: Cambridge University Press.

Whitelaw, E., and D. I. K. Martin. 2001. Retrotransposons as epigenetic mediators of phenotypic variation in mammals. *Nature Genetics* 27: 361–365.

Whiten, A., and R. Ham. 1992. On the nature and evolution of imitation in the animal kingdom: Reappraisal of a century of research. *Advances in the Study of Behavior* 21: 239–283.

Whiten, A., J. Goodall, W. C. McGrew, T. Nishida, V. Reynolds, Y. Sugiyama, C. E. G. Tutin, R. W. Wrangham, and C. Boesch. 2001. Charting cultural variation in chimpanzees. *Behaviour* 138: 1481–1516.

Wilkins, A. S. 2002a. Interview with Ernst Mayr. *BioEssays* 24: 960–973.

Wilkins, A. S. 2002b. *The Evolution of Developmental Pathways*. Sunderland, MA: Sinauer.

Wilkins, J. F., and D. Haig. 2003. What good is genomic imprinting: The function of parent-specific gene expression. *Nature Reviews Genetics* 4: 359–368.

Williams, G. C. 1966. *Adaptation and Natural Selection: A Critique of Some Current Evolutionary Thought*. Princeton, NJ: Princeton University Press.

Wilson, D. S. 1983. The group selection controversy: History and current status. *Annual Review of Ecology and Systematics* 14: 159–187.

Wolff, G. L., R. L. Kodell, S. R. Moore, and C. A. Cooney. 1998. Maternal epigenetics and methyl supplements affect *agouti* gene expression in A^{vy}/a mice. *FASEB Journal* 12: 949–957.

Wray, A., ed. 2002. *The Transition to Language*. Oxford: Oxford University Press.

Wright, B. E. 2000. A biochemical mechanism for nonrandom mutations and evolution. *Journal of Bacteriology* 182: 2993–3001.

Wright, B. E., A. Longacre, and J. M. Reimers. 1999. Hypermutation in derepressed operons of *Escherichia coli* K12. *Proceedings of the National Academy of Sciences of the United States of America* 96: 5089–5094.

Wright, S. 1968–1978. *Evolution and the Genetics of Populations*, 4 vols. Chicago: University of Chicago Press.

Wu, C.-t., and J. R. Morris. 2001. Genes, genetics, and epigenetics: A correspondence. *Science* 293: 1103–1105.

Wynne-Edwards, V. C. 1962. *Animal Dispersion in Relation to Social Behaviour*. Edinburgh: Oliver & Boyd.

Yeiser, B., E. D. Pepper, M. F. Goodman, and S. E. Finkel. 2002. SOS-induced DNA polymerases enhance long-term survival and evolutionary fitness. *Proceedings of the National Academy of Sciences of the United States of America* 99: 8737–8741.

Yoder, J. A., C. P. Walsh, and T. H. Bestor. 1997. Cytosine methylation and the ecology of intragenomic parasites. *Trends in Genetics* 13: 335–340.

Yokomori, N., R. Moore, and M. Negishi. 1995. Sexually dimorphic DNA methylation in the promotor of the *Slp* (sex-limited protein) gene in mouse liver. *Proceedings of the National Academy of Sciences of the United States of America* 92: 1302–1306.

Zhivotovsky, L. A. 2002. A model of the early evolution of soma-to-germline feedback. *Journal of Theoretical Biology* 216: 51–57.

Ziman, J., ed. 2000. *Technological Innovation as an Evolutionary Process.* Cambridge, UK: Cambridge University Press.

Index

Notes are indexed only when they significantly amplify the main text. They are identified by N preceding the page number to which they refer, and can be found between pages 385 and 416.

Acquired characters. *See* Inheritance of acquired characters
Adoption, 187–188. *See also* Foster parents
Agency, 208–209, 224, 231. *See also* Reproducer
Aging, 130
Agriculture, 237, 296–297, 363, 366–367, N237
Aisner, R., 170
Alarm calls, 34–35, 54, 159, 194, 196
Alleles 24–25. *See also* Gene
  metastable, N332
Alternative splicing, 65–67
Altruism, 34–5, 187
Amplification of DNA, 69–70
Animal traditions. *See also* Behavioral inheritance; Cultural evolution; Lifestyle evolution
  in birds and whales, 172–173
  in black rats, 170–171
  in chimpanzees, 177, 183, 313, 368
  complexity of, 178, 180
  and conservation practices, 190, 370
  and cumulative evolution, 176–180, 220, 335
  in insects, 189–190

  in macaques, 178–180, 183, 188, 207, 220
  origins of, 333–336
  prevalence of, 177
  stability of, 179, 182–185
  in tarbutniks, 158–159
  in tits, 169–170
Antirrhinum, 141
Aphids, 83–84
Arabidopsis, 268–269, 270
Ascaris, 68–69, 83, N68
Asexual reproduction
  and epigenetic inheritance, 114–118, 138, 148, 250, 275, 277–278, 346
  and sexual reproduction, compared, 82–84, 103, 346
Asimov, Isaac, 78
Assimilate-stretch principle, 290–292, 308–309, 309–310
Avital, E., 156, 291

Bacillus, 138
Bacteria, 237, 328–329
  epigenetic inheritance in, 119, 138, 332
  genetic systems of, 30, 85, 233
  mutation in, 79–80, 88, 93–98, 105, 106, 322–323, 345–347, 364

Baldwin, J. M., 289, 293
Baldwin effect, 289–290, 310–311,
  N289. *See also* Genetic assimilation
Bateson, P. P. G., 186
Bateson, W., 23–24
Beaver dam, 237
Behavioral imprinting
in ducks, 218
filial, 167–168
on habitat, 181, 184–185
on helpers and adopters, 186–187
in insects, 190, 240, 367
on parasite's hosts, 185, 240
sexual, 168, 181, 185–6
song, 172–173, 185
Behavioral inheritance
through behavior affecting substances,
  161–165, 182, 189–190, 334, 367
through imitation, 172–175
through nonimitative social learning,
  157–161, 166–172, 181, 189–190
origins of, 333–336
stability of, 181–184
Belyaev, D. K., 259–260, 272, 347–348
Biotechnology, 58, 136–137, 366, 382.
  See also Genetic engineering, by
  humans
Birds. *See also* Bowerbirds; Ducks;
  Mauritius kestrels; Ostriches;
  Peacocks; Tits
alarm calls of, 34–35
helpers in, 186–187
imitation by, 172–174 (*see also*
  Parrots)
imprinting in, 167, 173, 181, 185–187
parasitic, 185
seed caching by, 183
songs of, 54, 161, 172–173, 185, 202,
  218, 288, 368
Blackmore, S., 207–208, 211
Black rats, 170–171, 176, 178, 237
Bloom, P., 314
Bonobos, 304, 337–339, 349–351

Boojums, 382–383
Bowerbirds, 169
Bowler, P. J., 22–23
Boyd, R., 205
Brassica nigra, 98–99
BSE (bovine spongiform
  encephalopathy; mad cow disease),
  123, 125
Buss, D. M., 216
Butler, S., 21
Butterflies, 83, 190, 240

Caenorhabditis elegans, 76, 133, 135,
  144, 331, 344–345
Cairns, J., 79–80, 88
Canalization, 65. *See also* Genetic
  assimilation
of behavioral traits, 290, 311–312
of morphological features, 44,
  262–265, 267–268, 272–273, 282
and plasticity, 78, 312, 377
Cancer, 86, 372
diagnosis and treatment, 248, 364–365
epigenetic aspects of, 130, 144, 148,
  248, 366
Cannibalism, 124
Caporale, L. H., 101
Carroll, Lewis, 382
Carrot juice, 163–164, 181
Cassirer, E., 193–194
Categorization, 172, 175, 291–292,
  305–307, 308–309, 314–316
Cattle
bovine spongiform encephalopathy
  (BSE), 123, 125
domestication and dairying, 293–296
Cavalier-Smith, T., 122–3, 328–329
Cavalli-Sforza, L. L., 205
Cell division. *See* Mitosis; Meiosis
Cell heredity. *See* Epigenetic
  inheritance
Cell membranes, 122–123, 147, 151,
  328–329

Cell memory, 113–114, 137, 251–254, 325–326, 344–345. *See also* Epigenetic inheritance
erasing, 138–139, 149, 253–254
Cell organelles, 32, 33, 329
Central dogma, 31, 104–105, 152–153, N31
Chaperones, molecular, 266–270, 273, 347
Characters, definition of, 390. *See also* Genes, and characters; Traits as units of evolution
Chickens, 66–67, 153
Chimpanzees, 124, 173, 193, 337
cultural traditions in, 177, 183, 313, 368
genomes of, 76, 83
and language, 302, 304, 337–339, 349–351
Cholesterol, 61, 72
Chomsky, N., 218, 298–303, 313–316
Chromatin diminution, 68–69
Chromatin marking, 126–132. *See also* Genomic imprinting
Chromatin marks, 128, 134, 137, 145, 270, 366, 371
through DNA methylation, 128–131, 138, 140, 142, 143, 247–248, 329–332, 344–345, 366
environmental influences on, 144
evolutionary origins of, 329–332, 352
and genetic change, 247–250, 344–345
through histone modification, 131–132
persistence of, 138–144, 150–151, 256–257, 332, N332
through proteins, 131, 138, 270, 331–332, 345
transgenerational transmission of, 137–144, 269–270
Chromatin structure, 126–128, 246, 270, 329, 351–352. *See also* Chromatin marks

Chromosomes
bacterial, 30, 85, 138
elimination of, 70, 139, 256
inactivation of, 138, 256–257, 279–280
and Mendelian genes, 27–28, 29, 81, 85
number of, 5, 81, 83, 259
origins of, 342, 343, 351–352
polyploid, 69
polytene, 69–70
sex, 256–258, 278–280
structure of, 126–128, 246, 351–352
Weismann's view of, 16–17, 18–19, 68–69
Ciliates, 121–122, 138, 328
CJD (Creutzfeldt-Jakob disease), 123, 125, 266, 365
Clones, 82, 114, 250. *See also* Asexual reproduction
Cloning, 59, 150, 366–367
Cnemidophorus uniparens, 83
Cockroaches, 190
Coen, E., 142
Coevolution, N293
of genes and culture, 293–298
of genes and language, 306–310, 315–316
Cone stripping behavior, 170–171, 176, 178
Conservation practices, 168, 190, 369–370, 371–372
Coprophagy, 164
Copying processes, 234
content insensitive, 55, 210, 226
content sensitive, 55, 174–175, 209, 211 (*see also* Reconstruction processes)
Coronary artery disease, 60–62, 71–72
Correns, Carl, 24
Cortical inheritance, 44, 121–122, 328. *See also* Structural inheritance
Cosmides, L., 216, 231

Courtship behavior, 159–160, 168, 202, 230
Cows, 123, 125, 293–296, 379
Creationists, 9, 373
Crick, F. H. C., 30, 31, 48–49, 152–153
Crossing-over, 81, 84–85, 103
Crouse, H. V., 139
Cultural evolution
 in chimpanzees, 177, 183, 313, 368
 definition of, 160, 177, 205, 227–228
 in humans, 178, 204–223, 224, 226–229, 293–298, 312–313, 372, 374 (see also Language, evolution of)
 in Japanese macaques, 178–180, 183, 188
 Lamarckian view of, 189, 219–223, 228–229
 of literacy, 215
 of mathematics, 218–219, 231
 of music, 245–246
 in nonhuman animals, 159–160, 174–175, 176–180, 188–189, 205, 220, 313, 335–336
 rate of, 189
 and speciation, 159, 184–185
 in tarbutniks, 157–160
Culture, definition of, 160, 205
Cyanobacteria, 237
Cytoplasmic inheritance, 29–30, 32, 33, 45

Dairy farming, 293–296
Dance transmission, 174, 202, 204, 224, 291
Daphnia, 83
Darwin, C., 141, 145, 373
 and hereditary variation, 9, 13–16, 23, 141
 and the inheritance of acquired characters, 14–16, 311
 and niche construction, 176, 239
 pangenesis theory, 13–16, 22, N14
 and sexual selection, 288
 and the tangled bank, 10–11, 239
 and the theory of natural selection, 9–11, 239
Darwinism
 Darwin's, 10–16, 39
 four-dimensional, 1–2, 353, 355–356, 378–379
 gene-centered, 1, 4, 30–39, 41–42, 45
 Modern Synthesis, 24–34, 39, 42–45, 378
 social, 372
 transformations of, 39–40, 355–356
 universal, 11–12, 40–41
 Weismann's, 17–21, 39
Dawkins, R., 40, 373
 meme concept, 206–208
 recipe and cake metaphor, 33, 70, 209, 361
 replicator-vehicle concept, 36–37, 41–42, 102, 189, 207–209, 375–376
 selfish gene theory, 35–38 (see also Selfish genes)
Deacon, T. W. 193
Deforestation, 296–297, 368–369
Development. See also Canalization; Networks, genetic-developmental; Plasticity; Differentiation
 of behavior, 188, 218
 and cultural evolution, 211–212, 220–221
 Darwin's view of, 14–16
 evolution of, 251–254, 256–258, 277
 genomic changes during, 68–71, 78, 85, 88, 99–101, 102, 106, 139
 and heritable variation (see Instructive and selective processes; Lamarckian processes; Replicator-vehicle concept)
 Gould's view of, 38, 374
 in the Modern Synthesis, 27–28, 29, 44–45
 Weismann's view of, 16–19, 68
Developmental systems theory, 374
de Vries, H., 23, 24, 141–142

Dialects, 161, 172, 368
Diamond, J., 193, 237
Dicer, 133–135
Diet. *See* Food preference transmission;
  Maternal effects, on diet; Nutrition
Differentiation, 113, 117, 251–254, 330
Directed mutation. *See* Mutation,
  directed
Diseases, 140, 355–356. *See also* CJD;
  Coronary artery disease; Kuru;
  Lactose intolerance; Sickle cell
  anemia; Tay-Sachs disease
  complex, 58, 60–62, 71–75
  monogenic, 56–58, 75
  new treatments for, 75, 136–137, 248,
  364–365, 366, 367
DNA, 5, 320, 342, 351–352
  changes during development, 68–71,
  78, 85, 88, 99–101, 106, 152
  code and translation, 30–31, 49–52,
  53, 57, 271
  foreign, 139, 140, 329–330
  as information, 52–55, 67
  junk, 32, 52
  methylation (*see* Methylation, DNA)
  modular organization of, 55
  noncoding, 32–33, 52
  repair, 82, 86–87, 94, 103, 248,
  322–323
  repeated sequences in, 95–96, 98–99,
  135, 259, 323, 328, 330, 331–332
  replication, 33, 48–49, 55, 86, 130,
  323
  structure, 30, 48–49
Dobzhansky, T., 29, 227, 265
Dogs, 83, 175, 259, 282–283
Dolly, 59, 150, 367
Dolphins, 172, 173, 351
Domestication, 237, 241, 259–260,
  282–283, 293–294, 347–348
Donald, M., 224
Donkey, 139
Dor, D., 305–310, 315

Dosage compensation, 256–257,
  279–280
Dover, G., 362, N332
Drosophila (fruit fly), 27, 44–45, 76, 83,
  153, 262, 311
  chromosomes of, 69–70
  courtship dance of, 202
  DNA methylation in, 131, 331,
  344–345
  heat stress effects on, 104, 264–265
  selection experiments with, 264–265,
  267–268, 269–270, 273, 274
Ducks, 167, 218
Durham, W. H., 293–296
Dyslexia, 215

Earthworms, 83, 239–240, 241
Easter Island, 369
Educated guess. *See* Evolution, of the
  educated guess
EIS (epigenetic inheritance system). *See*
  Epigenetic inheritance
Endangered species, 168, 190, 369–371
Environment
  dual role of in evolution, 233, 276,
  281–282, 285 (*see also* Inheritance of
  acquired characters; Instructive and
  selective processes)
  fluctuating, 290, 312, 324–326, 347
  human impact on, 241, 296–297,
  368–369, 372
Epibeasts, 117–118
Epigenetic ecology, 371
Epigenetic epidemiology, 364–366
Epigenetic inheritance. *See also*
  Chromatin marks; Self-sustaining
  feedback loops; Structural
  inheritance; RNA interference
  and adaptive evolution, 114–118,
  144–145, 147–148, 151, 153–154,
  275–276, 359
  constraints on, 138–139, 148–150,
  152

Epigenetic inheritance (cont.)
and the evolution of multicellularity, 251–254
fidelity of, 150–152, N151
in fluctuating environments, 324–326
and heritability estimates, 362–364, N363
origins of, 324–333
in plants, 138, 140–142, 148–149, 249–250, 277–278, 358
transgenerational, 137–145, 148–149, 151–152, 358–359, 366, 371
and transmissibility, 150–151
types of, 119, 147
in unicellular organisms, 119, 138, 148, 324–325
Epigenetic landscapes, 63–65, 261–264, 265. *See also* Canalization; Genetic assimilation; Networks, genetic-developmental
Epigenetics, 114, 395, N261
Epimutations, 142, 248, 253, 359, 364–365
Escherichia coli, 97–98, 106, 322, 346–247
Ethical issues, 9, 74, 367, 371–372, 379–381. *See also* Eugenics
Eugenics, 30, 43, 230
Evolution
of categorization, 291–292
of complex organisms, 251–254, 324, 342, 343 (*see also* Evolutionary transitions)
of culture (*see* Cultural evolution)
Dobzhansky's definition of, 29
of the educated guess, 101–102, 103–104, 238, 249–250, 320–340, 345–347
of instincts, 261, 286–292, 311–312
of language, 173, 219, 223–4, 302–310, 312–317
of lifestyles, 158–9, 171, 176–180, 183, 220–221, 293–296, 539–540

of social learning, 334–336
rate of, 23, 144–145, 185, 188, 250, 260, 309, 347, 353, 356
of reproductive strategies, 82–85
Spencer's notion of, 21–22
Evolutionary byproducts, 94, 254–256, 258, 306, 320–321, 323, 327, 328, 329, 334, 335, 336, 241, 345–346, N254
Evolutionary progress, 21, 352–353
Evolutionary psychology, 206, 212–219, 221–222, 230–231, 380–381
Evolutionary transitions, 122, 146, 341–344, 351–352
Evolvability, 103, 345–348, 353
definition of N345
Exons, 66
Extraembryonic tissues, 257–258, 279–280

Fear response, 261, 286, 289–289
Feathers, 303
Feces, 163–164, 165, 334, 346–347
Feldman, M. W., 205, 362, 363
Finches, African parasitic, 185
Fire ant, 190
Fitch, W. T., 302–303, 313
Flax, 99
FLB (faculty of language in the broad sense), 302–303
FLN (faculty of language in the narrow sense), 302–303, 305, 306, 313–316, 338
Food preference transmission
in insects, 189–190, 240, 367
in mammals, 47, 71, 161–165, 178, 180–181, 182, 184, 334
Foster parents, 163, 165, 168, 170–171, 185, 187, 190
Foxes, 259–260, 272, 283, 348
Fruit flies. *See Drosophila*
Fungi, 30, 43, 49, 125, 133, 329. *See also Neurospora*

Gaia, 370
Gajdusek, D. C., 123–124, 125
Galton, F., 22–23, N14
Gametogenesis, 17, 25, 81
  and chromatin marks, 138–139, 145,
    149–151, 253–256 (*see also* Genomic
    imprinting)
Gánti, T., N320
Geldanamycin, 267, 269
Gemmules, 14–15, 20, 22, 39
Gene
  agouti, 142–144, 365
  APOE, 61–62, 71–72
  chicken Slowpoke (cSlo), 66–67
  crossveinless, 264
  cubitus interruptus, 262
  Dicer, 133–135
  Hsp90, 267–268, 270, 347
  knockouts, 63–65, 73, 130
  Krüppel, 269–270
  lactase, 293–295
  methyltransferase, 130
  novelty seeking, 59–60
  prion, 328
  sickle cell hemoglobin, 56–57,
    296–297
  Tay-Sachs, 57, 297
Gene regulation, 52–53. *See also*
  Chromatin marks; Networks, genetic-
  developmental; Self-sustaining
  feedback loops
Genes
  and characters, 6–7, 56–67, 71–74,
    76–77, 105–106, 281–282, 362–364,
    390
  contingency, 95–96
  cytoplasmic, 32, 33
  as DNA sequences, 7, 30–31, 33, 67
  dormant, 260, 272
  jumping (*see* Transposable elements)
  as Mendelian units, 24–27
  modifier, 43, 143
  mutases, 322

number of, 47, 75–76, N47
selfish, 34–38, 39, 77–78, 206, 207,
  280, 373 (*see also* Replicator-vehicle
  concept)
as units of information, 28, 49
Genetic accommodation, 312
Genetic assimilation, 245, 281–283
  of behavior, 283, 285, 289–292,
    311–312 (*see also* Baldwin effect)
  of language, 307–310, 314, 316, 340
  of morphological and physiological
    features, 262–265, 266–268, 273–274,
    275, 282
  simple model of, N264
Genetic astrology, 58–59, 67, 71–74,
  382
Genetic code, 49–50, 271, 342
Genetic determinism, 76–78
Genetic diseases, 56–58, 61–62, 71–75.
  See also Diseases
Genetic engineering
  by humans, 5, 62, 63, 75, 114, 133–5,
    136–7, 139–140, 333
  natural, 68–71, 78, 85, 86–87, 333
Genetic membranes, 122, 328
Genetic networks. *See* Networks,
  developmental-genetic
Genetic programs, 33
Genome, dynamic, 7, 88. *See also* DNA,
  changes during development;
  Genetic engineering, natural
Genome project, 5–6, 47, 371
Genomic conflict, 257–258, 278–280
Genomic defense systems
  methylation, 329–331, 344, 366
  RNAi, 135, 332–333, 366
Genomic imprinting, 139–140, 145,
  150, 254–258, 278–281, 346, 364,
  367, N254
Genotype-phenotype distinction, 28,
  33, 109–110, 207, 208–209, 245. *See
  also* Replicator-vehicle concept
Germinal selection, 19–20

Germ line
  continuity of, 18–19, 20, 68–69
  segregation of, 18, 31, 44, 45,
    148–149, 250, 253–254, 277
Germ plasm. *See* Germ line
Goethe, J. W. von, 141
Gottlieb, G., 218
Gould, S. J., 38, 229, 362, 373, 374,
  N38
Griesemer, J., 376, N376
Group selection, 20, 34–35, 37–38, 39,
  188, 345

Haemophilus influenzae, 94–96
Haig, D., 257, 278, 280
Hair, 320–321
Hamilton, W. D., 35, 37
Hares, 285
Hauser, M. D., 302–303, 313
Hearing, 66–67, 320
Heat shock effects, 98, 104, 264–265,
  274
Heat shock proteins, 266–270, 274, 347
Helpers, 186–187
Heredity. *See also* Information
    transmission; Inheritance systems
  Darwin's pangenesis theory, 14–16,
    22
  definitions of, 28, 353
  limited, 41, 121, 233–234
  in the Modern Synthesis, 24, 28–30,
    31
  nature of, 1, 10, 39, 42, 45, 109–111,
    344, 353, 355, 360, 375
  Weismann's theory of, 16–21, 68–9,
    360
Heritability, 362–364
Heterochromatin, 70
Hinny, 139
Hirschbein, L., 138
Histone code, 132
Histones, 126–128, 131–132, 329
Holliday, R., 113, 128, N130

Hormones, 145, 225, 259–260, 272,
  283
Horses, 68, 139
Hull, D. L., 229, 360–361, N36
Human behavior, genetic basis of, 1,
  59–60, 155, 206, 212–217, 219,
  230–231, 380–381
Human genetics, 56–62, 75. *See also*
  Diseases
Human Genome Project, 5–6, 47, 371
Hyenas, 286, 311

Ifcha Mistabra, 3
Imitation, 170, 172–175, 207–208, 211,
  226
Immelmann, K., 184
Immune system, 68, 88, 95, 102, 106,
  152, 323
Imprinting. *See* Behavioral imprinting;
  Genomic imprinting
Inbreeding, 186, 259, 269
Incest, 186
Information, 28, 52–56, 166, N52
  holistic, 121, 126, 165, 171, 200, 204,
    233–234
  latent, 174, 202, 225, 234–235
  modular, 55–56, 71, 121, 131,
    174–175, 203, 225, 234–235
Information transmission, 1–2,
  109–111, 158, 166, 176, 342–344. *See
  also* Inheritance systems
  through behavior-affecting substances,
    71, 165–166, 182, 334
  through cell structures, 126
  direction of, 31, 165, 172, 188, 204,
    233–235
  fidelity of, 86, 151, 183–184, 188, 221
    (*see also* Mutation; DNA repair)
  through DNA, 55–56, 71, 86
  through DNA methylation, 130–131
  through imitation, 173–175
  through nonimitative social learning,
    171–172

through RNAi, 136
through self-sustaining loops, 120–121
through symbolic communication,
   201–204, 224–226 (*see also* Memes)
systems compared, 71, 109–111,
   165–6, 174, 182, 184, 201–204, 225,
   233–238, N201
Inheritance of acquired characters, 7,
   13–15, 22–23, 38, 44–5, 106–107,
   355, 360–361. *See also* Lamarckian
   processes
through behavioral and symbolic
   transmission, 228–229
and the central dogma, 104–105
Darwin's views on, 13–16, 21, 311
Dawkins's views on, 36–37, 42
through genetic assimilation,
   260–265, 273–274, 286–292, 311
Hull's views on, 229, 360–361
Spencer's views on, 21–22, 228
Weismann's views on, 17–19, 20, 21,
   22, 36, 360
Inheritance systems, 2, 42, 107,
   109–111, 146–147, 233–238, 355–356
behavioral, 155–176, 180–191, 205,
   224
compared, 109–111, 144, 149–150,
   150–151,165–166, 182–184, 189,
   201–204, 214–216, 218–219, 225,
   233–238, 324–325, 362–364
epigenetic, 113–154, 358—359, 395
and evolutionary transitions,
   341–344, 351–352
genetic, 24–34, 47–52, 55–56, 71,
   79–107
interactions of genetic and behavioral,
   161, 177, 184, 239–240, 286–293,
   311–312, 334
interactions of genetic and epigenetic,
   133–136, 139, 140, 146, 240–241,
   245–283, 324–333, 344–345
interactions of genetic and symbolic,
   230–231, 293–298, 299–310,
   312–317, 340 (*see also* Evolutionary
   psychology)
origins of, 320–341 (*see also*
   Evolution, of the educated guess)
symbolic, 201–212, 219–222, 223–227
Innate responses, 217–218, 240, 261,
   286–292, 311–312
Innovations, N307. *See also* Variation
   generation
Insects. *See also* Aphids; Butterflies;
   Drosophila
behavioral transmission by, 189–190,
   240, 367
chromosomes of, 69–70, 139, 256
social, 20, 34, 190
warning colors of, 282
Instincts, 217–218, 240, 261, 286–292,
   311–312
Instructive and selective processes, 1,
   102, 147–148, 189, 227–229,
   235–237, 344, 356–357, 374
Intelligence and symbolic
   communication, 213, 215, 231, 301,
   304–306, 316, 240, 350–351
Interactors, N36
Introns, 66–67

Jablonka, E., 2, 156, 213, 291, 306–310,
   315, 362
Japanese quails, 186
Jaynus thought experiment, 114–118,
   137, 144, 151, 156, 349, 352, 358
Jews, 17, 162, 178, 181, 202, 221,
   297
Johannsen, W., 28, 36, 44
Jumping genes, 33, 44, 94, 135, 247,
   249–250, 330, 332, 346
Juniper berries, 162–164
Just so stories, 380–381

Kammerer, P., N22
Kanzi, 337–339, 340, 349–351
Kauffman, S. A., 327

Keller, E. F., 67
Kin selection, 35, 37
Kisdi, E., 362
Koshima island, 178–179
Kuru, 123–124, 266, 365

Lactose intolerance, 293–296
Lamarck, J. B., 13, 311, N13
Lamarckian processes, 21–23, 235–238,
  273–274, 286, 353, 357, 360–362. *See
  also* Inheritance of acquired
  characters; Instructive and selective
  processes; Neo-Lamarckism
 in behavioral and cultural evolution,
  189, 205, 221–223, 228–229,
  310–311
 in Darwin's theory, 13, 15, 311
 and genomic change, 7, 102, 106–107,
  152, 319
 origins of, 320–353
Lamb, M. J., 2
Language
 use by bonobos, 304, 337–339, 340,
  349–351
 Chomsky's view of, 218, 298–303,
  313–315
 Dor's view of, 305–310, 315
 evolution of, 173, 219, 223–224,
  302–310, 312–317, 339–340, 352
 functionalists' view of, 303–305,
  316–317
 learning, 204, 299–301, 305, 316,
  337–338, 368
 module for, 216, 218, 299–300, 313,
  315, 338, 381
 as a symbolic system, 195–199
Learning, 155, 237. *See also* Language,
  learning; Selection, for learning
  abilities
 early, 162–169, 180–181, 182,
  184–185, 218
 by imitation, 170, 172–175, 207–208,
  211, 226

through intentional teaching, 55, 195,
  204, 224
 social, 157–160, 161–180, 188,
  189–190, 209–212, 333–336, 349
 by trial and error, 157, 165, 166, 170,
  209, 334–335
Lenneberg, E. H., 218
Levels of selection, 20, 34–40, 41–42,
  148, 342–344, N34
Lewontin, R. C., 74, 371, 374, 377
Lexigrams, 338–340
Life
 definition of, 40–41, 370
 origins of, 320, 328, 332–333, 342,
  349, N320
Lifestyle evolution, 176–180, 183
 human, 220–221, 293–296, 339–340
 in Israeli black rats, 171, 176
 in Japanese macaques, 178–179, 183,
  220
 in tarbutniks, 158–159, 176
Linaria, 140–142, 148, 149, 358
Lindegren, C. C., 43
Lindquist, S, 266–268, 270–271, 273,
  328
Linkage, 81, 342, 351–352
Linnaeus, C., 140–141, 142
Lions, 286, 311
Literacy, 204, 213–215, 237, 316
Liu, J.-J., 328
Lodish, H. F., 58
Lorenz, K., 167
Lovelock, J., 370
Lu Hsun, 383, N383
Lyon, M. F., 256
Lysenko, T. D., 43, 44, 259

Macaques, 178–180, 183, 188, 207,
  220, N178
McClintock, B., 88–89, 94, 249, 260
Maize, 44, 249
Malaria, 296–297
Margulis, L., 369

Maternal effects, 139, 143, 152, 349, 365–366, 377. *See also* Genomic imprinting
on behavior, 145–146, 210, N165
on diet, 71, 162–165, 170–171, 189, 190
Mathematics, 200–201, 218–219, 231
Mauritius kestrels, 181
Maynard Smith, J., 40, 41, 121, 373
his airplane analogy, 33
and evolutionary transitions, 343–344, 351
his generalization of Darwinism, 11
and Lamarkism, 104
Mayr, E., 104, 229
Medawar, P. B., 229
Meiosis, 17, 81–82, 103
Membranome, 122
Memes, 37, 206–212, 219–220, 221–222, 224–226, 229–231, 380
Mendelian genetics, 24–30, 43–45, 81
Mendelian ratios, 25, 141
Methylation, DNA, 128–131, 142, 143–144, 148, 246–249, 364–365
as genomic defense, 140, 329–331, 333, 344, 366
and the evolution of marks, 329–332, 344–345
Mice, 44, 67, 130
imprinted genes in, 140, 258, 279
yellow, 142–144, 358–359, 364–365, 366
Midgley, M., 220
Midwife toad, N22, N266
Milk
in cultural evolution, 293–296
as a route of information transfer, 71, 152, 163, 165, 189
Milk-bottle-opening behavior, 169–170, 172, 209, 210–211, 335
Miller, G., 217
Mimetic culture, 224
Mitosis, 16, 68, 128

Mobile elements. *See* Jumping genes
Models, 378–379
of canalization, N63
of cortical inheritance N121
of cultural evolution, 205–206, 226, N205
of the evolution of adoption, N186
of the evolutionary consequences of sexual imprinting, N168
of genetic assimilation, 311, N264, N289, N298, N311
of language, N308
of regulatory networks, 327, N119
of the selection of epigenetic variants, 325, N324
of Steele's hypothesis, N152
Modern Synthesis, 24, 29, 31, 33, 42–45, 79, 227, 378
Modules, mental, 155, 212–219, 222, 230–231, 311–312, 380–381, N218. *See also* Language, module for
Molecular drive, N332
Mongolian gerbils, 145–146, 154, 161, 225–226
Monkeys, 181, 291–292. *See also* Macaques
Morgan, C. L., 289, N289
Morgan, T. H., 27–28, 30
Mosquitoes, 296–297
Moxon, E. R., 94, 96
Mule, 139
Music analogy, 109–111, 119, 146, 245–246, 260
Mutagens, 33, 86, 87
Mutation
causes of, 33–34, 70–71, 85–86, 94, 96, 103, 247–250, 320–323, 344–346
directed, 7, 79–80, 87–89, 99–101, 105–106, 325
discriminate, 96
hot spots for, 94–96, 323
induced, 92–94, 97–100, 249, 322–323, 364

Mutation (cont.)
  interpretive, 92–101, 105, 320–323, 364
  random, 7, 31,,33–34, 39, 70–71,
    79–80, 87–89
  rate of, 79, 87, 92–99, 142, 322, 347
  thought experiment, 89–92
  types of, compared, 89–101
Mutationism, 23–24, 27, 29

Neo-Lamarckism, 21–23, 289, 311, N21,
  N22. See also Lamarckian processes
Networks
  behavioral-ecological, 179–180, 183,
    239–242, 368–370
  genetic-developmental, 6, 7, 62–65,
    67, 72–74, 76–78, 261, 266, 273, 280,
    281, 327, 375 (see also Canalization,
    of morphological features; Genetic
    assimilation)
  symbolic-cultural, 200, 220–222
Neurospora, 43, 49
Neutralist-selectionist debate, 32–33, 65
  (see also Selective neutrality)
Niche construction, 176, 236–237, 241,
  285–286, 374, 377–378
  cultural, 292–298, 307–310, 368
Novelty seeking behavior, 59–60, 362
Nuclear monopoly, 29–30
Nucleosomes, 126–128, 131–132
Nursery rhymes, 55, 210–211, 299,
  N211
Nutrition, 143–144, 293–296, 365–366.
  See also Food preference transmission

Ohno, S., 256
Origins of life, 320, 328, 332–333, 342,
  349, N320
Osborn, H. F., 289, N289
Ostriches, 262–263
Oyama, S, 374, N52

Palmer, C. T., 380
Pangenesis, 14–16, 22, N14

Paramecium, 122, 138, 328
Parrots, 173, 174, 195–198, 286–288
Pasteur, Louis, 16
Pathogens, 94–96, 323
Payne, R. B., 185
Peacocks, 288
Peas, 24–25
Peloric variants, 141–142, 149
Petunias, 133
Phenotype. See Genotype-phenotype
  distinction; Plasticity; Replicator-
  vehicle concept
Pictures as symbols, 199–201, 204, 209,
  210
Pinker, S., 298, 314
Plants See also Antirrhinum, Arabidopsis;
  Linaria, Maize; Peas; Petunias
  epigenetic inheritance in, 133, 138,
    142, 148–149, 249–250, 277–278,
    359
  germ line of, 44, 138, 149, 250,
    277–278
  RNAi in, 133, 134, 135
  stress responses in, 94, 98–99, 249,
    346
Plasticity, 62, 77–78, 266, 312, 356, 377
  definition of, N62
  temporally extended, 348–349, 377
Podospora, 125
Polyploidy, 69, N145
Polyteny, 69–70
Potato washing habit, 178–180, 183,
  188
Prions
  in Aplysia, 126
  and the central dogma, 153
  of mammals, 124–125, 151, 266, 328
  of yeast and Podospora, 125, 270–272,
    273, 275, 328, 347
Progress, evolutionary, 21, 352–353
Proteins
  conformation of, 124–126, 266–270,
    273, 328

structure of, 30–32, 49, 56–57, 61
synthesis of, 30–31, 50–51, 57, 65–67, 119–120, 271–272
Prusiner, S. B., 124
[PSI⁺], 271–272, 275, 328, 347
Pugh, J. E., 113, 128
Punishment, 222–223
Pure lines, 28

Queitsch, C., 268, 269

Rabbits, 22, 162–163, 285, 293
Radman, M., 322
Rain forests, 368–369
Rape, 380–381, N380
Rats, 163, 173. *See also* Black rats
Rechav, G., 213
Recombination, 81–82, 84–85, 103–104, 247
Reconstruction processes. *See also* Copying processes, content sensitive
    behavioral, 165, 171–172, 188–189
    epigenetic, 151–152
    fidelity of, 151, 179, 182–185, 221
    symbolic, 209–212, 220–221, 226–227
Recursion, 299, 302–304, 315–316
Reductionism, 374–375
Replication. *See* Copying processes, content insensitive; DNA replication
Replicator-vehicle concept, 36–37, 41–42, 70, 102, 189, 206–207, 208–212, 226, 229, 375–376, N36
Reproducers, 376, N376
Reproduction. *See* Asexual reproduction; Sexual reproduction
Reverse transcription, 152
Reverse translation, 31, 104–105, 152
Ribosomes, 50–51, 98
Richerson P. J., 205
Riggs, A. D., 113, 128
RNA, 50, 343
    messenger (mRNA), 30, 50–51, 56–57, 66–67, 133–134, 152, 271–272

ribosomal (rRNA), 50, 98–99
small interfering (siRNA), 133–134, 152, 366
transfer (tRNA), 50–52
RNA interference (RNAi), 132–137, 144, 149, 152, 365, 366
    evolution of, 332–333
RNA splicing, 65–67
RNA world, 332–333, 376
Robinson Crusoe, 194–197, 286–287
Robots, 40–41, 58, 76, 78, 146
Ruden, D. M., 270, 273
Rutherford, S. L., 266–268

Sangster, T. A., 268
Savage-Rumbaugh, E. S., 337, 350
Schmalhausen, I. I., 265, 272, 377
Sciara, 139
Science and society, 9, 43–45, 259, 367, 371–373, 379–383. *See also* Eugenics; Evolutionary psychology; Sociobiology
Scrapie, 123, 124, 153
Seagulls, 286
Seed caching, 183
Selection. *See also Drosophila*, selection experiments; Instructive and selective processes
    of behavioral variants, 158–161, 196, 259–260
    of epigenetic variants, 148–149, 151, 270, 273, 275–276, 324–325
    of human sexual partners, 216–217, 219, 288, 380–381
    leading to instincts, 286–292, 311–312
    leading to mental modules, 155, 212–213, 215–217, 311–312, 380–381
    for learning abilities, 286–290, 308, 335–336
    levels of, 20, 34–40, 41–42, 148–149, 239, 342–344
    for mutability, 96, 322–323, 346–347
    organic, 289

Selection (cont.)
sexual, 216–217, 230, 287–288, 291
Selective neutrality, 32, 65, 73, 280. *See also* Canalization
Selfish DNA, 249
Selfish genes, 34–38, 39, 77–78, 206, 207, 373
Self-sustaining feedback loops, 119–121, 137,151, 154, 325–327
Sex chromosomes, 256–258, 278–280
Sexual reproduction, 14–15, 17, 18–19, 20, 29, 80–85, 101, 103–104, 346
Sexual selection, 216–217, 230, 287–288, 291
Shapiro, J. A., 70–71, 78, N88
Shaw, George Bernard, 21
Sickle cell anemia, 56–57, 296–297
Sign language, 305
Signs and symbols, 194–201, 205, 340, 349–351
Silver foxes, 259–260, 272, 283, 348
Skin thickening, 261, 262–263
Slugs, 300
Snails, 83, 96
Snakes, 96, 261, 286
Snarks, 382–383
Social learning. *See* Learning, social
Sociobiology, 78, 372, 380–381. *See also* Evolutionary psychology
Soft inheritance, 7, 229, 361. *See also* Inheritance of acquired characters; Lamarckian processes
Solenopsis invicta, 190
Somatic selection hypothesis, 152
Sonneborn, 122
SOS response, 322–323
Spalding, D., 167, 195, 286–288
Speciation, 141, 356
resulting from epigenetic differences, 145, 347–348
resulting from behavioral differences, 159–160, 184–185, N184
Spencer, H., 21–22, 228, 372–373

Sperber, D., 206
Splicing, 65–67
Starlings, 161, 173
Steele, E. J., 152
Sterelny, K., 375–376, N218
Stress responses, 345–348, 356
involving a change in the mode of reproduction, 83–84
involving epigenetic change, 249–250, 259–260, 266–268, 272–275
involving genomic change, 88–89, 92–94, 97–101, 103–104, 249–250, 321–323, 364
Structural inheritance, 44, 121–126, 137, 147, 151, 270–272, 327–329
Symbolic communication, 191, 193–201. *See also* Language
by bonobos, 337–339, 349–351
as an inheritance system, 201–204, 221, 225–226, 229, 233–238, 355–356 (*see also* Memes)
nonlinguistic, 199–200, 223–224, 307, 340
origins of, 336–337, 339–340, 343, 350–351
Symbols. *See* Signs and symbols
Szathmáry, E., 41, 121, 343–344, 351

Tarbutniks, 156–161, 176, 349, 352, 358
Tay-Sachs disease, 57, 297
Teleology, 319. *See also* Evolutionary progress
Templeton, A. R., 60
Terkel, J., 170
Thalassemia, 75
Thalidomide, 365
Thornhill, R, 380
Thought experiment
about behavioral evolution, 156–160
about epigenetic inheritance, 114–118
about the evolution of instincts, 286–287

about the evolution of literacy, 213–215

about the evolution of mathematical reasoning, 218–219

about nonrandom mutation, 89–92

comparing genetic and nongenetic inheritance, 109–111, 245–246

comparing signs and symbols, 194–199

Tits, 169–170, 172, 209, 211, 335

Toadflax. *See Linaria*

Tobacco plants, 134

Tooby, J., 216, 231

Traits as units of evolution, 41, 77, 281–282, 356, 376

Transcription, 50–52, 119–121, 128–132

Transgenes, 139, 140

Translatability of symbols, 203

Translation of DNA information, 30–31, 50–51, 271–272. *See also* Reverse translation

Transposable elements, 33, 44, 94, 135, 247, 249–250, 330, 332, 346

Transposon sequences, 142, 143, 365

True, H. L., 271

Tschermak, E. von, 24

Twins, 268, 363, 365

Unicellular organisms, 138, 148, 251, 324–326, 342. *See also* Bacteria; Ciliates; Yeast

Units of evolution/selection, 20, 34–40, 41, 76–77, 206, 227, 280, 342–344, 356, 376. *See also* Memes

Universal Darwinism, 11–12

Universal grammar, 298–303, 313–314

Uterine environment, 145–146, 154, 162–163, 182, 225–226

Variation, 1–2, 10, 11–12, 319

behaviorally transmitted, 156–176, 189

continuous, 23, 27

Darwin's explanation of, 13–16

genetic, 25–30, 55–56, 215

heritable epigenetic, 114–145, 147–152, 270, 275, 359

hidden, 32, 260, 262, 265, 267–269, 271–274, 276, 290, 291, 312, 347

Johannsen's explanation of, 28

origin of, 102–103, 235–238, 319, 356, 357, 360–361

Weismann's explanation of, 19–20

Variation generation, 235–238, 319, N307

behavioral, 165–166, 171–172, 174–176

cultural, 203, 210–212, 220, 231, 236–237, 307–308

epigenetic, 132, 138, 144–145, 234, 236, 258–260, 270–274, 359, 364–365

through mutation, 7, 29, 70–71, 80, 85–103, 105–106, 247–250

through sex, 19–20, 29, 81–85, 87, 101, 103–104

Virchow, R., 16

Viruses, 33, 85, 124, 135, 137, 152, 329, 332

Vitamin D, 295

Waddington, C. H., 63–65, 78, 261–266, 268, 272, 275, 289, 377

Warning colors, 282

Water fleas, 83

Watson, J. D., 30, 48–49

Weatherall, D. J., 75

Weismann, A., 16–21, 22, 24–25, 31, 34, 36, 68, 229, 360

West-Eberhard, M. J., 312, 377, N62, N281, N298

Whales, 161, 172–173, 194, 205, 299, 351

Whitelaw, E., 142–143

Wolves, 224, 283

Wright, B. E., 97–98, 322
Wright, S., 32, 119

X chromosomes, 256–257, 258,
    278–280

Y chromosomes, 256, 258, 278–280
Yeast, 83, 125, 138, 270–272, 273, 275,
    328

Zuckerman, Solly, 245